Springer Series on
Atoms+Plasmas

5

Editor: Günter Ecker

Springer Series on
Atoms+Plasmas

Editors: G. Ecker   P. Lambropoulos   H. Walther

Managing Editor: H. K. V. Lotsch

Volume 1   **Polarized Electrons**   2nd Edition
By J. Kessler

Volume 2   **Multiphoton Processes**
Editors: P. Lambropoulos and S. J. Smith

Volume 3   **Atomic Many-Body Theory**   2nd Edition
By I. Lindgren and J. Morrison

Volume 4   **Elementary Processes in Hydrogen-Helium Plasmas**
By R. K. Janev, W. D. Langer, K. Evans, Jr. and D. E. Post, Jr.

Volume 5   **Pulsed Electrical Discharge in Vacuum**
By G. A. Mesyats and D. I. Proskurovsky

Volume 6   **Atomic and Molecular Spectroscopy**
By S. Svanberg

G. A. Mesyats   D. I. Proskurovsky

Pulsed Electrical Discharge in Vacuum

With 120 Figures

Springer-Verlag Berlin Heidelberg New York
London Paris Tokyo Hong Kong

Professor Dr. Gennady A. Mesyats

USSR Academy of Sciences, Ural Division
Institute of Electrophysics, 91, Pervomaiskaya Street
SU – 620219 Sverdlovsk, USSR

Dr. Dimitri Proskurovsky

USSR Academy of Sciences, Siberian Division
Institute of High Current Electronics, 4, Akademichesky Avenue
SU – 634055 Tomsk, USSR

Series Editors:

Professor Dr. Günter Ecker

Ruhr-Universität Bochum, Institut für Theoretische Physik, Lehrstuhl I, Universitätsstraße 150
D-4630 Bochum-Querenburg, Fed. Rep. of Germany

Professor Peter Lambropoulos, Ph. D.

University of Crete, P.O. Box 470, Iraklion, Crete, Greece, and
Department of Physics, University of Southern California, University Park
Los Angeles, CA 90089-0484, USA

Professor Dr. Herbert Walther

Sektion Physik der Universität München, Am Coulombwall 1
D-8046 Garching/München, Fed. Rep. of Germany

Managing Editor: Dr. Helmut K. V. Lotsch

Springer-Verlag, Tiergartenstrasse 17
D-6900 Heidelberg, Fed. Rep. of Germany

ISBN-13: 978-3-642-83700-5   e-ISBN-13: 978-3-642-83698-5
DOI: 10.1007/978-3-642-83698-5

Library of Congress Cataloging-in-Publication Data. Mesiats, G.A. (Gennadiĭ Andreevich) [Impul'snyĭ élektricheskiĭ razriad v vakuume. English] Pulsed electrical discharge in vacuum / G.A. Mesyats, D.I. Proskurovsky. p. cm.–(Springer series on atoms + plasmas ; 5) Translation of: Impul'snyĭ élektricheskiĭ razriad v vakuume. ISBN-13: 978-3-642-83700-5 1. Electric discharges through gases. 2. Breakdown (Electricity) 3. Vacuum. I. Proskurovskiĭ, D.I. (Dmitriĭ Il'ich) II. Title. III. Series. QC711.M5413 1989 537.5'3–dc 19   88-35986

The text was prepared using the PS™ Technical Word Processor

2154/3150-543210 – Printed on acid-free paper

Preface

This monograph is intended to provide the reader with an up-to-date description of advances in the physics of electrical breakdown and discharges in vacuum. These achievements are the results of studies of the fast processes which form the essence of the discharge phenomenon. The need to carry out these investigations was dictated by the lack of experimental data both on the kinetics of the breakdown of vacuum electrical insulation and on the nature of the nonstationary phenomena associated with spark and arc discharges in vacuum. For many years this lack of information was caused by experimental difficulties due to the short-term and microscopic nature of the phenomena under investigation. Such problems could be solved only by using equipment able to provide high temporal and spatial resolution. Relevant equipment, which is a spin-off from the high-voltage nanosecond-pulse technology, has been developed at the Institute of High Current Electronics, USSR Academy of Sciences Siberian Division (Tomsk) and employed in the experiments for the investigation of the discharge processes.

We believe that the discovery of the explosive electron emission (EEE) phenomenon and the demonstration of the fundamental role played by microexplosions and EEE in the development of vacuum breakdown is the most important result of the investigations carried out at our Institute. The discovery of the explosive electron emission provided a better understanding of the physics of electrical breakdown and discharges in vacuum. It stimulated extensive experimental and theoretical studies aimed at analyzing the processes which take place at the electrodes and in the discharge gap, accounting for their nonstationary nature. The investigation into the fast processes of electrical breakdown and discharges in vacuum facilitated not only an understanding of the physics of the discharge, but provided also solutions to a number of new scientific and engineering problems related mainly to high-power pulse electronics. However, none of the applications are discussed in this monograph due to limitations on space.

The results to be presented were obtained during the last two decades at the Institute of High Current Electronics (Tomsk). In writing this book we used results of our colleagues and co-workers S.P. Bugaev,

E.A. Litvinov, G.P. Bazhenov, R.B. Baksht, V.P. Rotshtein, Ya.Ya. Yurike, E.B. Yankelevich, N.A. Ratakhin, V.F. Puchkarev, and others. At the same time, some research groups in the USSR, USA, GDR, Great Britain and other countries were engaged in investigating the same problems. Among these we would like to mention the research groups of G.N. Fursey (Leningrad, USSR) and B. Jüttner (Berlin, GDR), which have benefited from our experience of the high-voltage nanosecond-pulse methods and have subsequently remained in close scientific contact with us. Their results as well as those obtained by other research groups will be included.

In this monograph we do not go into the details of all the phenomena and processes peculiar to vacuum discharges. Therefore we suggest that readers consult additional literature.

We believe that the present book will be useful to specialists in physical electronics, plasma physics, electrical discharges, as well as in high-voltage switching and accelerator engineering.

We would like to thank H.C. Miller (General Electric, USA) for copies of his bibliographies on vacuum electrical discharge, which enabled us to compile the references for the English edition of our book.

The authors are most grateful to V.N. Romanenko for her efforts in editing and preparing the manuscript and to T.K. Cherkashina for translating the book into English. We are also grateful to V.A. Maksimenko and N.Ye. Timoshenko who typed the manuscript, as well as to V.A. Sazhenova and V.I. Sorokin who prepared the illustrations.

Special thanks are due to G. Ecker who was so kind to read the manuscript with great care, for his extremely useful discussions and comments. Lastly, we would like to acknowledge the assistance of H.K.V. Lotsch in enabling us to incorporate the many manuscript improvements.

Sverdlovsk, July 1988   G.A. Mesyats
Tomsk   D.I. Proskurovsky

Contents

1. Introduction 1

2. Review of Vacuum Breakdown and Discharge Studies 5
   2.1 The Electrode Surface in a Vacuum Discharge 5
        2.1.1 Preparation of Electrodes 5
        2.1.2 Determination of Micropoint Parameters 7
        2.1.3 Effect of Emission from Non-metallic Inclusions ... 9
   2.2 Vacuum Insulation, Properties and Breakdown 11
        2.2.1 Prebreakdown Phenomena 12
        2.2.2 Microdischarges 13
        2.2.3 The Breakdown Voltage 14
   2.3 Kinetics of Vacuum Electrical Breakdown 15
        2.3.1 Characteristic Times of Breakdown 15
        2.3.2 Role of Electrodes in the Development
               of Breakdown 18
        2.3.3 X-Ray Pulse at Breakdown 19
   2.4 Field Electron Emission
        to Vacuum Breakdown Transition 21
   2.5 Hypotheses on Vacuum Breakdown Initiation 23
        2.5.1 Physical Processes Leading to Vacuum Breakdown .. 21
        2.5.2 Cathode-Initiated Breakdown 24
        2.5.3 Anode-Initiated Breakdown 26
        2.5.4 Comparison between Cathode and
               Anode Mechanisms for Breakdown Initiation 27
        2.5.5 Microparticle-Initiated Breakdown 31
   2.6 Spark Stage of Vacuum Breakdown 33
   2.7 The Discharge Arc Stage. The Cathode Spot 35
        2.7.1 Physical Properties of the Cathode Spot 36
        2.7.2 Cathode-Spot Models 40

3. Experimental Equipment and Techniques 43
   3.1 Electrical Measurement Techniques 43
        3.1.1 High-Voltage, Nanosecond Pulse Generators 44
        3.1.2 Current and Voltage Pulse Recording 46

3.2 Diagnostics of the Radiation
     that Accompanies Breakdown 47
     3.2.1 Electro-optical Recording of the Light Emission ... 48
     3.2.2 Photoelectrical Recording of the Light Emission ... 50
     3.2.3 Spectral Investigation
            of the Discharge Plasma Radiation 52
     3.2.4 X-Radiation Recording 53
3.3 Vacuum Equipment 54
3.4 Preparation and Examination of Electrode Surfaces 55

4. Pulsed Nanosecond Breakdown of Vacuum Gaps 58
4.1 Time Characteristics of the Pulsed Vacuum Breakdown ... 58
     4.1.1 The Influence of Electrode Conditioning 60
     4.1.2 The Influence of the Vacuum 60
4.2 Study of Light Emission at Pulsed Breakdown 62
     4.2.1 Single-Shot Investigations 62
     4.2.2 The Continuous-Operation Regime 64
     4.2.3 Comparison with Other Data 66
4.3 Electrode Erosion Studies 70
     4.3.1 Cathode Erosion 70
     4.3.2 The Tracer Method 70
     4.3.3 Anode Erosion 71
4.4 Nature of the Discharge Current at Breakdown 72
4.5 Mechanism of Pulsed Breakdown of Vacuum Gaps 75
     4.5.1 The Role of the Cathode 75
     4.5.2 The Cathode Plasma and the Electron Current 76
     4.5.3 Anode Phenomena 77

5. Cathode Processes in a Pulsed Vacuum Discharge 79
5.1 EEE Initiation by High-Density FEE Current 79
     5.1.1 Experimental Conditions 80
     5.1.2 Description of EEE Current 82
     5.1.3 The Point Explosion Delay Time 84
     5.1.4 Calculation of the Emitter Heating 85
     5.1.5 The Vacuum Discharge Delay Time 87
5.2 Erosion of Point Cathodes 89
     5.2.1 The Fast Current Rise 89
     5.2.2 The Slow Current Rise 92
     5.2.3 The Point Erosion Rate 93
     5.2.4 Erosion Due to Joule Heating 93
     5.2.5 Comparison with Experiment 96
5.3 EEE Current Density Measurements 96
     5.3.1 Current Density of a Point Cathode 96
     5.3.2 Current Density from a Massive Cathode 97

    5.3.3 Measurements Based on Erosion 99
    5.3.4 Experimental Data 102
  5.4 Microstructure of the Cathode Surface 104
    5.4.1 Erosion Traces in SEM 104
    5.4.2 The Field Enhancement Factor 107
  5.5 The Contribution of Droplet Ejection
     to Cathode Erosion 110
  5.6 Pressure in the Emission Zone 113
  5.7 Formation of Cathode Microstructure 114

6. Cathode Flare Plasma 118
  6.1 Velocity of CF Plasma Expansion 118
    6.1.1 The Grounded Grid and Collector Method 119
    6.1.2 The Photoelectric Method 120
    6.1.3 The Transverse Magnetic Field Method 120
    6.1.4 The Method of the Anode Erosion Mark 121
  6.2 CF Plasma Parameters 121
    6.2.1 CF Plasma Density 121
    6.2.2 CF Plasma Composition and Temperature 124
  6.3 EEE Current Effect on the Dynamics
     of the Plasma Light Emission 126
  6.4 A Model for CF Plasma Expansion 128
    6.4.1 The Adiabatic Model 128
    6.4.2 MHD Calculation 130
    6.4.3 The Model of an Ideal Plasma 132

7. Current Passage in the Spark Stage of Breakdown 136
  7.1 Electron Emission from CF Plasma into Vacuum 137
  7.2 Electron Emission from CF Plasma,
     Experimental Studies 139
  7.3 Current-Voltage Characteristics of a Single-CF Diode . . . 144
  7.4 Dynamics of the CF Electron Emission Boundary 146
  7.5 CF Plasma Potential Distribution and
     Plasma Emissive Properties 147
    7.5.1 Probe Measurements of the CF Plasma Potential . . 148
    7.5.2 The Nature of the Instability of CF Emission 151
  7.6 Spark Current Between Broad-Area Electrodes 154
    7.6.1 Calculation of the Spark Current Rise 155
    7.6.2 The Role of Cathode and Anode Flares 156

8. Formation of New Emission Centers on the Cathode 159
  8.1 Mechanisms of New EC Formation Under the Plasma . . . 159
    8.1.1 Mechanism of the Explosion of Micropoints 161
    8.1.2 Mechanism of the Explosion of the Liquid Neck . . 163

8.1.3 Mechanism of the Breakdown of
Non-metallic Inclusions 165
8.2 New EC Formation and Operation
Under Cathode Plasma 166
    8.2.1   Experiments Without Application
        of a Magnetic Field 166
    8.2.2   Effect of Transverse Magnetic Field
        on New EC Formation 169
    8.2.3   Results and Discussion 172
8.3 "Screening" Effect and Electron Beam Structure
in a Diode 175
    8.3.1   "Screening" Effect 175
    8.3.2   Influence of Neighbouring CFs
        on the Electron Beam Structure in the Diode ... 178

9. **Anode Processes in the Spark Stage
of Vacuum Breakdown** 181
9.1 Anode Heat Conditions 181
    9.1.1   Power Density Deposited at the Anode 181
    9.1.2   The Anode Temperature 183
9.2 Surface Structure of the Anode in the Discharge Zone .. 186
    9.2.1   Summary of Previous Work 186
    9.2.2   Metallographic Studies 187
    9.2.3   Electron-Microscopic Studies 190
    9.2.4   Mechanisms of the Anode Surface Damage 191
9.3 Formation of Anode Flares 192
    9.3.1   Conditions for AF Formation, Its Composition
        and Temperature 192
    9.3.2   The Expansion Velocity of AF 195
9.4 X-Radiation Generated at the Anode 198
    9.4.1   X-Radiation on Discharging a Line 199
    9.4.2   X-Radiation on Discharging a Capacitor 200

10. **Fast Processes at DC Breakdown of Vacuum Gaps** 203
10.1 Electrical Study of DC Breakdown 204
    10.1.1 Electric Circuit 204
    10.1.2 Prebreakdown Current and Breakdown Voltage . 205
    10.1.3 The Current Rise Time at Breakdown 207
    10.1.4 X-Radiation and Electrode Erosion at Breakdown 210
10.2 Optical Studies 210
    10.2.1 Determination of the Time
        of Appearance of Light 210
    10.2.2 Electro-optical Breakdown Studies 211
10.3 Comparison with Results of Other Investigations 212
10.4 EEE Initiation at DC Breakdown 219

10.4.1 EEE Initiation under Pure Conditions 219
10.4.2 EEE Initiation and the Total Voltage Effect . . . 221
10.4.3 Criteria for Vacuum Breakdown
and EEE Initiation 225

11. **Nonstationary Processes
in the Vacuum Arc Cathode Spot** 227
11.1 The Motion of Vacuum Arc Cathode Spots 227
11.1.1 The Effect of Surface Condition 227
11.1.2 The Influence of a Magnetic Field 232
11.1.3 Spontaneous Formation of Cathode Spots
in Pulsed Arc Discharges 234
11.2 Response of the Vacuum Arc to Current Transients . . . 235
11.2.1 Experimental Equipment and Technique 236
11.2.2 Results 238
11.3 Vacuum Arcs at Threshold Currents 241
11.3.1 The Threshold Current of a Vacuum Arc 241
11.3.2 Cathode Spot Current Density 244
11.4 Numerical Simulation of Processes in an
Explosive Emission Center 246
11.5 Explosive Electron Emission and
the Vacuum Arc Cathode Spot 252

12. **Pulsed Electrical Discharge in Vacuum at
Cryogenic Electrode Temperatures** 255
12.1 Field Electron Emission
at Low Cathode Temperatures 255
12.1.1 Effect of Superconductivity on FEE Current . . . 255
12.1.2 The Nottingham Effect and Superconductivity . 256
12.1.3 Other Emission Effects 258
12.2 Field Emission Current
Preceding the Explosion of a Point 259
12.3 Characteristics of the Vacuum Discharge
at Cryogenic Temperatures 261
12.3.1 Experimental Conditions 261
12.3.2 Experimental Results 261
12.4 Vacuum Discharge Between Electrodes
Made of High-Temperature Superconductors 263
12.4.1 General Notions 263
12.4.2 FEE from High-Temperature Superconducting
Cathodes 264
12.4.3 Vacuum Discharge 265

References 269

Subject Index 291

1. Introduction

Progress in modern physics, and electrical and electronic technology depends on the utilization of unique properties of the vacuum. A high vacuum is required for a variety of electrophysical processes (vacuum electronic and cathode-ray devices, charged-particle accelerators and separators, generators of electromagnetic radiation, etc.). Such systems utilize the insulating properties of vacuum and its ability to restore its electric strength after discharge. In vacuum capacitors, electrostatic generators and vacuum switches a high vacuum serves as an excellent insulator. Vacuum switches and spark gaps are able to switch heavy currents for short durations. They are notable for their small size, high stability, reliability, and noiseless operation. During the last two decades special interest has been shown in applications such as the creation of nanosecond and microsecond high-power, high-current pulsed charged-particle accelerators used in inertial confinement fusion experiments, the generation of high-power X-ray and microwave pulses, the gas lasers, etc.. Developing such high-power pulse devices requires knowledge not only of the pulsed electric strength for vacuum insulation, but also of the characteristic breakdown times, the kinetics and related phenomena.

In the early sixties, when we began to investigate the electrical breakdown of vacuum gaps using high-voltage nanosecond pulses, there was an apparent lack of information about the mechanism of vacuum breakdown. The majority of the hypotheses which existed at that time were based on the results of experiments carried out with insufficient temporal and spatial resolution. It turned out to be difficult to identify the most important facts and to describe all the processes of vacuum breakdown from the time of the application of the electric field up to the transition to the arc discharge. Therefore, we suggested that nanosecond-pulse technique, together with up-to-date methods for recording fast processes, could provide new information about the mechanisms involved.

Indeed the experiments that we have carried out later on have shown that the irreversible failure of vacuum insulation during the nanosecond pulse is caused by an explosive-like appearance of plasma microblobs named cathode flares. All further processes occurring in the gap (growth of breakdown current, X-ray flash, cathode and anode erosion, anode flare) are related to the operation of cathode flares. As a result of investigations into the time behavior of most typical processes at the electrodes and in the vacuum gap, it has been established that during the period of the cathode flare operation intense electron emission from the cathode takes place. It is caused by the explosive transition from the solid state to a dense plasma of the cathode material due to the heating of localized cathode regions by the emission current. This phenomenon was named "explosive electron emission". In addition, we were able to distinguish several subsequent steps in vacuum pulsed breakdown, each characterized by certain physical processes. Most importantly, the breakdown initiation which is associated with the explosion of micropoints under the action of a high-density field emission current, as suggested by earlier investigators, was confirmed.

It has been of interest to apply the method for recording fast processes to the study of the kinetics of the dc vacuum breakdown and to compare this phenomenon with the pulsed discharge. We have established that the fast stage of the dc breakdown develops similarly to a pulsed breakdown, i.e., the discharge is due to the initiation of an explosive electron emission at the cathode. Its distinctive feature is that for a slow voltage rise, explosive emission can be induced by migration, adsorption, desorption and ionization processes or by transport of material particles from one electrode to another.

Our results and conclusions do not contradict the results of numerous preceding works but do provide a rather more substantial explanation. Many of the phenomena previously found are put in perspective to the development of vacuum breakdown and discharge.

Thus the study of fast processes has provided a great deal of understanding of the physics of electrical breakdown and discharge in vacuum. These studies have forced a radical change in the assumptions about mechanisms of vacuum breakdown, sparking and arcing in vacuum. Investigations of the explosive electron emission have provided a great stimulus toward a revision of existing concepts for the vacuum arc cathode spot.

The topics of this monograph have been the subject of reviews and numerous articles in journals. They were also discussed at Int'l Symposia on Discharges and Electrical Insulation in Vacuum (ISDEIV) and at Int'l Conferences on Phenomena in Ionized Gases (ICPIG) [1.1-12]. This monograph aims to summarize our own results as well as those of others, placing emphasis on the techniques for investigating fast processes in vacuum discharges.

2

The book comprises twelve chapters. A general review of studies on vacuum breakdown and spark and arc discharges in vacuum is given in Chapter 2. Particular attention is paid to the dynamics of the breakdown development. It is shown that, in spite of the widespread use of vacuum electrical insulation and vacuum discharges, the phenomena associated with this mechanism remain a subject for discussion.

The third chapter is devoted to a description of the apparatus and methods for investigating discharge phenomena of nanosecond duration with high sensitivity and the necessary spatial and temporal resolution.

Presented in Chapter 4 are the data on the breakdown of vacuum gaps by high voltage pulses with a nanosecond rise time. A regular sequence of pulsed breakdown stages is shown to exist and the processes occurring in these stages are discussed. The discussion of the breakdown mechanism clearly demonstrates that the explosion of cathode micropoints plays a dominant role and is accompanied by dense plasma generation and a simultaneous initiation of intense electron emission. This emission is referred to here as "explosive electron emission".

Comprehensive investigations of the cathode processes in the spark stage of vacuum breakdown support the existence of the explosive emission phenomenon. They are described in the fifth chapter along with experimental data which have enabled us to establish the explosive nature of cathode erosion and to determine the temporal and spatial scales of the explosive emission cycles. The data also helped us to analyze the current density at the metal-to-plasma phase transition and to acquire knowledge of the dynamics of the cathode surface structure in vacuum discharges.

Chapter 6 presents the measurements of the parameters of the cathode flare plasma, the nonstationary plasma appearing at the explosion of micropoints. The data on the plasma expansion velocity, temperature and composition are summarized. Also discussed is the influence of the explosive emission current on the dynamics of the plasma radiation and the pattern of plasma expansion.

The seventh chapter is devoted to the study of the emissive properties of the cathode flare plasma and the mechanism of the current growth during the vacuum breakdown spark stage. Current-voltage characteristics of vacuum diodes with a single cathode flare are described. These characteristics are now widely used in designing high-current pulsed diodes with explosively emitting cathodes. The data are classified according to the characteristic features of the transition from stable to unstable electron emission. The reasons for such a transition are considered.

Chapter 8 describes the results of investigations into the formation and the "motion" dynamics of explosive emission centers on the cathode. The spatial zones associated with the formation of new centers are defined and the mechanisms leading to the formation of such centers are analyzed. The influence of the electron space charge and that of the localization of the explosive emission on the structure of the electron flow in the diode gap are discussed.

In the ninth chapter the anode processes taking place in the breakdown spark stage are investigated. Successive phases of anode destruction are described, and the conditions for their appearance are determined. The formation of an anode plasma from a desorbed gas and anode material is considered, and the anode plasma parameters, together with the plasma expansion model, are described. Parameters of X-ray pulses emitted by the anode in the breakdown spark stage are presented.

The kinetics of dc vacuum breakdown is treated in the tenth chapter. During these investigations we, for the first time, were able to determine, within nanosecond accuracy, the start of the fast breakdown stage during which the gap is filled up with conducting medium. It is established that the start of dc breakdown is inseparably linked with the initiation of explosive electron emission.

The eleventh chapter is devoted to a study of a number of nonstationary processes occurring in vacuum-arc cathode spots. The influences of the cathode surface condition and the applied magnetic field on the motion of cathode spots are discussed. Results for the time lag of cathode spot emission and the nonstationary processes at the threshold arc currents are described. Also presented are some recent results of numerical simulation of the processes occurring in a single emission center. The general analysis of the nonstationary processes taking place in cathode spots demonstrates that the explosive emission processes and microexplosions play a fundamental role in the behavior of nonstationary cathode spots of spark and arc discharges.

The twelfth and concluding chapter presents a review of the results available in the literature, and reports on new data on vacuum discharges with electrodes cooled to cryogenic temperatures. Field electron emission from such electrodes, critical values of the field emission current density and the development of vacuum breakdown are discussed. The chapter concludes with an analysis of results on the field electron emission and pulsed vacuum breakdown with electrodes made of high-temperature superconducting materials.

2. Review of Vacuum Breakdown and Discharge Studies

The electric strength of vacuum insulation, the laws of nature and the physics of electrical breakdown and discharges in vacuum have been studied extensively. It seems imposssible to give equal attention to all these investigations and to discuss comprehensively all the problems concerning the vacuum discharge phenomenon in a single book. In preparing the present chapter we concentrated on those aspects which enabled us to better understand the nature of fast processes that occur in vacuum discharges.

2.1 The Electrode Surface in a Vacuum Discharge

The surface condition of the electrodes and its influence on vacuum breakdown has extensively been discussed for many years. There is a good deal of information in the literature [2.1-7], so that we can restrict ourselves to a brief outline.

2.1.1 Preparation of Electrodes

The condition of an electrode surface depends on a number of factors. It has been established that the determining factors are the treatment techniques employed in the preparation of electrodes, the methods of conditioning the electrodes in the discharge chamber and the conditions under which the discharge gap is subsequently operated.

When treating an electrode surface mechanically, the crystal grains undergo destruction and deformation. The thickness of the destroyed layer depends on the type of metal and the techniques of treatment; it may be a few or several tens of micrometers [2.3, 7]. In this layer, individual grains are bound much more loosely than in the original metal. It may also contain particles of abrasive materials and chemical reaction products. The presence of the latter is stimulated by the high temperatures produced by friction.

To remove the destroyed layer electrochemical polishing is usually employed. Electrochemical polishing dissolves surface protrusions, giving a smooth electrode surface. The surface finish after such treatment is directly related to the purity of the electrode material. For instance, after electrochemical polishing of aluminum monocrystals the size of irregularities left on the surface is about 10 nm or less [2.8]. However, electrochemical polishing usually causes the formation of a thin oxide layer which can be removed only by conditioning of the electrodes in vacuum.

Conditioning in vacuum usually starts by heating the electrode up to high temperatures. Degassing and diffusion of impurities towards the surface along the grain boundaries cause a change in the surface microrelief. This change is due to the fact that the system seeks to minimize its surface energy, and the surface rearranges into crystalline faces with a lower surface tension. Such rearrangement produces parallel protrusions on the surface with steps of height up to several micrometers. The appearance of the electrode microirregularities is also assisted by metal evaporation and condensation, recrystallization, phase transformation, anisotropy of the coefficients of expansion, etc. The rate of cooling is also important. Rapid cooling results in the surface becoming rougher [2.1, 3, 5].

High-temperature heating does not remove the surface contaminations completely. Electrode treatment in a glow discharge is an efficient method of removing oxides and other stable compounds [2.3, 7, 9-12] and results at the same time in the sputtering of micropoints by ion bombardment.

Prolonged application of voltage produces a considerable change in the surface microrelief. Factors such as the type of metal, impurities on the surface, surface temperature and electric field initiate such processes as surface diffusion, surface rearrangement in an electric field, evaporation, field desorption, migration of impurities, etc. Moreover, several investigators observed the growth of chaotic microconglomerations, microprotrusions, as well as the extraction of micropoints. A detailed analysis of these processes exceeds the scope of the present book.

The microrelief of the electrode surface significantly changes in the process of electrical breakdown in vacuum. During breakdown the discharge current concentrates in localized areas of electrodes. The energy released from these areas is spent on heating, melting and erosion of electrode material. Thus, the process of cathode erosion gives rise to microcraters and micropoints which eject, at a high velocity,

6

microdrops of liquid metal. Melting and evaporation of isolated anode areas result in the formation of localized zones of destruction and in transfer of anode material to the cathode.

From the above it is clear that to obtain smooth electrodes is neither a simple nor, in some cases, an economical task.

2.1.2 Determination of Micropoint Parameters

To allow a quantitative characterization of surface roughness of the cathode, the factor of electric field enhancement at the micropoints, β, is introduced. It represents the ratio of the true value of the electric field at the protrusion apex to its average macroscopic value $E_{av} = V/d$, V being the gap voltage, and d the electrode separation.

For simple protrusion geometries some correlations between the factor β and the irregularity parameters can be found [2.1, 13-17]. For the useful range of values one can employ some simple approximate relationships between β and h/r_e, h and r_e being the microprotrusion height and the tip radius, respectively. For instance, for a semiellipsoid on a plane with $7 \leq h/r_e \leq 100$ we have [2.1]

$$\beta = ah/r_e + 1 , \tag{2.1}$$

with $1 \geq a \geq 0.4$; for a cylinder with a spherical tip [2.13]

$$\beta = h/r_e + 2 , \tag{2.2}$$

for a cone of height h with a spherical tip of radius r_e and taper angle θ in the range $5° \leq \theta \leq 10°$ [2.14]

$$\beta = h/2r_e + 5 . \tag{2.3}$$

The micropoint parameters are frequently determined from measurements of prebreakdown field emission current versus applied voltage. For high electric fields and low temperatures (<300K) the current from a point is described by the Fowler-Nordheim equation [2.18]

$$j = \frac{1.54 \cdot 10^{-6} \ E^2}{\phi} \exp\left[\frac{-6.83 \cdot 10^7 \phi^{3/2} \ \theta(y)}{E}\right] , \quad \text{where} \tag{2.4}$$

$$y \simeq 3.8 \cdot 10^{-4} \ \sqrt{E/\phi} , \tag{2.5}$$

$$\theta(y) \simeq 0.956 - 1.06y^2 , \qquad (2.6)$$

j is the current density [A/cm^2], E is the electric field at the emitter tip [V/cm], ϕ is the work function [eV], and $\theta(y)$ is Nordheim's function.

It is commonly assumed that the main contribution to the pre-breakdown current is made by the electron emission from a few most efficiently operating micropoints on the cathode. They provide an effective field enhancement factor β and an effective emitting area S_e. Taking them into account, (2.4) can be reduced to the form

$$\log(i/V^2) = -2.84 \frac{d\phi^{3/2}}{\beta V} + \frac{4.6}{\phi^{1/2}} - 5.81 + \log(\beta^2 S_e/\phi d^2) , \qquad (2.7)$$

where i is the current in the gap. From (2.7) it follows that the graph of $\log(i/V^2) = f(1/V)$ is a straight line with the slope

$$m_{FN} = \frac{d\log(i/V^2)}{d(1/V)} = -2.84 \cdot 10^7 \frac{d\phi^{3/2}}{\beta} , \qquad \text{hence} \qquad (2.8)$$

$$\beta = \frac{2.84 \cdot 10^7 \, d\phi^{3/2}}{m_{FN}} . \qquad (2.9)$$

Thus, by specifying the work function ϕ one can determine the field enhancement factor β from the slope of the straight line and the emitting area S_e from the y intercept.

Some typical curves are shown in Fig.2.1 [2.19]. The β values found from these plots are in the range 10 to 100, and may reach several hundreds. Thus, we have experimental proof that (2.7) holds in the presence of microprotrusions. This is also confirmed by the calculation in [2.20,21] where the β value as determined for the total current is shown to be close to the enhancement factor for the sharpest protrusion. The emitting area is larger than the area of any individual protrusion, but it is less than the total area of the emitters. Hence, it can be seen that the estimates of β and S_e obtained for the total current make it possible to characterize some of the sharpest protrusions present on the cathode surface. Various methods of cathode surface treatment give different β values. One can check the change in the surface properties by measuring Fowler-Nordheim curves [2.22-25].

However, the method described above for checking electrode surfaces, although relatively simple and easy to handle, does involve some uncertainty. To characterize the emissive properties of an individual

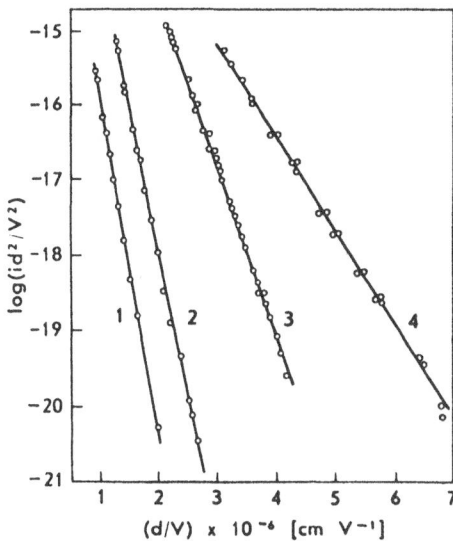

Fig.2.1. Typical Fowler-Nordheim plots of the field emission current between parallel tungsten electrodes for gap spacings 0.005 (*1*), 0.025 (*2*), 0.102 (*3*), and 0.406 cm (*4*) [2.19]

cathode section one needs to know three parameters: the field enhancement factor β, the emitting area S_e, and the work function ϕ. The Fowler-Nordheim graphs enable one to determine only two parameters provided that the third one, typically the work function, may be chosen arbitrarily. Therefore experiments which eliminate this uncertainty are of particular interest. For example, the surface can be studied using alternative methods in order to evaluate the condition of the cathode independently. *Little* et al. [2.26,27] searched for emission sites using a luminescent anode. In an overwhelming majority of cases microprotrusions were found at emission sites. Their profiles were examined with a shadow electron microscope. The field enhancement factor, according to estimates of *Little* et al., reaches 100 and more. Protrusions have been observed in copper, nickel, tungsten, aluminum, tantalum and stainless steel electrodes. Moreover, it has been shown that the emission current is only weakly dependent on an increase in temperature or on the action of light on the cathode. Hence, one can believe that there are no regions with an abnormally low work function.

2.1.3 Effect of Emission from Non-metallic Inclusions

Cox et al. [2.24,28] developed a method for measuring the emission current from a single emission center on a plane electrode. The parameters of the emitting area were determined with the use of the Fowler-

Nordheim equation as well as from measurements of the electron current distribution at the anode. The electrodes were mounted inside the scanning electron microscope (SEM) chamber allowing the site of emission to be shown and photographed in situ. It is shown that in all cases the emission current versus electric field obeys the Fowler-Nordheim equation, although the existence of emission centers is not always due to the presence of surface points. The emission current is also strongly affected by the presence of non-metallic inclusions.

Experimental studies using various techniques [2.25,29-32] have confirmed that when the Fowler-Nordheim equation predicts abnormally high values of β ($\beta > 10^2$), the non-metallic inclusions present on the cathode surface play the part of emission centers.

To explain these experimental findings, *Latham* [2.33] has suggested a model based on the "non-metallic" nature of the prebreakdown emission. It is assumed that a microinclusion possesses good dielectrical properties, so that the external field can penetrate through it to the metal cathode. Then, at a certain field strength, electrons can tunnel through the barrier at the metal-dielectric boundary into the conduction band of the dielectric. These electrons cause the formation of avalanches. The holes produced in this way moved towards the metal surface, aiding the consequent tunnelling of electrons from the metal. The tunnelling electrons, while passing through the dielectric at the bottom of the conduction band, rapidly thermalize. They then enter the region with a strong field near the vacuum-dielectric interface, where they gain energies of up to several electron-volts before passing through the potential barrier.

The model suggested by *Latham* enables a better understanding of the experimental effects observed to accompany the emission from non-metallic inclusions. It also provides an alternative way of showing linearity of the Fowler-Nordheim plots for the emission current from such inclusions. Furthermore, the model explains the abnormally high value of the field enhancement factor β as calculated from these characteristics. It is assumed that the prebreakdown current is provided by the emission of hot electrons obeying the Maxwell-Boltzmann statistics. By analogy with the Richardson-Dashman relationship, the hot electron current is described by the expression

$$j = AT_e^2 \exp(-e\chi/kT_e) , \qquad (2.10)$$

where T_e is the increased electron temperature, χ is the height of the potential barrier at the metal-dielectric interface (the electron affinity of the dielectric). The electron temperature can be found from the relation $e\Delta V = (3/2)kT_e$, hence

10

$$T_e = \frac{2e\Delta V}{3k} \qquad \text{and} \qquad (2.11)$$

$$\Delta V \simeq \frac{1}{\epsilon} \frac{\Delta d}{d} V , \qquad (2.12)$$

where k is Boltzmann's constant, e is the electronic charge, Δd is the thickness of the microinclusions, and ϵ is the relative dielectric constant of the insulator. Substituting (2.11, 12) into (2.10) results in

$$j = A \left(\frac{2e\Delta d}{3k\epsilon d} \right) V^2 \exp\left(- \frac{3ed}{2\Delta d} \frac{\chi}{V} \right) . \qquad (2.13)$$

Equation (2.13) gives a straight line in the coordinates $\log(i/V^2)$ – $1/V$ with the slope

$$m_L = - \frac{3\chi\epsilon d}{2\Delta d} . \qquad (2.14)$$

Substituting (2.14) into (2.9) gives a formal expression for the factor

$$\beta = \frac{1.89 \cdot 10^7 \ \phi^{2/3} \ \Delta d}{\chi\epsilon} . \qquad (2.15)$$

In practice, β values derived from (2.15) are in the range 10^2-10^3. Recent experiments [2.32] with microparticles artificially deposited on the cathode confirmed the existence of high β values.

Thus, when trying to estimate the parameters of the cathode emission centers from the current-voltage characteristics of the prebreakdown current, one should always keep in mind the experimental conditions.

2.2 Vacuum Insulation, Properties and Breakdown

Nowadays there is a great deal of experimental material on the electrical properties of vacuum. The effect of gap width, electrode material and temperature, electrode configuration, methods of treatment of the electrode surface, and vacuum conditions on the voltage hold-off capacity have been investigated. Prebreakdown phenomena and the effect of electrode conditioning by discharges and prebreakdown currents have been studied. It is well known [2.1,3-6] that the nature of the breakdown of vacuum insulation depends both on the type of the applied voltage (ac, dc, or pulsed) and its magnitude, and on the electrode condition. For instance, with a dc voltage, three types of insulation failure can be observed: the appearance of a prebreakdown (dark) current, the occurrence of microdischarges, and the direct breakdown, i.e., a discharge with an abrupt drop in the gap voltage.

2.2.1 Prebreakdown Phenomena

When an increasing voltage is applied across a vacuum gap, a noticeable current appears between the electrodes beginning from a certain level, called the dark or prebreakdown current. In the majority of vacuum breakdown studies, the investigation into the behavior of prebreakdown current was one of the important aspects. For a long time its nature was a subject of much discussion, the reason being that the dependence of the prebreakdown current i on the electric field at the cathode, E_{av}, when presented in a Richardson-Shottky plot [$\log(i) = f(E_{av}^{1/2})$] or in a Fowler-Nordheim plot [$\log(i/E_{av}^2) = f(1/E_{av})$], often yielded straight lines [2.30,31]. This uncertainty was probably due to insufficient purity of the vacuum and the difficulties in controlling such important parameters as the cathode surface microstructure, the properties of the surface layers, the composition of absorbed gases, etc. Today, due to progress in vacuum engineering and to improvements in the techniques for surface examination, a more complete knowledge of the behavior of the prebreakdown current can be achieved. It has been established that the prebreakdown current, under UHV (ultra-high vacuum) conditions with carefully cleaned electrodes, is virtually purely electronic. The ion fraction grows as both the electrodes become more and more contaminated. Under "pure" conditions, the prebreakdown current increases exponentially with the applied voltage, independently of the cathode temperature. This fact indicates unambiguously that it is a field emission current. The emission centers are micropoints on the cathode surface. At an initial increase of the voltage across virgin electrodes, the current, even in ultra-high vacuum, can be due to the electron emission from the cathode areas with a low work function (usually these are regions where non-metallic microinclusions are present [2.25,29], which are destroyed by breakdowns). Introduction of water, acetone, air, oxygen and nitrogen vapors into the discharge chamber up to partial pressures of the order of 10^{-5} to 10^{-4} Pa considerably suppresses the electron emission from the cathode. The reduction in the emission is due to gas adsorption at the electrode surface and an increase of the work function [2.36]. Pre-breakdown current instabilities are caused, to a large extent, by the migration of the adsorbed molecules to the tips of the surface protrusions and their subsequent desorption, which is intensified by ion bombardment of the cathode [2.36,37].

In a number of investigations the material transfer from one electrode to another was observed at voltages much lower than the breakdown voltages. Two aspects of this phenomenon should be pointed out: the matter is transported preferentially from the anode to the cathode, and the matter transport occurs in the form of polyatomic particles. The mechanism of particle formation is not known, but there are some indications that these particles are only weakly attached to the electrode [2.1-3,5,6].

2.2.2 Microdischarges

In some cases the breakdown is preceded by microdischarges, i.e., low-power prebreakdown current pulses of duration up to 10^{-1} s and amplitude up to 0.01 A. They show up on application of a voltage greater than 20 to 30 kV across a vacuum gap with contaminated electrodes ($d \geq 1$ mm). The repetition rate of the microdischarges increases with the applied voltage and decreases as the electrodes become progressively cleaner. In mass-spectroscopy studies it was established that the microdischarge current, in addition to its electron component, contains positive and negative ions of the adsorbed gases. In the literature there is a prevailing opinion that the self-extinction of microdischarges is caused by a reciprocal secondary emission of positive and negative ions, while electrons seem to play no significant role in the appearance and maintenance of microdischarges. The electrons are a product of γ-processes at the cathode [2.1-6]. It should be noted that data exist which indicate the existence of some other kind of microdischarge. It has been observed [2.36,37] that the microdischarges which do not give rise to vacuum gap breakdown are caused by the destruction of local clusters of atoms adsorbed on cathode protrusions which, under the action of the applied electric field, migrate toward the point tips. A considerable portion of energy necessary for surface migration is provided by the cathode. The possibility of the existence of another form of microdischarge has been discussed in [2.38,39]. It is assumed that under the action of field emission current the microdischarges can be associated with explosive destruction of submicroscopic protrusions present on the cathode where the total amount of the protrusion material is insufficient for a breakdown to develop. *Tomaschke* and *Alpert* [2.39] did, in fact, observe, for unclean electrodes, the formation of a "fringe" consisting of a thin "hair" of about 0.1 μm height on the cathode, the modulation level of the prebreakdown current approaching unity. *Tatarinova* [2.40] related the appearance of microdischarges to the occurrence of a discharge in the gas desorbed from the surface of cathode micropores.

2.2.3 The Breakdown Voltage

An important characteristic is the breakdown voltage. It should be noted that the concept of breakdown voltage in itself is, to a certain degree, conditional, since it can be interpreted in a number of different ways (e.g., the voltage that a gap can support at a constant value of dark current or at which a certain number of visible breakdowns occur per unit time) [2.1-6]. Moreover, experimental conditions are often so different (and uncontrollable in many aspects) that it makes sense to consider only some trends to serve as a rough quantitative guide. An experimentally obtained relationship of the breakdown voltage V_{br} on the electrode separation d has the following form:

$$V_{br} = \text{const} \cdot d^{\alpha_1} . \tag{2.16}$$

Maitland [2.41], having processed a great number of experimental data from various sources, found that the most probable value of the parameter α_1 is about 0.6-0.7. The value $\alpha_1 = 1$ would correspond to the breakdown condition at a constant electric field strength $E_{br} = V_{br}/d$ independent of d. This is usually the case only for relatively narrow gaps (d≤1mm) and voltages not above 30kV. For d ≥ 1mm $\alpha_1 \rightarrow$ 0.6-0.7, i.e., as the vacuum gap width is increased, the breakdown electric field E_{br} decreases. Under such conditions, the breakdown of a gap is characterized by an absolute value of breakdown voltage rather than by the E_{br} value. In the literature on vacuum breakdown, this phenomenon is called a "total voltage" effect.

It has been established in numerous experiments that the breakdown voltage is strongly dependent on the electrode material. An observed trend is that V_{br} increases with the electrode material's melting point (with the exception of graphite). There is evidence that V_{br} increases with the mechanical strength of the electrode material [2.42]. The anode material has frequently been shown [2.1-6] to have a domi-

14

nant effect on the breakdown voltage. Most probably this is due to the fact that anode material is preferentially transported across the gap.

The breakdown voltage is significantly affected by the duration of the applied voltage and the rate of voltage rise (Sect.2.3). A number of other factors are known to influence the breakdown voltage, e.g., the conditioning of the gap by successive breakdowns, vacuum conditions, parameters of the electric circuitry to which the gap is connected, and the electrode curvature, area and temperature. These factors are discussed below.

2.3 Kinetics of Vacuum Electrical Breakdown

In a vacuum electrical breakdown the low-conductivity state of the vacuum gap, which is due to the existence of the dark current, changes into the high-conductivity state characteristic of an arc discharge. This transition occurs exceptionally rapidly due to the avalanche-like accumulation of conducting medium in the gap. Unfortunately, in the overwhelming majority of experiments carried out up to the early 1960s the temporal resolution of the experimental equipment was inadequate in coping with the duration of such short processes. This triggered off a great number of sophisticated hypotheses which were not verified by experiments with sufficient temporal and spatial resolution. Therefore, in order to understand the breakdown mechanism it makes sense to concentrate on the studies utilizing pulsed techniques.

2.3.1 Characteristic Times of Breakdown

Owing to the sluggishness of the electrical breakdown development in any medium, the discharge does not develop immediately on application of a voltage pulse with a short rise time, but with a certain time lag commonly named the delay time t_d. Then, for a certain period, usually named the commutation time t_c, the transition to a low-voltage arc discharge occurs. Thus, by investigating the pulsed breakdown one is able to determine the time of the development of the phenomenon. The data from various diagnostic methods enabled us to study the kinetic development. Let us review the data available at the beginning of our study.

The first publication on oscillographic studies of the pulsed vacuum breakdown is thought to be that by *Hull* and *Burger* (1928) [2.43], which revealed that 10^{-6} to 10^{-5} s after the application of an electric field of about $5 \cdot 10^5$ V/cm there occurs a "purely electronic" discharge which turns into an arc discharge after 10^{-7} s. Approximately the same result was initially obtained by *Snoddy* [2.44]. In later publications, information appeared about the use of pulses the length of which was estimated to be 10^{-7} to 10^{-6} s [2.45]. It was established that pulsed breakdown fields are stronger than dc ones, but in the work mentioned above no systematic data have been given.

Considerably more studies, in which microsecond pulses were used, were carried out in the 1950s to the early 1960s [2.46-55]. It was shown that the breakdown voltage increases with the duration of the applied voltage. The pulse to dc breakdown voltage ratio is called the pulse factor K_p. Using pulses with a 12 μs rise time, investigations arrived at values of K_p in the range of 1.3 to 1.6. With pulses of 300 kV and pulse times between 10^{-4} and 10^{-7} s [2.53], K_p was less than 1.7. *Tarasova* and *Kalinin* [2.53] found that there was no dependence of K_p on the gap width. In [2.47, 50, 52] no mention was made of the pulse factor and the dc breakdown voltage.

The above-mentioned studies did not contribute any essential information about the breakdown delay time t_d and the gap voltage decay time. According to the estimates by *Boyle* et al. [2.48], t_d at breakdown across 10^{-4} to 10^{-3} cm gaps amounts to 10^{-7} to 10^{-6} s, the voltage decay time being less than 10^{-8} s. *Slivkov* [2.47] found t_d of about 0.5 μs on the application of a pulse with a 1.5 μs rise time to a 1 mm gap. Approximately the same result was obtained by *Tarasova* and *Kalinin* [2.53]. As a rule, the time t_d was not associated with the pulse factor K_p. *Leader* [2.52] estimated the decay time for a 0.3 mm gap to be of the order of 10^{-7} s. It should be emphasized that there was no attempt to relate the oscillograms for breakdown to the physical phenomena occurring in the vacuum gap.

The first systematic studies of vacuum breakdown time-voltage characteristics for the duration of 10^{-7} to 10^{-5} s of the applied voltage were carried out in the early 1960s by *Kalyatskii* and *Kassirov* [2.57, 58] (commercial vacuum; steel, copper, aluminum, lead, and graphite electrodes). Breakdowns were triggered by varying the onset time of the pulses. For gaps with d = 1.5, 1.0, and 0.5 mm the value of K_p was discovered to increase as the pulse duration was decreased from 3 down to 0.2 μs. *Olendzskaya* and *Salman* [2.59] investigated time-voltage characteristics of the breakdown between plane steel electrodes of 100 mm in diameter with separations d of 5 to 15 mm at pulsed voltages of up to 500 kV. As the rate of voltage rise was increased, the breakdown voltage also increased, while the breakdown delay time decreased (Fig.2.2.). The results of [2.59] agree qualitatively with those of [2.57,58] as well as with recent results [2.60,61].

It should be noted that both pulsed and dc breakdown voltages were low in [2.57,58]. In the authors' opinion, this is due to a comparatively low resistance of the discharge circuitry. In dc breakdown studies the circuitry resistance limiting the discharge current is usually chosen to be ~10^5 Ω, while to obtain rapidly rising short pulses, it is necessary to use low circuit resistances and hence high operating currents. In [2.58] and also in a number of others [2.47,53], the breakdown voltage was shown to increase with circuit resistance. We, however, believe that the low breakdown voltages found in [2.47,53, 58] are due to the combined effect of the high-current discharge regime and poor vacuum conditions in the experiments. The commercial vacuum conditions could also significantly affect the breakdown time characteristics, which makes it difficult to interpret the results from a physical point of view. Moreover, the oscillographic methods themselves provide no essential basis for an unambiguous conclusion about the breakdown mechanism. In view of this, we shall now discuss some results published on studies of the breakdown development where different methods have been used.

It is important to note that the study of the development of vacuum breakdown with the use of steep pulses can only be qualitative, since the processes which contribute to the initiation of breakdown are most likely dependent on the rate of rise of the applied voltage and its peak value. More attractive is the use of rectangular pulses.

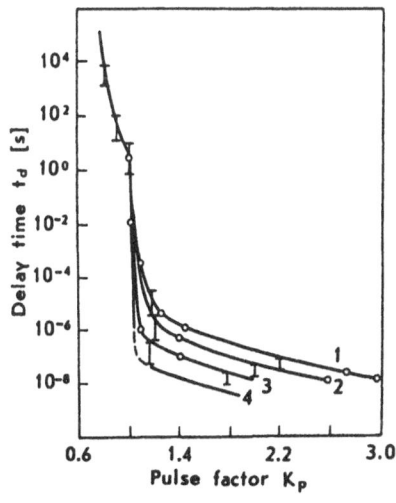

Fig.2.2. Breakdown delay time as a function of the pulse factor for steel electrodes with gap spacing d = 15 (1), 10 (2), 5 (3), and 1 mm (4) [2.59]

As already shown in earlier studies, transient processes in a breakdown across ~1 mm gaps can proceed with a duration of ~10^{-1} s. Therefore, it is preferable to use voltage pulses with much shorter rise times, e.g., 10^{-9} s.

2.3.2 Role of Electrodes in the Development of Breakdown

A characteristic feature of the vacuum electrical breakdown is the fact that the conducting medium necessary for its development can only be supplied to the gap by the electrodes. Therefore, a principal problem is to elucidate the role played by the electrodes in the initiation of the breakdown and in its subsequent development. The formation of vapor medium from the electrode material in the presence of a charged particle flux should be accompanied by the excitation and ionization of the medium, and consequently by light emission of the vapor in the gap. Thus, studying the light emission dynamics is of principal importance.

Snoddy [2.44] is thought to have been the first to make an attempt to scan the light flash at pulsed breakdown with the use of a rapidly rotating mirror. An analysis of photographs of the scanned image of the luminous region for a breakdown between copper electrodes showed that at first an anode spot glow appears in (1 to 4)$\cdot10^{-7}$ s and after that, with a (1 to 2)$\cdot10^{-7}$ s delay, a cathode glow that lasts until the discharge terminates. In *Snoddy's* opinion, the breakdown proceeds in two stages: the "pure electronic discharge" with a duration of less than $5\cdot10^{-7}$ s and following this a low-voltage discharge in the copper vapor. After discharges, craters could be observed on the anode. According to *Snoddy's* estimates, the anode temperature in the discharge reached about 2900 K. *Snoddy* assumed that damage to the anode was caused by the electron beam action.

A more systematic study of pulsed breakdown glows using the same technique was carried out by *Chiles* [2.62]. He established that the glow usually starts at the anode and propagates toward the cathode with an approximately constant velocity. The glow from the cathode side appears before the anode glow arrives at the cathode. In *Chiles'* opinion, the heating of the anode is caused by field emission current, while the cathode spot occurs later as a result of heating the cathode surface by ions extracted from the anode vapor. *Chiles* believed that anode vapor velocities of (5 to 9)$\cdot10^{5}$ cm/s could not be produced by thermal processes only, because such velocities should correspond to a vapor temperature of about $2\cdot10^{5}$ K. In his opinion, the vapor leaves the anode surface in an ionized state, the ions recombine at a distance of $(2 \text{ to } 5)\cdot10^{-2}$ mm from the anode surface and then pass through the gap in the form of excited neutrals. The results of this work were used for several decades as a forcible argument for the validity of the "anode" mechanism of the vacuum breakdown initiation and development.

Thirty years later, the work of *Chatterton* et al. [2.63] appeared, which presented results for spatial patterns of the glow development in the initial stage of a dc breakdown across a 0.075 mm vacuum gap formed by tungsten electrodes of 1 mm in diameter. The experimental technique was as follows. The light that appeared in the gap passed through the optical delay line at the input of the Kerr electrooptical shutter. It was projected from the shutter output onto the input of an image optoelectronic amplifier with a 10^4 gain; the amplified image was photographed. At the same time, a dc voltage was applied to the gap under examination and to the shutter plates. The gap separation was chosen sufficiently small so that the voltage decay time during breakdown, which determines the speed of the Kerr shutter, would be not more than a few nanoseconds. The time from the beginning of the breakdown until the operation of the shutter could be varied in the range of 0 to 30 ns by varying the length of the optical delay line. *Chatterton* et al. established that the glow always occurred at the anode side and expanded into the gap at a velocity of the order of 10^6 cm/s. At the cathode no glow was observed. In the opinion of the present authors, the results obtained are in good agreement with the calculations of *Chatterton* [2.14], which predict that under such experimental conditions the anode should be significantly heated by the action of prebreakdown field emission currents.

Thus, by the time we obtained our experimental evidence that the dominant role was played by cathode processes in vacuum breakdown initiation and development, many results were already available which provided every reason to consider the anode mechanism of breakdown initiation to be plausible.

2.3.3 X-Ray Pulse at Breakdown

In 1897 *Wood* [2.64] found that in a vacuum electrical discharge, X-rays are generated, which, in his opinion, is due to the electron bombardment of the anode. The possibilities for producing high-power pulsed X-ray tubes, based on the vacuum breakdown phenomenon attracted attention from the early 1930s [2.65]. First pictures of an X-ray focus were taken by *Chiles* using a camera-obscura [2.62]. In the mid-1950s, recording the time behavior of the intensity of an X-ray pulse of duration less than 10^{-6} s became feasible [2.65]. By this time, X-ray tubes of a tremendous pulse power had been developed [2.66]; but the mechanism of their operation remained unknown, particularly with respect to the process starting with the electron emission from the cathode up to the state where the gap is filled with a conducting medium.

In the early studies [2.62] as well as in the investigations of the 1960s [2.65] it was commonly accepted that, under the action of a field emission current, heating and vaporization of the anode material occurs and the filling of the gap with the anode vapor partially ionized by electron flow is just the cause for the transition to low-voltage arc discharge. This concept was supported by the fact that there is a delay in the appearance of the cathode glow with respect to the start of the anode-vapor glow. The proof that the currents that cause vaporization of anode material are of field emission nature, is not convincing. This can be inferred from the experiment of *Jamet* with a three-electrode pulse X-ray tube of inverted design (a needle anode is mounted in line with the tube axis) [2.67]. Although the discharge in the tube was initiated by igniting a spark at the cathode, the cathode spot could not be seen on the electro-optical-transducer recording device sooner than 500 to 600 ns after the beginning of the breakdown, when the current in the main gap had reached several hundred amperes. Apparently, this current was not due to the field emission but it was rather the current extracted from the plasma of the igniting spark. Having analyzed the data of [2.67], we came to the conclusion that the absence of an image of the igniting spark glow in several first frames is due to a specific feature of the tube design (the cathode is "screened" by the igniting electrode). After a certain time, when the cathode spots "jump" over onto the igniting electrode, conditions are suitable for a rapid increase of the gap conductance. Just at this moment the gap voltage falls abruptly together with the intensity of the X-rays. *Jamet* [2.67] was able to provide valid arguments for the idea that the transition to arc discharge, under the conditions of his experiment, is due to filling of the gap with conducting medium from both the anode and the cathode sides.

Preferential filling of the gap with anode vapor and a delay of the cathode glow appearance was revealed by *Tsukerman* et al. [2.68] using a photochronograph with a mirror sweep in studying the operation of a super-high-power pulsed X-ray tube of the diode type. However, the question about the nature of the cathode emission in the initial stage of discharge before the appearance of the cathode glow still remained open. In this time interval, according to our estimates, the discharge current could reach several kiloamperes.

In view of the available experimental data on the generation of X-rays at breakdown, we come to the conclusion that the observations discussed above are insufficient to gain an understanding of the temporal mechanism of breakdown.

2.4 Field Electron Emission to Vacuum Breakdown Transition

As already noted, field electron emission can play a decisive role in the occurrence of breakdown between electrodes of large area. In this respect, it is of great interest to discuss the investigations of the vacuum electrical breakdown which treat the cathode as a classical point field electron emitter. Since the emitting area of such a cathode is usually the surface of a single microcrystal, a number of circumstances of principal importance arise which facilitate the study of breakdown phenomena. *Dyke* and co-workers were the first who took advantage of this along with ultra-high-vacuum technology, pulse technique and field emission microscopy [2.51,69,70]. They established that the transition to breakdown is preceded by two characteristic processes. The first is a spontaneous rise of the emission current during the voltage pulse, i.e., at a gradual increase of the amplitude the rectangular pulses of field emission currents are distorted into rising ones. This phenomenon could be repeatedly reproduced in a single test. However, once the voltage was increased by about 1% more, breakdown inevitably occurred, and the current increased by more than two orders of magnitude (up to 50 to 100 A) during a time estimated as $5 \cdot 10^{-8}$ s. The other characteristic phenomenon is the appearance of a bright luminous ring around the emission image of a clean tungsten emitter. For all the points tested with different tip radii and taper angles the same condition was satisfied: breakdown started on reaching a field emission current density in the range $j_{br} = 4 \cdot 10^7$ to $7 \cdot 10^7$ A/cm^2, which corresponded to the electric field $E_0 = 6.5 \cdot 10^7$ to $7 \cdot 10^7$ V/cm at the emitter tip. The breakdown caused melting of the emitter tip, resulting in an increase of the tip radius by nearly two orders of magnitude.

It has been shown convincingliy that the occurrence of the breakdown in this case is not related to the bombardment of the cathode by ions produced at the anode, because the time required for the ions to traverse the interelectrode gap is longer than the voltage pulse duration [2.51,69,70]. Using two emitters placed closely together, the current densities of which differed by a factor of two, it was demonstrated that the bombardment by residual gas ions and by ions formed as a result of the destruction of one of the emitters is not accompanied by a breakdown between the anode and the other emitter, the emissive properties of which remain unchanged.

In this way *Dyke* and co-workers established that the reasons for the occurrence of breakdown in a field electron emitter-anode system are entirely dependent on the processes which take place in the emitter itself, the decisive factor in this case being the current density which is unambiguously related to the electric field strength. *Dyke* and co-workers attributed the discovered phenomena to the effect of heating of the point by the self-emission current. At a sufficiently high current density the Joule heating of a tip results in the transition of pure field emission to field-assisted thermionic emission. Increasing the current promotes a further increase in temperature of the emitter tip and so forth. The process develops in an avalanche-like manner. The setting-in of vaporization of the cathode material in the presence of electron current leads to the production of ions and to the compensation of the electron space charge. This causes a further increase in current and hence the development of breakdown.

A theoretical analysis of the rise in the emitter temperature by *Dyke* et al. [2.69,70] involves Joule's heating as well as thermal losses due to heat conduction and radiation. Despite the fact that neither the calorimetric effect (Nottingham effect) nor a temperature dependence of the cathode material resistivity are taken into account, the analysis predicts that heating the emitter tip up to 3300 K in a pulse requires current densities close to those obtained in experiment. *Gor'kov* et al. [2.71] carried out a more detailed "thermal" calculation for a field emitter taking into account the temperature coefficient of resistance. These calculations showed a good agreement with the experimental results obtained by the present authors.

Nevertheless, the nature of the sharp current burst at a transition to breakdown still remained unclear. This drew the attention of *Alpert* et al. [2.19]. In their opinion, it is difficult to explain such a rapid increase of current by the setting-in of neutralization of the space charge by ions produced by vaporization of the emitter tip, since no abrupt change of the vapor pressure occurs at the melting point of tungsten. Moreover, for different metals, the values of vapor pressure at the melting point may differ by ten orders of magnitude, so the nature of the transition to breakdown should be considerably different for field-emission cathodes made of different materials. Another disadvantage of treatment of the transition to breakdown suggested by *Dyke*'s group is that in essence they have not described the influence of ions on the abrupt current rise.

Studies of principal importance in this respect are those carried out by *Sokol'skaya* and *Fursey* [2.73] who showed experimentally that the abrupt transition to breakdown cannot be explained by the avalanche growth mentioned above. By carefully approaching the critical state, they were able to locate a quasistationary section in the emission current waveform which showed that the spontaneous current rise had almost ceased. A further increase in the current transit time nevertheless resulted in the breakdown. This result made doubtful the hypothesis of emitter disintegration due to the above mentioned avalanche, since there was practically no increase of current before the breakdown. It was suggested that the transition to vacuum breakdown is directly related to a process analogous to electrical breakdown in semiconductors.

The breakdown phenomenon in a vacuum gap with a cathode of the field emitter type can be explained qualitatively only by the processes that develop at the cathode itself. However, as shown above, in experiments with plane electrodes (and even in a system with a tapered cathode as in a pulse X-ray tube [2.68]) the anode glow was always recorded earlier than the cathode one. A thorough theoretical analysis of the electrode processes is necessary in defining the conditions under which one or the other mechanism of vacuum breakdown initiation may play a predominant role.

2.5 Hypotheses on Vacuum Breakdown Initiation

2.5.1 Physical Processes Leading to Vacuum Breakdown

There are several hypotheses on the mechanism of the initiation of vacuum breakdown. They result from a variety of physical processes which occur on the application of a high voltage to the electrodes, but are also due to insufficient understanding of the breakdown in vacuum. The most important physical processes which can lead to a breakdown, deduced from experiment [2.3, 5], are listed below.

Field electron emission. The field emission current, when flowing through a point on the cathode, causes it to heat up, then melt and vaporize, thus in the end leading to breakdown. On the other hand, breakdown can be initiated by electrons accelerated in the vacuum gap, which transfer their energy to a section of the anode surface, causing its heating and vaporization.

Secondary emission of electrons, ions, and photons. In this case, breakdown can occur if the process turns out to be cummulative.

Metal or impurity particles which are weakly attached to electrodes can become detached under the action of the voltage applied across the gap, and on impact with the opposite electrode, they can create the conditions for breakdown (heating and vaporization of a particle, deformation of the electrode surface). Breakdown can occur, when the particle is approaching an electrode, due to the occurrence of an igniting discharge between the electrode and the particle. Moreover, vaporization of the particle in flight under the influence of fast electrons can also create the conditions for a breakdown to occur.

The effect of the electric field ponderomotive forces can result in a change in the electrode surface microrelief (the formation of efficiently operating micropoints on the cathode or the detachment of material pieces from electrodes).

Non-metallic inclusions and films existing on the cathode can become efficiently operating emission centers, and their breakdown can ignite the discharge.

Gas desorption from the electrode surface can promote the occurrence of a gas discharge which provides a medium for the breakdown.

We mainly concentrate on the physical processes that accompany field electron emission, since, in our opinion, all the remaining processes lead in the end to the enhancement of the electric field at the cathode, the intensification of field electron emission, and the breakdown of the gap.

2.5.2 Cathode-Initiated Breakdown

When we began our studies there was no serious objection to the feasibility of the field-electron-emission mechanism of the breakdown of millimeter-wide vacuum gaps. It was considered proven that electron emission occurs from local sites on the cathode surface that have the form of points at which the electric field can be enhanced some tens or even hundreds of times. However, the role played by field emission itself was understood in different ways. Some investigators showed preference to the "anode" initiation mechanism of breakdown by electron microbeams, while others considered the avalanche-like vaporization of cathode micropoints under the action of field emission current to be the reason for the breakdown.

One of the most serious investigations which reported experimental evidence for the "cathode"-initiated breakdown mechanism is an article by *Alpert* et al. [2.19]. First of all, they turned their attention to an early investigation of *Boyle* et al. [2.48] carried out under UHV conditions of clean tungsten electrodes separated by $d = 10^{-4}$ to 10^{-3} cm. Proceeding from the results of this work, it was shown that if the micropoint electric field enhancement factor β is taken into account, the local breakdown field $E_0 = \beta V_{br}/d$ appears to be independent of the gap spacing and equal to the E_0 value found in [2.69, 70] for the case of breakdown in an electrode system with a tungsten field emitter.

To verify the hypothesis for gaps wider than those used in [2.48] *Alpert* and his co-workers carried out experiments with tungsten electrodes of large area under "pure" vacuum conditions having gap spacings ranging from 0.005 to 0.635 cm. For each gap width, after conditioning by breakdowns, the prebreakdown current-voltage characteristic was measured. A plot of $\ln(i/V^2)$ against $f(1/V)$ revealed a straight line (Fig.2.1) with a slope of $2.96 \cdot 10^7 \phi^{3/2}$ V/E. From the slope the breakdown electric field E_0 could be determined provided that V_{br} was known. *Alpert* and his co-workers have reported on the correlation of their data with data found in the literature (Fig.2.3). From this figure it is clear that as the gap spacing is varied by five orders of magnitude, the breakdown electric field E_0 remains constant within the limits of experimental accurracy and that it is independent of electrode geometry. For tungsten $E_0 = (6.5\pm1)\cdot10^7$ V/cm. Using similar techniques, the E_0 values for a variety of electrode materials were determined in [2.74, 75], observing that E_0 in the range of $(5$ to $11)\cdot10^7$ V/cm.

A very important consequence is that the breakdown voltage is directly related to the prebreakdown field emission current, and can be predicted from its measurements. *Alpert* et al. [2.19] believed that the nonlinear function $V_{br}(d)$ (the "total voltage" effect) revealed by many investigators is due to the increase of the field enhancement factor β

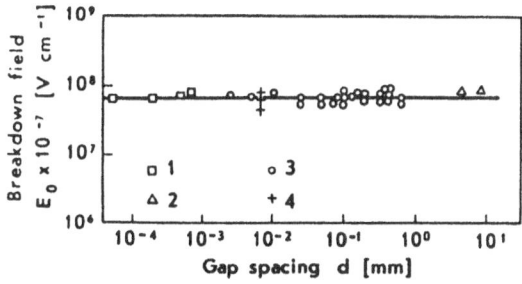

Fig.2.3. Local breakdown field at the cathode as a function of the gap spacing for tungsten electrodes. The data are taken from [2.48] (*1*), [2.69] (*2*), [2.72] for cathode radii of 1.75 cm (*3*) and 0.0035 cm (*4*).

with the gap spacing d. A significant increase of β in the range of d = 10^{-4} to 10^{-3} cm could be well understood, since the gap width is comparable with the height of micropoints on the cathode, while for higher d values the increase of β is attributed to effects of macrogeometry (d becomes comparable with the electrode dimensions or with their radii of curvature). Similar assumptions have also been made in a number of other works [2.3, 6, 14, 16, 17]. Thus under "pure" experimental conditions, the hypothesis of cathode-initiated breakdown has enabled a satisfactory correlation of the prebreakdown current with the conditions of breakdown initiation. Upon reaching E_0 at the point tips, an avalanche-like destruction of the emitters and a release of conducting medium occurs, which in the end results in breakdown [2.19, 75].

2.5.3 Anode-Initiated Breakdown

In our opinion, the main shortcoming of the hypothesis of cathode-initiated breakdown developed in the mid 1960s is that the proponants of this hypothesis confined themselves to an explanation of the initial explosive-like release of conducting medium, but gave no interpretation for the process of breakdown development itself until its transition to a vacuum arc. In a system with a field-emission cathode [2.51, 73] the breakdown development is quite feasible, since the discharge current flowing through a single point could result in release of a large amount of conducting medium. On the other hand, in a system with plane electrodes, where micropoints of extremely small volume were observed on the cathode surface, it was rather difficult to explain the transition to arc only by the release of conducting medium due to the explosion of the micropoints which were present at the start. Moreover, the nature of the secondary processes occurring on the cathode, which could provide the release of additional conducting medium, was entirely obscure. This led to a widespread hypothesis of an anode initiation in which the field electron emission was considered to provide the electron supply only. The following experimentally established facts formed the background for the popularity of this hypothesis:

    (i) the preferential transfer of the anode material to the cathode in the breakdown stage;

    (ii) the dominant influence of the anode material on the breakdown voltage; and

    (iii) the primary appearance of the glow at the anode.

Semyonov [2.76] was the first to suggest in 1929 that breakdown depends on heating the anode by electrons followed by gas and vapor release. The ideas of this hypothesis were carried further in [2.48, 77].

The hypothesis was developed most completely in the early 1960s by *Maitland* [2.36] who suggested that breakdown of the gap starts as the power density deposited to the anode by the field-emission electron beam reaches some critical value. To support his conclusions *Maitland* reported the results of his own work on the anode erosion at breakdown of vacuum gaps by rectangular voltage pulses with the duration $t_p = 4.5$ μs. He observed that after discharge a great number (10^4-10^5) of craters about 5 μm in diameter appear on the anode surface. In his opinion, each crater is formed as a result of the action of a single electron microbeam, all 10^4-10^5 beams acting simultaneously and each carrying a current of the order of 10^{-4} A. The bombardment of the anode by such beams with a power density of $\simeq 10^8$ W/cm^2 proceeds during a time interval of 10^{-10}-10^{-8} s, after which evaporation of the the anode material starts and a crater forms. After a certain time the heat evolution at the anode decreases abruptly because of dissipation of electron energy in the vaporization of the anode. The vapor fills up the gap, ions form and electron avalanches occur in it, the gap conductance increases and the discharge changes into an arc. *Maitland* attributed the light flash at the anode preceding the appearance of the cathode glow to the primary appearance of the ionized vapor cloud at the anode.

2.5.4 Comparison between Cathode and Anode Mechanisms for Breakdown Initiation

Thus, by the mid 1960s, the opinion had been formed that the breakdown mechanisms based on the leading role of field electron emission could be confirmed by experimental findings. Attempts were made to find the limits of both the cathode and the anode mechanisms [2.3, 14, 78, 79].

A generalized theoretical analysis of the vacuum breakdown initiated by field electron emission was first made by *Chatterton* [2.14]. He suggested that the attainment of a melting point at one of the electrodes should be considered as a criterion for breakdown initiation. The temperatures of the emitter tip and the anode spot were calculated for a steady-state case, the heat conductivity and the electrical resistivity being assumed to be independent of temperature. The cathode heat balance involves two possible sources each taken separately: Joule's heating and Nottingham's heating. It is also assumed that the removal of heat is due to the heat conduction. For materials with a melting point T_m which is lower than the inversion temperature

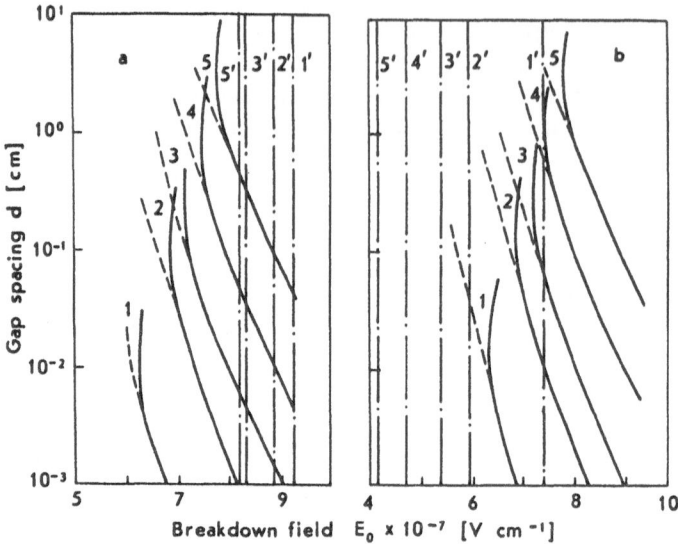

Fig.2.4a,b. Theoretical variation of the breakdown field with gap spacing for conical (a) and cylindrical (b) protrusions with $r_e = 5\cdot10^{-6}$ cm and $\beta = 10$ (1,1'), 40 (2,2'), 70 (3,3'), 150 (4,4'), and 300 (5,5'). Solid and dash curves correspond to anode initiated breakdown; dash-dot curves to cathode initiated breakdown [2.14]

$$T^* = 5.32\cdot10^{-5}\ E\phi^{-1/2}, \tag{2.17}$$

at which the Nottingham effect changes from heating to cooling, the emissive heating can play a decisive part. For materials with $T_m > T^*$ emissive cooling of the micropoint takes place, but its contribution is low compared to the Joule heating. The relations between the cathode temperature T_c and the field emission current density (and, hence, the electric field E_0) have been obtained. In order to find the anode temperature the beam radius R_a at the anode is determined by solving the problem of motion of a boundary electron from an emitting micropoint of radius r_e. This gives the relation between the anode temperature and the local field E_0 at the cathode. Shown in Fig.2.4 are the calculated E_0 values versus gap spacing, at which T_m is attained in a steady-state case in a localized region of the cathode or anode surface (tungsten). The vertical lines indicate the critical cathode microfield necessary for the heating of micropoints that is independent of the gap width and is only defined by the value of the field enhancement factor β. The curves represent the dependence of E_0 on the gap width for the same cathode protrusions, for the case of heating the anode up to T_m. The nonlinearity of the curves is due to a nonlinearity in the relation-

28

ship between the penetration depth of electrons into the anode and their energy, as determined by the Widdington formula.

Having analyzed his results, *Chatterton* concluded that there apparently exists a certain value of field enhancement factor β^* for every concrete condition, at which the probability for the attainment of the melting point becomes equal for both electrodes. The β^* value can be found by, respectively, plotting the functions $\beta(E_{0min})$ and $\beta(E_0)$ for anode and cathode from Fig.2.4. The intersection point of these curves determines β^*. For $\beta < \beta^*$ primary melting occurs on the anode, while for $\beta > \beta^*$ the cathode protrusion melts first. Since the microprotrusions observed on the cathode are mainly conical, and since studying the breakdown glow has shown that the anode vapor appears earlier than that of the cathode, *Chatterton* arrived at the conclusion that there is good agreement between theory and experiment, which confirms the anode-initiated breakdown mechanism. At the same time, with the β values found in the experiments of *Alpert* et al. [2.72], the theoretical values of E_0 based on both the "cathode" and the "anode" mechanisms are not much different from those obtained experimentally. Therefore, in *Chatterton's* opinion, more detailed studies under carefully controlled conditions are necessary in order to reveal the range of validity for each of the two mechanisms.

In a theoretical study by *Charbonnier* et al. [2.78], the non-stationary nature of thermal processes at the anode is taken into account. Two characteristic time constants of heat propagation are involved: $t_1 = x^2 \rho c/(2\lambda)$ and $t_2 = R_a^2 \rho c/(2\lambda)$, where ρ, c and λ are the density, the heat capacity, and the heat conductivity of the anode material, respectively; x is the penetration depth of electrons into the anode. If the duration t_p of the voltage pulse applied to the gap is much smaller than t_1 and hence t_2, the heat conduction loss could be neglected and the anode temperature could be assumed to be the same all over the volume $\pi R_a^2 x$. If $t_p \gg t_2$, the temperature in the center of the anode spot can readily be found from the equation for the stationary heat conduction for a surface heat source. The temperature of the cathode tip is determined by taking into account the Joule heat release and the Nottingham heat. It is assumed that the criterion for the occurrence of breakdown is the attainment on either electrode of the temperature T_{-2} corresponding to the saturated vapor pressure of the order of 10^{-2} Pa, although no explanation is offered why increasing the pressure up to such a value should result in breakdown. In [2.78] a rather simple expression for β^* was given by

$$\beta^* = \sqrt{r_e j_{br} E_0/(4q_a)} \,, \tag{2.18}$$

29

where the numerator and the denominator designate the power densities required to reach T_{-2} at the cathode and anode, respectively. It is obvious that as t_p decreases, q_a increases, while β^* decreases.

Charbounier et al. [2.79] verified their theorical predictions under exceptionally pure experimental conditions using dc and pulsed (2-$100\,\mu s$) voltages of amplitude up to 30 kV for tungsten, molybdenum, copper, and aluminum cathodes. With $\beta < \beta^*$ they observed, using a telescope device, localized "hot" spots on the anode just before breakdown, the breakdown beginning when a power density of $q_a = (3$ to $4)\cdot 10^4$ W/cm² was reached at the anode. With $\beta > \beta^*$ they found cathode "hot" spots. In this case, the calculated value of the field emission current necessary for heating the protrusion tip up to T_{-2} agreed satisfactorily with the experimental value of the prebreakdown current.[1]

Some remarks should be made on the works described in [2.71, 72]: (1) the equation of the cathode heat balance was solved for a stationary case; (2) the attainment of T_{-2} at one of the electrodes does not yet explain how the breakdown develops subsequently; (3) the experiments (d<1 mm) were carried out with insufficient temporal resolution, so it can not be stated unambiguously which of the electrodes is responsible for the evolution of conducting medium at a given stage of breakdown; (4) the experimentally determined power density at the anode responsible for the occurrence of the primary anode instability turns out to be one or two orders of magnitude less than the power density necessary for the attainment of the melting point on the anode measured in other investigations [2.80, 81], although the experimental conditions in these two studies were not significantly different.

Utzumi [2.80] considered whether the thermal instability primarily appears at the cathode or at the anode. The same topic has been pursued in several works by *Slivkov* [2.1, 3], who not only made progress on the principles which govern the attainment at the electrode of some critical temperature, but also made an attempt to ascertain how the anode thermal instability could lead to the subsequent transition to breakdown. He based his consideration on the BKG (Boyle-Kisliuk-Germer) hypothesis [2.48] that the spontaneous increase of the field emission current from a cathode protrusion might be due to the growing space charge of positive ions produced by ionization of the anode material vapor. Moreover, he assumed that the anode vapor density decreases proportionally to the squared distance from the anode and that the velocity of positive ions moving toward the cathode is deter-

---

[1] For the case of aluminum electrodes, Bennette et al. [2.79] predicted theoretically and showed experimentally the realization of the particle detachment mechanism.

mined not only by the electric field but also by the recharging of the ions in the anode vapor. Later on, *Slivkov* considered another possibility for the "anode" mechanism that manifests itself, which is associated with the occurrence of a self-maintained discharge in the vapor cloud formed near the anode as a result of vaporization of the anode material under the action of the field electron beam [2.3]. He also took into account the possibility of heating the emitting protrusion by positive ion bombardment as well as the possibility of stretching a protrusion from a molten spot on the anode under the action of electric field.

A comparison of the field emission currents from the protrusion for the three above-mentioned mechanisms shows that the difference between them is less than an order of magnitude. If one takes into account that the calculations are done using a number of simplifying assumptions and bearing in mind that we assume only a stationary case, it is easy to understand how difficult it is to estimate which of the mechanisms considered is the most probable. *Slivkov* [2.3] showed that the difficulties are doubled due to the fact that the experimentally measured prebreakdown currents represent a sum of currents from a great number of emission centers and, still more important, that they are obtained using time-lag methods which are incapable of recording currents 10^{-7} to 10^{-8} s prior to breakdown. This is a particularly important time interval when prebreakdown phenomena become more intense.

2.5.5 Microparticle-Initiated Breakdown

For a long time, one of the most widely spread hypotheses of the mechanism of vacuum breakdown initiation was based on the impact of a particle on the electrode. Such an idea was first suggested in 1952 by *Cranberg* [2.82], who assumed that for the breakdown to occur, a certain minimum energy should be released at the impact of a particle on the electrode. By the beginning of the 1960s this hypothesis had been generally accepted due to a number of reasons: (a) it was based on simple physical considerations as well as on the experimental observation that microparticles of electrode material are present on the electrode surfaces; (b) for the first time a hypothesis provided the explanation of the "total voltage" effect; (c) the experiments with artificial introduction of microparticles into the gap provided the function $V_{br}(d)$ which was the same as the one predicted by the hypothesis ($\alpha_1 \simeq 0.5$); d) the change of V_{br} depending on the electrode material could be explained. The greatest difficulties appeared when it was attempted to apply this hypothesis to the pulsed breakdown of vacuum gaps with a pulse duration less than the possible microparticle transit time. In addition, it was shown experimentally [2.83] that the particles artificially introduced into the gap were able to initiate breakdown when starting from the anode and impacting onto the cathode.

Davis and *Biondi* [2.84, 85] developed the idea of the initiating role of a liquid metal drop which has broken off from the anode. The main point of their model is the following:

The electrons emitted by a microprotrusion on the cathode, being accelerated in the gap, transfer their energy to a region of the anode surface and heat it up to a certain critical temperature. This temperature is sufficient for the electric field force to be in excess of the force of mechanical cohesion of the anode material. As a result, a protrusion starts growing on the anode surface. As the heat removal reduces, the heating and the further growth of the protrusion intensify, a molten charged particle breaks away and moves towards the cathode. In its flight the particle is bombarded by electrons from the same cathode microprotrusion and evaporates as a result of heating. Electron avalanches which develop in the vapor form a plasma cloud and cause an increase of the prebreakdown current, thus resulting in breakdown.

Having analyzed the possibilities of the avalanche-like electron multiplication in the vapor, *Davies* and *Biondi* came to the conclusion that the conditions for the development of vapor breakdown could be satisfied only in a 10^{-3} cm thick vapor layer adjoining the microparticle, i.e., the breakdown in the vapor cloud resulted in the formation of a plasmoid with a characteristic size of 10^{-3} cm. According to the estimates of *Davis* and *Biondi*, the time of formation of such a plasma cloud was 10^{-8} s, i.e., it was much less than the time of flight and the time of evaporation of the microparticle. No detailed consideration was given to the mechanism of the ionization processes occurring in the gap after the formation of microplasma because of the absence of experimental data relating to the moment when the gap voltage started to decay.

Summarizing the discussion on the hypotheses about the mechanism of vacuum breakdown initiation, we should point out that their common principal limitation is that all these hypotheses were restricted to the nature of the event of primary evolution of conducting (or vapor) medium and did not treat the relation between this event and the further development of the breakdown. Since this relation was not understood, there was in fact no clear understanding about the secondary processes taking place in the subsequent stages of breakdown when conditions for heavy discharge currents could be created. In order to convince ourselves of this statement, let us review the works on the vacuum breakdown spark stage available by the early 1960s.

2.6 Spark Stage of Vacuum Breakdown

As a result of the vacuum breakdown initiation the discharge goes over into a new phase - the spark stage, the high-current, high-voltage phase of a self-maintained discharge with a falling current-voltage characteristic, which finally develops into a vacuum arc if the power source provides a current which exceeds the arc threshold current.

First attempts to analyze the process of current rise and voltage fall in a vacuum spark discharge (in low-pressure X-ray tubes) were made in 1938 by *Steenbeck* [2.86]. He assumed the existence of an efficient electron source on the cathode. The electrons, being accelerated in the gap, ionize the residual gas. The accumulation of positive ions results in a decrease of the electronic space charge. The electron current increases until the mean volume densities of electron and ion charges become equal. Measurements of *Kingdon* and *Tanis* [2.87] indicated a disagreement between the theory [2.86] and experiment for the range of low pressures. Beginning from a certain pressure (about $4 \cdot 10^{-2}$ Pa) the measured currents turned out to be higher than those predicted theoretically and were, in fact, pressure independent.

An attempt to modify the theory [2.86] was undertaken by *Fünfer* [2.88], who took a more complete account of some elementary processes. According to *Fünfer's* data , at a pressure of $4 \cdot 10^{-3}$ Pa, a discharge capacitor capacitance C = $5 \cdot 10^3$ pF, and V_0 = 40 kV, the predicted pulse current amplitude and duration were, respectively, 2 A and 30 μs, while the experimental values were 50 A and 1.5 μs. Thus, it was not possible to interpret the observed experimental values of the spark discharge current amplitude and pulse duration of the basis on the current enhancement due to gas pressure.

Flynn [2.89] suggested a theory for the current rise in a gap with an igniting electrode mounting on the cathode, according to which the accumulation of conducting medium in the gap occurred due to the motion of the cathode plasma generated by the igniting device rather than due to the ionization of residual gas. It was assumed that electrons were extracted from the front surface of the expanding plasma cloud and accelerated towards the anode, the electron current density being self-space-charge limited. Practically all the interelectrode voltage was supposed to be applied between the plasma front and the anode. As a rule, experimentally obtained current and voltage waveforms, current pulse amplitude and duration, were in good agreement with the calculated parameters. It was shown experimentally that the current pulse duration increased, in accordance with the theoretical predictions, in direct proportion to the vacuum gap width d and was independent of residual gas pressure in the range of $5 \cdot 10^{-3}$ to $3 \cdot 10^{-1}$ Pa [2.89].

M. and A. Goldman [2.77] calculated the growth of the vacuum spark current proceeding from the assumption that the current plasma was generated by the anode and that the current growth was associated with shortening of the gap by the anode plasma. It was inherently assumed that the cathode had an unlimited emissivity, although the mechanism of this emission was not specified. Slivkov [2.1,3] analyzed the development of an auxiliarily ignited spark discharge and showed that when the gap is shortened by the anode plasma its amount should be much greater. His experiment on determining the minimum ignition energy stored at both the cathode and the anode showed that in the latter case the energy required should be four orders of magnitude greater than in the former one [2.1,3]. However, if one takes into account that in a number of experiments on vacuum breakdown the anode plasma generation and motion was observed to be preferential, it could be well understood why the hypotheses of Slivkov and Goldman were more popular than that of Flynn [2.89].

It should be noted that all the results of spark discharge studies available in the literature were obtained for tubes with artificial ignition at the cathode or anode. It has turned out that the cathode initiation of spark discharge can be achieved more easily. A high plasma density in the igniting spark is of great importance in this case. However, this still does not prove that in a spontaneous vacuum breakdown the plasma generated at the cathode fills up the gap, since the amount of cathode plasma capable of initiating the breakdown is several orders of magnitude less than that necessary to shorten the gap. Therefore, in Slivkov's opinion [2.1,3], in a spontaneous breakdown without ignition, the necessary amount of plasma is generated as a result of some secondary processes. Using the data of Chiles [2.62], he arrived at the conclusion that the plasma shortening the gap is initiated only at the anode.

The development of the pulsed breakdown spark stage due to explosive vaporization of material from local anode sites irradiated by electron microbeams was also considered in an early work by Kassirov and Mesyats [2.90].

In conclusion, at the time when the research program was formulated, the mechanism of the phenomena occurring in the spark stage of vacuum breakdown was not understood unambiguously. The situation which prevailed allowed rather forcible arguments to be drawn in favor of either of the hypotheses. Therefore, the problem was to study in more detail the process of filling of the gap with conducting medium and to elucidate the leading mechanism of the electron emission from the cathode which provides for the existence of a high-current discharge.

2.7 The Discharge Arc Stage. The Cathode Spot

Electrical breakdown of a vacuum gap with a sufficient power of the supply source terminates in the form of an arc. The vacuum breakdown phenomenon is used, among others, to initiate the vacuum arc [2.91-93]. The arc form of discharge is characterized by the following physical features: (i) a low discharge voltage comparable in magnitude with the ionization potential of the cathode material; (ii) a high current density in the region of the discharge-cathode bridging; (iii) high concentrations of charged particles in the near-cathode region. In the case of vacuum arc, the cathode is a single source to deliver particles into the gap. Thus, the peculiarity of the arc form is completely determined by the processes occurring at the cathode and in the region adjoining its surface, i.e., in a so-called "cathode spot".

Since our aim was to investigate the vacuum breakdown phenomenon including the vacuum arc stage, the two following points should be emphasized. First, the study of this phenomenon had to answer unambiguously the following questions: at which breakdown stage the cathode spot arises and what are the processes responsible for its formation, how it develops, and when the discharge does acquire the features of a true arc discharge. As shown above, the available literature on the experimental and the theoretical studies of the development of spark discharge (without artificial ignition at the cathode) and vacuum breakdown did not answer these questions completely. Secondly, the investigation into the development of vacuum breakdown resulted in some findings about the processes which occur in the cathode spot. As will be shown below, these investigations made it possible to draw a number of fundamental conclusions concerning the nonstationary nature of the cathode spot processes.

In this section we do not attempt to elucidate entirely the status of cathode spot investigations. We only briefly discuss the physical properties of vacuum arc cathode spots and the theoretical ideas on the cathode spot processes, restricting ourselves to the data of [2.91-96].

2.7.1 Physical Properties of the Cathode Spot

The cathode spot is a small brightly luminous region at the cathode surface through which the passage of current between the cathode and the arc column occurs. The cathode surface within the spot is heated up to a high temperature which can be in excess of the boiling point. Above the cathode surface, inside the spot, a near-cathode plasma exists. The main practical difficulties in experimentally studying cathode spots are associated with their rapid movement (with a velocity up to the order of 10^4 cm/s) and the extremely small size of the spot ($\sim 10^{-3}$ cm). In recent years it was possible to establish [2.91,94] that the rapidly moving cathode spots occur initially on all metal cathodes (the first-type spots). They exist independently of each other and cause insignificant erosion on the surface (according to the estimates of [2.94], the erosion rate for copper is about $4.5 \cdot 10^{-7}$ g/C). *Lyubimov* and *Rakhovsky* [2.94] believed that the main volume of experimental information gained in recent years seems to be related to this type of spot. Some time after the discharge ignition (50-500 μs at $i \simeq 70$-100 A), individual spots also appear (the second-type spots). They are larger and "move" much more slowy than the first-type spots. The number of second-type spots increases with time and in the end only these spots remain on the cathode surface. They produce significantly more pronounced erosion ($\sim 10^{-4}$ g/C). In the opinion of *Rakhovsky* and coworkers [2.92,94], the first- and second-type spots are different in nature. However, we believe that the question of the distinction in the physical nature of these types of spot needs further investigation.

A cathode spot consists of several actively emitting sections. *Kesaev* [2.91] introduced the concept of the cathode spot cells, i.e., individual microspots which conduct current and reproduce the vapor-plasma medium. *Lyubimov* and *Rakhovsky* [2.94] gave a more accurate description of the cathode spot fragments. At a certain current exceeding a threshold value, some cells (fragments) are observed to disappear spontaneously and new ones form. As a result, the cathode spot moves in a chaotic manner. A number of attempts are known to have been made to explain the division and the chaotic motion of the cathode spots [2.91,94,95]. One of the hypotheses is based on the so-called "principle of maximum magnetic field" [2.91].

It has been established experimentally that a vacuum arc discharge is intrinsically unstable, therefore there always exists a non-zero probability of it being spontaneously extinguished. That arcs undergo a spontaneous extinction is confirmed by the appearance of spikes on voltage waveforms. These are indications of regenerating voltage pulses and subsequent breakdowns, faciliated by the existence of decaying plasma and heated cathode microareas. The average arcing time can be much in excess of the lifetime of a single cell (fragment) of the cathode spot.

One of the most debatable points in the studies of the arc phenomenon is the value of the cathode spot current density. The background for such a discussion is well understood, since the spot current density determines the mechanism of emission and energy evolution at the cathode as well as the nature of the near-cathode processes. A direct measurement of the cathode spot current density is impossible. Therefore, it has to be judged by some secondary effects which accompany the cathode spot operation, i.e., by the plasma glow or by the erosion trace left by the spot on the cathode surface. *Kesaev* [2.91] has noted truly that all these methods "in their very principle are capable of predicting only overestimated values of the emitting area to be determined". It should be noted that most of the investigators measured current density in the range 10^5 to 10^8 A/cm.

As a result of the cathode spot operation the cathode surface is eroded. Based on the results of some experiments with film cathodes, *Keasev* [2.91] showed that all the processes responsible for the erosion occur in the 1 to 2 μm thick metal surface layer. The erosion of the cathode due to temperature influences, as characterized by the occurrence of molten craters about 1 μm in size, was finally established in the period from the late 1960s to the early 1970s [2.97,98]. These results referred mostly to pulsed discharges where the arc channel was artificially displaced under the action of an external magnetic field. A great deal of this work was devoted to investigations into the mechanism of cathode erosion in a vacuum arc discharge. The erosion rate and the phase composition of erosion products were measured for a great variety of metals. The information gained in cathode erosion studies during 1963 to 1975 was given in a review by *Daalder* [2.99]. The data for copper cathodes, which were investigated in most detail, are taken from this work and reproduced in Fig. 2.5. The erosion rate was often considered to be a function of the discharge current, but as can be seen from the figure, the measurements covering a rather wide range of currents give somewhat contradictory results. This suggests that the erosion rate seems to be dependent not only on the arc current, but also on some other parameters.

For a long time it was believed that the main part of the erosion products is in the form of vapor [2.92]. Experience has shown, however [2.99,100,102,110,111], that this state should be called plasma rather than vapor as the cathode spot generates plasma rather than vapor, and ions are the mass background of the former. *Kimblin* [2.102] introduced the concept of "ion erosion" of the cathode. Moreover, *Tuma* et al. have

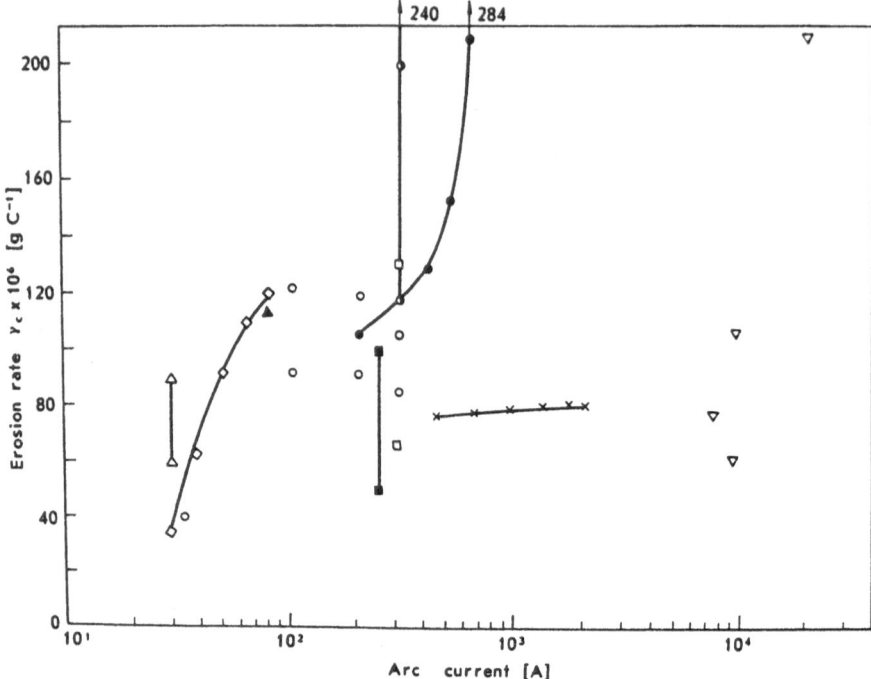

Fig.2.5. Data on the erosion rate of copper cathodes obtained by a number of investigators [2.99]: □ [2.100], ◐ [2.101], ▲ [2.102], • [2.103], ◇ [2.104], ○ [2.105], ▽ [2.106], △ [2.107], x [2.108], ■ [2.109]

shown that the true vapor fraction is not above 1% in the case of a copper arc, the rest of the erosion products consisting of ions and drops [2.111].

Tanberg (1930) [2.112] was the first to arrive at the conclusion that in the operation of the vacuum arc on a copper cathode a vapor jet ejects from the region of the cathode spot with a velocity of the order of 10^6 cm/s. According to *Tanberg*'s measurements, the specific force acting on the cathode is about $1.7 \cdot 10^{-4}$ N/A. Since then, the velocity of plasma jets in arcs operating on various cathodes has been measured using many different techniques. These measurements have confirmed *Tanberg*'s estimates: A cathode does indeed eject a plasma jet with a velocity of about $1 \cdot 10^6$ cm/s. Most surprising in this respect is the fact that this velocity is several times greater than the velocity which ions can acquire under the action of the applied voltage and it is directed opposite to the external field. Several attempts to explain this paradox are known [2.94, 100, 110].

One more finding is of interest, viz. the presence in the cathodic jets of multiply-charged ions of cathode material. It has been esta-

38

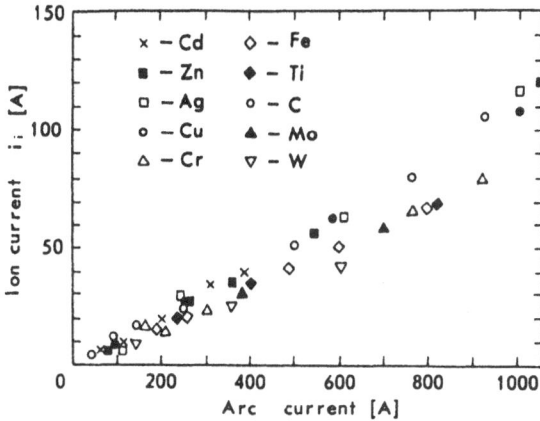

Fig.2.6. Ion current amplitude as a function of the arc current for various cathode materials [2.102]

blished [2.100,110] that in the case of low-boiling-point metals (Cd, Zn, Pb) singly-charged ions prevail and the ionization degree of the cathodic jet is 15-25%. For higher boiling points of cathode material the degree of ionization in the jets is greater (up to 30-50% for Cu, Al, Ni, and Ag and up to 70-90% for C, Mo, and W. Doubly-, triply-, and even quadruply-charged ions exist in the plasma. However, it has been established [2.102] that the plasma jet ion current is independent of the cathode material (Fig.2.6.) and amounts to 8-10% of the arc current. *Daalder* [2.99] has shown that for every cathode material the "ion erosion rate" is a constant independent of current, while the total erosion rate of the cathode increases with the current due to an increase of the neutral fraction (mostly the drop fraction).

Highly important parameters characterizing a cathode spot are the concentration and the temperature of particles in the spot. Because of extremely small cathode spot dimensions, the measurement of these parameters is mainly limited to the spectral method [2.91,92]. Measurements for copper cathodes [2.94] have shown that in the spots of the first type the charged particle concentration is about $6 \cdot 10^{17}$ cm^{-3} and it weakly changes with the distance from the cathode. Electron temperature measurements based on the determination of the ratio of ion line strengths gave the value $T_e \simeq 2$ eV, while those using the ratio of atomic line strengths resulted in $T_e \simeq 0.8$ eV. *Lyubimov* and *Rakhovsky* [2.94] expressed the opinion that such a disagreement can be attributed to the fact that the measurements in the two cases were at different regions of the cathode spot (internal and external, respectively). Almost the same values of the cathode spot plasma parameters are obtained for spots of the second type [2.94].

2.7.2 Cathode Spot Models

The presence of the near cathode potential drop is one of the principal factors leading to the existence of a cathode spot. It is commonly considered that the high cathode temperature is maintained due to the energy deposited by ions coming from the near-cathode plasma, although at the current density of the order of 10^8 A/cm² and higher, Joule's heating of the current contraction region at the cathode can provide a significant contribution [2.91, 113-118]. The energy released at the cathode is spent to compensate for the losses due to heat conduction, evaporation of atoms and electrons (electron emission) and radiation. It is assumed [2.91-96] that there is a positive-space-charge layer of thickness of 10^{-5} to 10^{-6} cm (of the order of the ion mean free path) over the metal surface across which the cathode voltage drops (10-20 V) and thus a high electric field is created at the cathode surface (10^6 to 10^7 V/cm). Furthermore, it is postulated that the ions accelerated in the space charge layer transfer to the cathode surface not only their kinetic energy, but also their neutralization energy. Due to the strong electric field and high cathode temperature, electron emission occurs from the cathode simultaneously with evaporation of atoms. The electrons, having acquired an energy in the near-cathode layer, spend it to ionize the vapor and heat the plasma being formed. The electron flow relaxation zone is $\sim 10^{-4}$ cm in size. It is assumed that the electric field in this zone is much lower than in the near-cathode layer. The beam relaxation zone extends into the arc column.

Lee and *Greenwood* [2.119] were the first to make an attempt not only to evaluate the spot current density, but by using a set of equations describing all the processes in the spot, to also determine its parameters (the electron current density and fraction, the electric field at the cathode, the cathode temperature, and the spot radius) for known values of the cathode voltage drop, work function, heat conductivity and evaporation constants in *Dashman*'s formula. Further development of the ideas of *Lee* and *Greenwood* is associated with the works of *Goloveiko* [2.120], *Beilis* et al. [2.121, 122] and *Ecker* [2.96, 123, 124]. In [2.114, 115] the transport processes in the near-cathode region are described using all the equations of motion and energy balance for heavy particles and electrons in the three-component cathode plasma state. To describe the cathode processes a generalized equation of emission and an equation of energy balance at the cathode are used. The system of equations is closed, so that the spot current density, the electron current fraction, the plasma electron temperature and density can be determined, whereas the spot current, the cathode drop, the erosion rate, as well as the cathode material constants are known. Calculations were carried out for a quasi-stationary cathode spot with a current varying from 100 to 500 A. The results have shown that as the current is increased in this range, the current density, the ionization degree, the electron temperature, and the cathode temperature fall from $1.4 \cdot 10^5$ down to $2.2 \cdot 10^4$ A/cm^2, from 0.2 down to 0.7, from 1.6 down to 1.2 eV, and from 3800 down to 3600 K, respectively.

The next step in this direction was made by *Nemchinsky* [2.125, 126] who was able to make the system of equations not only mathematically but also physically closed by relating the arc steady state to the minimum voltage drop across it. The calculation he carried out for a stationary arc on a copper cathode gave the following results: current density $5 \cdot 10^5$ A/cm^2, cathode temperature 3940 K, plasma temperature $1.6 \cdot 10^4$ K, ion current fraction 0.36, cathode drop 12.3 V, arc total voltage drop 20.1 V. From the calculation it also followed that the minimum voltage drop (19.5V) corresponded to a current of 140 A.

Another approach to the determination of the main cathode spot parameters has been developed in the work of *Ecker* [2.96, 123, 124, 131-133]. He proceeded from the fact that the exact values of the cathode spot parameters as well as the exact equations governing these parameters are unknown. Therefore, he used only limiting inequalities, which make it possible to outline the range of parameters at which the cathode spot can exist. Thus, the electron emission current density should be sufficient to provide for the near-cathode plasma ionization, that is $j_e V_c > j_i V_i$ (j_i is the ion current density, V_c is the ionization potential), but it should not be greater than the space-charge-limited current density, i.e., $j_e \leq j_i (M_i/M_e)^{1/2}$. By analogy, either all neutrals evaporating from the cathode are ionized and then return back to the cathode or no single particle is ionized. In addition, an approximate equality describing the energy balance at the cathode is used. All the above-described conditions are expressed in the form of characteristic curves on the cathode temperature (T_c) - cathode current density (j) coordinates. *Ecker* therefore plotted so-called "diagrams of existence" for a cathode spot. It turned out, for instance, that for a spot resting on a smooth surface of a copper cathode two models can be applied: the "0"-mode with $T_c \simeq 4000$ K and $j \simeq 10^6$ A/cm^2 and the "1"-mode with $T_c \simeq 5000$ K and $j \simeq 10^8$ A/cm^2. The "0"-mode can exist only if the spot current is in excess of a certain minimum value, which principally agrees with well-known experimental findings [2.91]. It is possible to believe that the "0"-mode corresponds to slowly moving spots, for which a quasi-stationary description of processes is well suitable. Meanwhile, there is a reasonable agreement between the spot parameters obtained in the calculation [2.121, 122, 125, 126] and those predicted by *Ecker* using the "0"-mode. As to the "1"-mode, the applicability of the nonstationary description is not so obvious in this case. For example, in [2.127, 128] it was shown that the existence of stationary spots on cathodes made of refractory metals is impossible. Therefore, the rapidly moving cathode spots should be described using nonstationary models. A nonstationary cathode model based on the stepwise movement of the cathode foot corresponding to the velocity of the spot motion was first analyzed on the basis of his existence diagram by *Ecker* [2.134]. It was found that essentially only the energy characteristics are affected by the motion of the spot. This results in an increase of the critical currents in comparison to those for the stationary spot.

Other nonstationary models of the cathode spot suppose, as a rule, that some explosive processes occur at the cathode. Thus, in the 1950s up to the 1960s the opinion was repeatedly expressed that the cathode spot can be treated as an element of an exploding wire [2.113, 115-117]. *Fursey* and *Vorontsov-Vel'yaminov* [2.129] suggested a qualitative model of the vacuum arc initiation based on the explosion of micropoints available on the cathode. Later on, the "explosive" model of the cathode spot was discussed in [2.94, 129-133]. However, a considerable advance in the understanding of the nonstationary spot processes became possible only due to a success in the study of explosive electron emission, and the cathode and near-cathode plasma phenomena it causes.

3. Experimental Equipment and Techniques

Analysis of results of the study of the electrical breakdown and spark discharge phenomena in vacuum, presented in Chap.2, enables one to draw some conclusions about the requirements that should be satisfied in order for the experimental equipment and techniques to be capable of providing unambiguous inference about the mechanism and nature of the processes under investigation. The first requirement is a high temporal resolution ($\sim 10^{-9}$s). It is obvious that methods of high-voltage nanosecond-pulse engineering should be employed. The second requirement is a high spatial resolution of the equipment. Indeed, as shown above, despite a large electrode surface the failure of vacuum insulation and the development of breakdown are related to some processes occurring in localized regions, often 10^{-5} to 10^{-4}cm in diameter. Therefore, optical and SEM methods for examining electrode surfaces can be expected to give relevant information. To obtain information on the kinetics of the breakdown development it is necessary that the spatial resolution available in recording the light emission accompanying a breakdown should be much greater than the electrode separation. In fact, the spatial resolution should be better than 10^{-2} cm. The third requirement is to use methods of high sensitivity, since the problem involves the recording of various types of low-voltage signals. This requirement necessitates the use of highly sensible recorders. The fourth requirement, an excellent vacuum together with clean electrodes, could be fulfilled with the use of UHV equipment and special methods for the preparation of electrodes.

3.1 Electrical Measurement Techniques

The electrical methods of investigation include measuring the pre-breakdown current, the breakdown voltage (both dc and pulsed), as well as the breakdown delay time and the commutation time by recording the discharge current and voltage waveforms. To measure these quantities, high-voltage dc and pulse generators and corresponding voltage- and current-recording techniques are necessary. We outline only a few of the types of equipment widely used in electrical measurements.

3.1.1 High-Voltage, Nanosecond Pulse Generators

To produce rectangular high-voltage pulses with a high rise rate, pulse voltage generators are used in which an accumulator cell discharges into a load through a fast switch. The discharge circuit has low stray parameters [3.1,2]. The role of the switch is most commonly played by a spark gap with compressed gas as working medium. High-voltage rf cables are used as pulse-forming and transmitting circuit elements.

A circuit producing high-voltage nanosecond pulses, which involves a line transmitting a pulse to the "test" vacuum gap, is represented in Fig.3.1. The pulse-forming line L_1 is charged through a charge resistor R_0 from a high-voltage stabilized dc source up to a voltage V_0. On operation of the spark gap SG, a rectangular voltage pulse with amplitude $V_0/2$ and duration $t_p = 2l/v_w$ is passed through the transmission line L_2 (l is the length of the pulse-forming line, the velocity of the electromagnetic wave propagating along the pulse-forming line is defined as $v_w = c/(\epsilon\mu)^{1/2}$, where c is the velocity of light, ϵ and μ are the permittivity and permeability of the line dielectric, respectively). In this case, the current pulse amplitude i_{max} is $V_0/(2\rho_w)$, where ρ_w is the wave resistance of the line. The duration of the current rise with no stray inductance and capacitance in the discharge circuit is determined by the commutation time of the spark gap.

The above-described generator scheme has an essential disadvantage. Since the resistance of the vacuum gap changes during breakdown, part of the energy of the travelling wave reflects from the gap. As a result of repeated reflections from the end of line L_1 it is not a single voltage pulse which acts on the gap, but several reflected ones. Therefore this scheme was mainly used in studying time characteristics of the breakdown phenomenon. However, in a number of investigations, particularly those concerned with electrode erosion, it was necessary to protect the gap from reflected pulses. It was also important that the discharge could be interrupted in any stage of its development. We

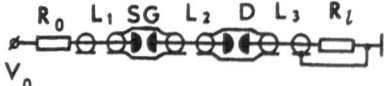

$$R_0 \quad L_1 \; SG \; L_2 \quad D \; L_3 \quad R_l$$

V_0

Fig.3.1. Rectangular-pulse forming circuit

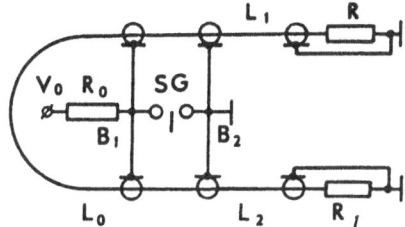

Fig.3.2. Rectangular-pulse forming circuit without reflection from the load

therefore designed a generator which can produce, without reflection, a pulse into a load with any value of the resistance. The schematic principle used was described in [3.3]

A modification of this scheme using coaxial cables is shown in Fig.3.2. A pulse-forming line L_0 is connected through the transmission line L_1 to the resistor R with wave resistance ρ and through the transmission line L_2 to the load R_1, in general unmatched with the cable's wave resistance. The cable braiding at the ends of L_0 and the braidings at the beginnings of L_1 and L_2 are connected to the electrodes of the spark gap SG with the low-industance busbars B_1 and B_2. Busbar B_1 and the cable braiding of L_0 are connected through the charge resistor R_0 to the high-voltage source; busbar B_2, together with the cable braidings of L_1 and L_2, are grounded. On operation of spark gap SG a voltage wave with amplitude $V_0/2$ propagates through the lines L_1 and L_2. The incident and reflected waves from the load are absorbed in the resistor R. A variation of the resistance of the load R_1 does not give rise to repeat voltage pulses on this load. If a vacuum gap is used as a load resistor, then as long as $R_1 \gg \rho_w$ the voltage gap amplitude $V \simeq V_0$. The duration of pulses produced by such a generator is determined by the length 1 of the pulse-forming line L_0 as $t_p = 1/v_w$.

The design of our generator was described in detail elsewhere [3.4]. The controlled spark gap is installed as an individual element of the generator switching unit, thus permitting one to renew it or to perform preventive overhaul. Typical voltage waveforms of the pulses formed by the generator are illustrated in Fig.3.3. It can be seen that

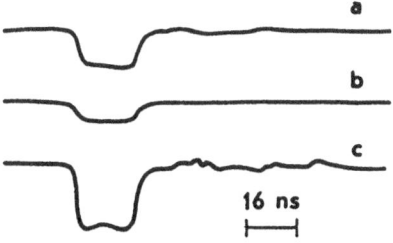

16 ns

Fig.3.3. Resistive-load voltage pulse waveforms: a) $R_e = 620\Omega$, $V_0 = 45\,kV$; b) 75Ω, $35\,kV$; c) 2.5Ω, $30\,kV$

the load resistance has practically no effect on the voltage waveform.

Besides the two 10-50 kV generators described, some other modifications based on the same principles were used, no special description is required here.

In experiments with voltages up to 200 kV, a small Arkad'yev-Marx generator was used. The output capacitance of the generator is $4 \cdot 10^{-10}$ F. The VPG (voltage pulse generator) was loaded directly by the vacuum gap. The pulse rise time was 3 to 5 ns, while the voltage pulse duration amounted to 30-80 ns depending on the load resistance. The pulse could be shortened. To do this, a cut-off spark gap was connected parallel to the vacuum gap. It operated, like the spark gaps of the VPG, at a nitrogen pressure of (3 to 5)$\cdot 10^5$ Pa.

3.1.2 Current and Voltage Pulse Recording

Nanosecond electrical signals were recorded using wide-band capacitive voltage dividers, low-ohmic current shunts, Rogowski coils, and high-speed oscilloscopes. A capacitive voltage divider has a coaxial configuration, so that its wave resistance is equal to the cable's wave resistance. A capacitor formed by the walls of the container and the internal metal-foil cylinder isolated from the container by a polyethylene film has a higher capacitance (a low-voltage arm), while a capacitor formed by the foil and the coaxial central conductor has a lower capacitance (a high-voltage arm). The divider has an upper-frequency limit of ~10^{-9} Hz and a lower-frequency limit of ~10^7 Hz. Thus it could be used for recording of signals with a duration not exceeding a few tens of nanoseconds and was mainly employed in the triggering of oscilloscopes.

Most widely used is the resistive voltage divider in which the cable's wave resistance is the high-voltage arm resistance, and low-ohmic resistors connected in parallel represent the low-voltage arm. The resistors are mounted on a metal disc and inserted into the break in the cable braiding. Owing to the availability of the "return wire" in the form of a metal disc, the divider self-inductance is significantly lowered.

When properly manufactured and mounted, such a divider allows undistorted pulses with a rise time smaller than 1 ns to be recorded. The resistive voltage divider connected in a line with distributed parameters is in fact a reverse current shunt. In our experiments, such resistive shunts were mounted directly in the discharge vacuum chamber to lower the stray parameters of the measuring circuit.

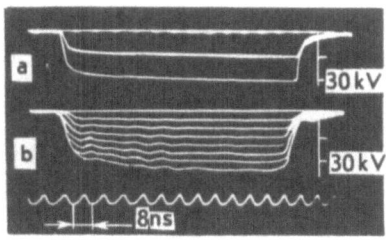

Fig.3.4a,b. Voltage (current) pulse waveforms recorded using the reverse current shunt (a) and the Rogowski coil (b)

When studying spark discharges with high values of di/dt there appeared to be a need to measure the current at localized regions on the cathode. It was important in the measurment process itself to produce no additional potential difference between these cathode regions. The same problem occurred in studying the vacuum arc discharge. From this point of view the Rogowski coil is suitable mostly as a current pickup. It is known (Chap.2 and Sect.3.5) that the Rogowski coil ensures that the input signal, within certain limits, is proportional to the measured current, but it introduces a negligible resistance (inversely proportional to the square of the number of the coil turns) into the current-carring circuit. Used in the experiment were miniature Rogowski coils with toroidal magnetic cores. The instrument resistance (input resistance of the oscilloscope) was 75 Ω, hence the resistance introduced in the circuit amounted to $7.5 \cdot 10^{-3}$ Ω with the number of turns equal to 100. Figure 3.4 shows some typical pulse voltage (current) waveforms recorded under the same conditions using a resistive reverse-current shunt and a Rogowski coil. To record the single nanosecond current and voltage pulses oscilloscopes with bandwidths of 2000, 350, and 50 MHz and sensitivies of 5, 10, and 10^{-3} V/mm, respectively, were used.

3.2 Diagnostics of the Radiation that Accompanies Breakdown

Vacuum electrical breakdown is known to be accompanied by visible, UV and X-radiation. In the preceding chapter we reviewed some studies of the spatial and temporal development of the glow, which had been carried out prior to our experiment, and pointed out their shortcomings. These investigations might be successful, if the newest equipment intended for the study of fast processes could be employed to detect the discharge radiation. First of all, let us mention the achievements in the development of photoelectric diagnostics of weaky luminous and fast-moving plasmoids that appeared by the mid 1960s as a result of the advent of high-speed electro-optical image converters (EOIC), destined for operation with nanosecond exposures [3.6], and nanosecond photomultiplier tubes (PMT) [3.7]. We shall briefly describe below the equipment for observation of the light emission and the radiation accompanying breakdown.

3.2.1 Electro-optical Recording of the Light Emission

An experimental set-up for electro-optical investigations in single-shot operation is shown in Fig.3.5 [3.8]. A voltage pulse with parameters V_0 $\leq$ 40 kV, t_p = 100 ns, and $t_1 \simeq$ 1 ns is produced by the three-channel generator 1. The pulse is fed to the vacuum chamber 2, to the anode of EOIC 3, and to the cut-off spark gap through the cables L_1, L_2, and L_3, respectively. The time of the frame exposure on EOIC 3 is determined by the difference in length of the cables L_2 and L_3, and amounts to about 3 ns. The time interval between the pulse arrivals at the vacuum chamber 2 and EOIC 3 can be changed by varying the length of cable L_1. Different phases of breakdown can thus be recorded. An image of the vacuum gap is projected onto the photocathode of EOIC 3, and from the screen of EOIC 3 it was projected onto the photocathode of the input stage of the image intensifier 4. The EOIC voltage pulse and breakdown current waveforms were recorded at the same time as the optical picture. The spread of the EOIC voltage pulse and the accuracy of synchronizing the breakdown oscillogram were about 0.1 ns. EOIC 3 was specially designed for nanosecond

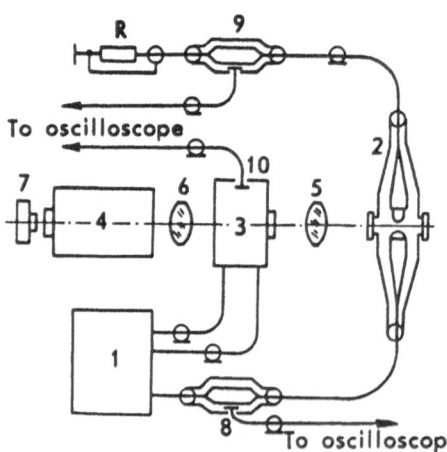

Fig.3.5. An experimental set-up to study light emission at breakdown with the use of a single-shot operating EOIC: (1) three-channel generator of high-voltage nanosecond pulses; (2) vacuum discharge chamber; (3) shutter EOIC; (4) electro-optical image intensifier; ($5,6$) lenses; (7) camera; (8-10) capacitive voltage dividers

operation. It was placed in a special metal package in order that the EOIC could be matched to cable L_2. A cut-off spark gap was built in the package to enable the anode to be deenergized during 0.5-0.7 ns. A multistage EOIC with magnetic focusing was used as an image intensifier *4*. To abate the noise of the intensifier first stages, we used the pulse power supply technique, in which an image is "remembered" and transmitted to the next stage. The total resolution of the electro-optical section was 12 line pairs per mm at an overall intensity gain of the order of 10^5.

A serious shortcoming of the photorecording system described is that only one shot per discharge can be obtained. In the early 1970s, super-high-speed electro-optical systems were developed in the USSR, in which specially designed grid-controlled EOIC are used. When operating in the frame regime (4 frames per screen), they provide a minimum exposure time as low as 10^{-8} s with an interval of $5 \cdot 10^{-8}$ s. Such systems could be useful in studying the breakdown of gaps a few centimeters wide. However, generally speaking, the use of the multiframe regime is awkward because the sharp rise of the glow brightness during breakdown would result in overloading of the EOIC at the last frames. Therefore photorecording of the glow which accompanies the breakdown in a continuous operation is of interest. The above-mentioned devices are also intended for operation in the frame sweeping regime, but the study of the breakdown of short vacuum gaps is awkward due to the existence of a dead time of the control circuits ($\geq 10^{-8}$s). This is a basic shortcoming if one attempts to study fast processes during a dc breakdown. With pulsed breakdown there is always an opportunity to anticipate the occurrence of breakdown and to "prepare" the recording instruments for this moment. Such an opportunity does not arise for dc breakdown. One has to time the start of the recording from the moment of appearance of the phenomenon under investigation. A great deal of information is lost, particularly at initial stages, due to the inevitable existence of the apparatus dead time.

Since the current rise at dc breakdown appears to be close to linear, the idea has occurred to use EOIC in the continuous-operation regime by controlling the timing plates with breakdown current [3.9]. In doing so, the synchronization would be accomplished automatically with the sweep duration always corresponding to the breakdown current rise time. This idea has been accomplished by using a single-chamber EOIC with electrostatic focusing. It has a pair of deflecting plates and a grid near the cathode which produces a cut-off in the EOIC with a voltage of 20 to 200 V [3.9]. A concept uitilized in the design of such a recording system is shown in Fig.3.6. The increasing

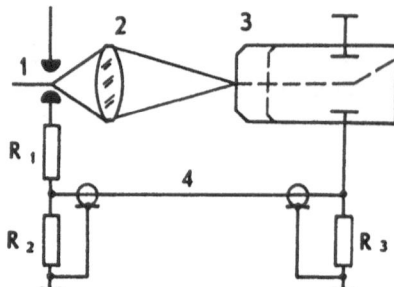

Fig.3.6. Schematic arrangement for recording light emission at dc breakdown: (1) vacuum gap; (2) lens; (3) EOIC, (4) cable (R_1-R_2): resistive voltage divider; R_3: matching resistor)

discharge current produces a voltage drop across the resistor R_2, which is applied to a deflecting plate of EOIC through cable 4. The cable is loaded by the resistor R_3 with the resistance ρ_w so that there is no reflection. Provided that $t_1 = t_2 + t_3$ (t_1 being the time of electromagnetic wave propagation through the cable, t_2 the time required for light to reach the photocathode, and t_3 the time of flight of electrons through the EOIC), the image is swept across the screen in synchronism with the current rise in the gap. This condition is satisfied when the cable length is 40 cm. Since we were only interested in the fast stage of breakdown until the onset of an arc, the EOIC was cut off via the control grid by a pulse from the shutter unit at the onset of arcing. The shutter unit was started by applying a voltage from the divider, which comprised a trigger circuit with a secondary-emission tube producing a voltage pulse with an amplitude of 100 V and a duration of more than 5 μs. These values were sufficient for a reliable cut-off of the EOIC. The shortcoming of this method is a significant image distortion due to the asymmetry of the deflection plate potentials in the EOIC. Nevertheless, this method appeared to be quite suitable to obtain a qualitative pattern of the breakdown development, particularly at dc voltages. With a preliminary calibration of the screen against the deflection voltage on the EOIC plates (with a step of 100V) it was possible by photographing a narrow slit to judge the space-time pattern of the glow development at breakdown within an accuracy of about 30%. Prior to each series of breakdowns, a photographic image of the gap was taken. The breakdown current and the deflection voltage waveforms were always recorded simultaneously with the electro-optical image of the light emission at breakdown.

3.2.2 Photoelectrical Recording of the Light Emission

When investigating the mechanism of vacuum breakdown there is a principal problem in establishing the place and the time of appearance

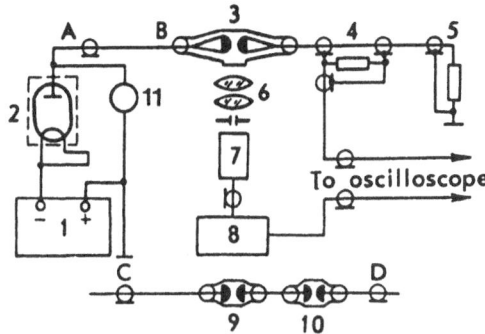

Fig.3.7. Schematic arrangement for recording the moment light appears in the gap at breakdown: (*1*) high-voltage source; (*2*) high-voltage, high-vacuum hot-cathode rectifier tube; (*3*) vacuum discharge chamber; (*4*) current shunt; (*5*) matching resistor; (*6*) lenses; (*7*) photoelectron multiplier; (*8*) wide-band amplifier; (*9*) spark gap; (*10*) spark gap; (*11*) kilovoltmeter

of glow in the gap. A nanosecond photomultiplier tube with a high current gain is useful in solving this difficulty [3.10]. A schematic representation of the experimental facility to indicate the instant of the appearance of light in any region of the gap using PMT is given in Fig.3.7. The glow is recorded by the photomultiplier tube *7*. By increasing the PMT supply voltage up to 2 kV as well as selecting an optimum dinode potential, and by connecting low-inductive capacitors between the last-stage dinodes, the current gain factor of the PMT could be increased up to 10^8 at a dark current of 10^{-6} A. The temporal resolution could be shortened to 3 ns. A signal from the PMT was applied to the input of the amplifier *8* with a bandwidth of 150 MHz and a voltage gain of 500. A signal from the amplifier output could be applied to the input of the oscillosocpe *9* instead of a signal from shunt *4*. The overall gain of the section appeared to be so high that the oscilloscope could record the pulses due to the emission of single electrons form the photocathode, and the amplitude of these weak pulses tended to saturate the amplifier. The interelectrode gap was projected onto the PMT photocathode using a pair of fast lenses *6*. A 0.1 mm wide spectrometry slit was positioned directly at the photocathode perpendicularly to the electrode axis. The slit enabled any section of the gap to be projected onto the photocathode. Precautions were taken to eliminate the possibility that the light scattered in the discharge chamber would get into the slit. The optical system collected about 3% of the light that appeared in the vacuum gap. Calculations have shown that for a single operating photoelectron to appear during a period equal to the PMT temporal resolution, it would be required to produce about $5 \cdot 10^2$ photons with wavelengths in the range detectable by this instrument (the quantum yield of the photocathode is about 10%)

Synchronizing the PMT signal with the onset of the breakdown current rise was accomplished by a reference spark which appeared on breaking down of an air microgap (about 0.1 mm) substituted for the "test" vacuum gap. In this case the breakdown delay time was 0.3 ns and the glow in the microgap could be assumed to appear simultaneously with the beginning of the spark current rise. The length of the signal-carrying cable was selected so that the light from the reference spark, initiated by the PMT signal, arrived at the oscilloscope input simultaneously with the beginning of the spark current rise. Thus in processing the oscillograms there was no need to take into account the signal delay time in both the PMT and the amplifier section. The interpretation of oscillograms was reduced to measuring the time interval between the onset of the PMT signal and the beginning of the breakdown current rise. By scanning the gap with the slit it was possible to obtain data on the kinetics of motion of the luminous medium.

3.2.3 Spectral Investigation of the Discharge Plasma Radiation

In a number of shots a monochromator or a spectrometer was placed between the vacuum chamber and the PMT with the aim of recording the appearance of any spectral lines of the discharge radiation.

Since the development of vacuum electrical breakdown is inseparably linked with the filling of the gap with a plasma conducting medium, the diagnosis of this plasma is of principal importance for understanding of the physics of discharge. A peculiarity of the plasma produced at breakdown is its spatial limitation, strong dynamic behavior, high gradients of density and temperature within the plasma, and its close attachment to the electrodes. Under these conditions, spectral methods are most suitable and informative for the diagnosis of plasma. These methods are well developed at present ([Ref.3.5, Chaps.5-7] and [3.11, 12]). The relative strengths of spectral lines is used to determine the plasma's electron temperature in the state of local thermo-dynamical equilibrium. To determine the electron concentration the method of Stark's line broadening as well as measuring the absolute strength of bremsstrahlung and recombination continua are employed.

Spectral studies were usually carried out with electrodes (mainly cathodes) of a limited area in order to localize the site of plasma generation. In some cases the whole discharge radiation spectrum was preliminarily obtained using a spectrograph. The availability of this information permitted any line to be easily identified with the use of a photoelectric spectrometer.

The photoelectric spectrometer involved a focusing optical section, a monochromator, a photomultiplier tube and a sensitive wide-band oscilloscope. If necessary, a wide-band preamplifier was placed between PMT and oscilloscope. The general scheme of the device differed from that shown in Fig.3.7 only in that a light-beam monochromator was placed between the optical section 6 and PMT 7. Signal synchronization was also accomplished employing a reference air spark. The light emitted by the plasma entered the monochromator input slit which was usually 100 μm wide. The spatial scanning was accomplished both along and across the vacuum gap. Special precautions were taken to screen the measuring circuits from any pulsed electromagnetic stray pick-ups.

In conclusion, it should be noted that at present laser methods have become widely employed in the diagnostics of the plasma generated in high-current vacuum diodes and self-magnetically-insulated vacuum lines [3.13]. The first results obtained by using these laser methods for the diagnosis of the vacuum breakdown plasma are described below.

3.2.4 X-Radiation Recording

For the recording of the X-radiation at breakdown we have used an arrangement schematically different only slightly from that shown in Fig.3.7 [3.14]. A photomultiplier was placed behind a plastic scintillator. The de-excitation time of the scintillator was less than 5 ns. The PMT output signal was applied directly to a high-speed oscilloscope input without amplification, since the X-ray pulse intensity was sufficiently high. Overloading the PMT (the maximum output current of the PMT should not be in excess of 0.8 A) can result in distortion of the pulse shape. The X-radiation was attenuated with the use of a set of metal foils placed in front of the scintillator.

A specific feature of the recording system is that both the breakdown current and the X-ray pulse were recorded on the same oscilloscope display (for a single sweep). In order to do this, the signal cable lengths were so chosen that the signal from PMT (through the connector of the matching load) and the breakdown current signal (through the input connector) would arrive at the deflecting system of the oscilloscope tube with a time difference greater than the X-ray pulse duration. We were able to use this method of recording because the X-radiation pulse turned out to be comparable in duration with the time of breakdown current rise, and there were no stray electromagnetic pick-ups. This method has the advantages that two well-synchronized signals may be recorded by a single-beam oscilloscope and that the data processing is simple.

3.3 Vacuum Equipment

The methods used to create high-vacuum conditions as well as the pressure of residual gases strongly affect the properties of the electrode surfaces and the conditions of breakdown of a vacuum gap. With this fact taken into account we have carried out experiments under both "commercial" and ultrahigh vacuum conditions. On the whole, three types of facilities were used. They differed in the means of producing and supporting the vacuum, and in the design of the discharge chamber. The "commercial" vacuum ($\simeq 2 \cdot 10^{-3}$ Pa) was obtained by conventional techniques. We have designed a special vacuum discharge chamber. In order to prevent distortion of pulses with nanosecond rise times, the chamber was manufactured as a coaxial conductor with a 75 Ω wave resistance. The chamber input and output were wave-resistance matched with special cone adapters which met the requirements of high-voltage insulation [3.1]. Insulators and seals were made up, respectively, of acrylic plastic and vacuum rubber. The chamber was supplied with branch pipes for pumping, electrode mounting and optical observations.

To eliminate the effect of organic contaminations on breakdown and to raise the degree of vacuum an installation was designed and made up for producing UHV oil-free vacuum. The vacuum system of the installation consists of three parts: fore vacuum, high vacuum, and ultrahigh vacuum. The forevacuum part consists of two zeolite sorption pumps. As a pressure of about 0.1 Pa is attained, the pumps are cut off and a getter pump starts to pump the system down to the pressure of $\simeq 7 \cdot 10^{-5}$ Pa. The UHV part made of stainless steel consists of a gate, a connecting tube, an operation chamber, and a getter ion pump cooled with liquid nitrogen; the seals are made of copper. After heating the discharge chamber for 20 hours and filling up the pump with liquid nitrogen, a pressure of about $4 \cdot 10^{-6}$ Pa may be steadily maintained.

The schematic of the vacuum chamber has been given elsewhere [3.1]. The glass insulators are brazed with connectors made of Kovar. Conical connectors made of acrylic plastic have to be removed whilst the chamber is being heated up. The chamber can hold a dc voltage of 60 kV and allow pulses with a 1 ns rise time to pass without distortion. A sylphon and a positioning mechanism are installed to vary the electrode separation.

Experience has shown that during the study of fast processes there is often no need to carry out experiments under high-vacuum conditions. Thus a vacuum installation which was simple to operate was designed, which can be called "intermediate" in view of the vacuum conditions it produces. It consists of a backing mechanical pump, a freezing trap, gates, a titanium getter ion pump, and a discharge chamber. The titanium pump is initated from a separate high-vacuum system. Since the volume to be pumped is not large, when the pressure of $6.5 \cdot 10^{-1}$ Pa is obtained, the baking pump is cut off and the titanium pump opened. The latter begins its operation without upsets and produces in the chamber space a vacuum of $6 \cdot 10^{-5}$ Pa in about 30 min. Oil vapor can get into the discharge chamber only during the process of pumping, which lasts a few minutes. To reduce the evolution of organic contaminants during the operation, all the seals in the high-vacuum part and the insulators in the discharge chamber have been made of Teflon.

The installation described was used particularly in a study of dc breakdown. Since in that experiment there was a significant power deposited onto the anode by the passage of the prebreakdown currents (up to hundreds of watts), the anode holder had to be cooled with water. To eliminate the leakage of charge through the water, the water was fed through long rubber hoses. A voltage divider with a resistance of 75 Ω, which served as a load, was built into the anode insultor. It consisted of three arms each made of a set of resistors connected in parallel. The voltages derived from the divider were intended to control the electro-optical image converter and to record the breakdown current.

A further description of vacuum conditions and of the discharge chamber units for a specific experiment will be given in the appropriate context.

3.4 Preparation and Examination of Electrode Surfaces

The choice of the form of electrodes, their dimensions and material, the technique of electrode preparation and the methods for analyzing the surface depend on the particular objective to be achieved. We discuss the most general aspects.

`Many electrode materials were used in vacuum breakdown studies, but most of the data were obtained for copper, molybdenum, tungsten, aluminum, lead, and graphite. Most frequently used in experiment were electrodes with a hemispherical operating surface (a hemisphere radius of 10 mm). This electrode geometry ensured a sufficiently uniform field in the gap ($d \leq 2mm$) and concentrated the sites of breakdown occurrence within a small region around the electrode axis. Both of these criteria are particularly important in optical investigations. In some cases the electrodes were pre-annealed in vacuum at a high temperature, and after annealing, the operating electrode surface was electropolished.

In the study described above a great deal of experiments were carried out with points having the geometry typical for field-electron emitters. The points were manufactured by electrochemical pickling of wire rods 0.1-0.8 mm in diameter [3.15]. This pickling permitted points with various taper angles θ and initial emitter tip radii r_θ = 0.1-0.5 μm to be produced. In some cases, to obtain better-quality surfaces, electrochemical smoothing of points (mostly made of copper) was carried out after electrochemical pickling. For some experiments, points (needles) with a tip radius R_e = 10-30 μm were required. Such radii were obtained by a complemental pickling of sharp points in polishing solution with their subsequent electropolishing.

The electrode surface was examined by optical as well as electron microscopy before and after exposing it to discharges. Examination of erosion traces left by discharges on the anode was made mostly in the optical microscope. The total magnification (of the microscope and the photographic enlarger) was determined by means of an objective micrometer. The error in determining the erosion trace dimensions depends on the total magnification, and it was never in excess of ± 1 μm throughout the experiment.

Points were photographed using a shadow optical microscope. Because of light diffraction at the tips of sharp points the magnification was not in excess of 200. The magnification scale in producing photographic prints was also determined using an objective micrometer.

The electrode separation was measured using a cathetometer with an accuracy of $\pm 10^{-2}$ mm. A cathetometer was also employed in observing and photographing the processes occurring in the gap. In the latter case the objective lens was replaced by a special mouthpiece with a camera.

To obtain shadow pictures of point cathodes with a greater magnification and better resolution an electron microscope was used instead of the optical one. An EOIC anode faceplate 2 cm in diameter

56

with a resolving power of about 30 pairs of lines per millimeter served as the microscope face.

The cathode surface microrelief was examined with the use of the JSM–U3 scanning electron microscope.

4. Pulsed Nanosecond Breakdown of Vacuum Gaps

One of our chief tasks was to study the electrical breakdown of vacuum gaps by application of rectangular voltage pulses of nanosecond rise time. This method has a number of advantages over the widely used technique of slowly varying voltages. Firstly, as mentioned above, it provides the opportunity of studying quantitatively the time characteristics of the development of breakdown. Secondly, owing to the short-term application of a voltage across the gap, it eliminates or significantly reduces the effect on breakdown of a number of slow processes (migration, diffusion, mechanical processes, adsorption). Thirdly, it makes it possibile to carry out experiments with stronger electric fields, which is of interest in itself.

The aims of our experiments were not only to investigate the breakdown time characteristics, but also to elucidate the space-time relationships of the light emission occurring in a vacuum gap at breakdown, the generation of X-radiation, the mechanisms of electrode erosion and material transfer from one electrode to another. The intention was that by comparing our results with the data of other investigators we would be able to find some general relationships, and could attempt to explain them.

4.1 Time Characteristics of the Pulsed Vacuum Breakdown

In order to obtain time characteristics of vacuum breakdown on a nanosecond time scale the discharge current and gap voltage waveforms were examined. In earlier investigations [4.1,2] only the gap voltage was recorded. However, since the discharge circuit has only the ohmic resistance R, the relation between the gap current i and the voltage V can be written as follows: $V = V_0 - iR$, i.e., in fact the current and voltage waveforms are mirror mappings of each other, and the time of decay of the gap voltage is equal to the current rise time (commutation time). Therefore, we shall denote the voltage decay time by t_c.

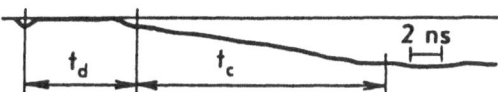

Fig.4.1. A current waveform recorded at vacuum breakdown with the use of a capacitive voltage divider with d = 0.5, V_0 = 50 kV

Figure 4.1 shows a typical waveform of breakdown current recorded with the use of a capacitive voltage divider connected to a transmission line L_3 (Fig.3.1). The current spike in the left part of the oscillogram corresponds to the moment of arrival of a voltage pulse at the vacuum gap, and is associated with the drift current flow through the gap. Then there is a section of insignificant current lasting during the breakdown delay time t_d, which is followed by a section of relatively monotonic rise of current up to a value limited by the discharge circuit resistance (corresponding to the commutation time t_c). The time t_c is usually measured between the 10% and 90% levels of the current peak value defined as V_0/R. The time t_d is conventionally determined from the apex of the capacitive current spike to the 10% level of the peak value. When studying the time characteristics, one hundred oscillograms per point were recorded, and average values of t_d and t_c were found. In tests with a vacuum of 10^{-6} Pa the interval between two pulses following each other was chosen to be equal to 5 s so that no gas monolayer could form on the electrode surface during this period (at such a vacuum the time of monolayer formation is about 100 s [4.3]). The same interval between two subsequent breakdowns was chosen when the tests were repeated with "commercial" vacuum. In our first experiments on nanosecond breakdowns [4.1,2], the times t_d and t_c were measured for copper, aluminum, steel, lead, and graphite electrodes in "commercial" vacuum with 0.1 to 1.0 mm gaps. The main results of this study are as follows:

1) At the gap overvoltage corresponding to $K_p > 1.1$ the breakdown delay time t_d is not in excess of several tens of nanoseconds;

2) the breakdown delay time decreases with an increase of the gap voltage;

3) the commutation time t_c is weakly dependent on the voltage peak value and increases with gap spacing;

4) the mean rate of breakdown current rise is of an order of magnitude of 10^{10} A/s; and

5) the times t_d and t_c are both independent of pressure in the range 10^{-3} to 10^{-2} Pa.

4.1.1 The Influence of Electrode Conditioning

Later on, a more detailed study of the time characteristics of vacuum breakdown was carried out [4.4,5]. As an example, let us show for copper electrodes separated by a 0.35 mm gap how the process of electrode conditioning with breakdowns affects the time characteristics and the breakdown voltage under the conditions of "commercial" (10^{-3} Pa) and oil-free (10^{-6} Pa) vacuum. In "commercial" vacuum, breakdown started to occur at a gap voltage of 40 kV and for about the first 200 breakdowns the time t_d was as a rule comparable with the pulse duration (no breakdown occurred). Then the time t_d gradually decreased. After about 1000 breakdowns the voltage applied across the gap could be decreased to 30 kV, breakdowns taking place with a delay $t_d <$ 100 ns.

A different picture was observed with an oil-free vacuum of 10^{-6} Pa. The first breakdown occurred at $V_0 = 32$ kV (the surface rearrangement and the diffusion of impurities into the electrode surface during prolonged heating of the vacuum chamber seemed to have an effect). After a short period, however, the time t_d began to lengthen periodically. and after 750 pulses no breakdown occurred at all. The voltage had to be increased up to 37 kV, but after about 100 pulses breakdowns ceased again. These results confirm the well-known fact that the presence of oil vapor lowers the gap breakdown voltage, thus shortening the breakdown delay time t_d.

Stable breakdowns of the gap occurred at voltages of 40 to 43 kV only, i.e., as a result of conditioning, the pulsed breakdown voltage increased by approximately 35% and the breakdown field reached about $1.25 \cdot 10^6$ V/cm. In the process of conditioning there was no noticeable dependence of the time t_c on the number of breakdowns and on the evacuation technique in both the vacuum systems. Similiar results on the effect of the conditioning process on the breakdown voltage and the time characteristics were obtained for aluminum and molybdenum electrodes.

4.1.2 The Influence of the Vacuum

Comparing the results of [4.4,5] with those of [4.1,2] the conclusion can be drawn that improving the vacuum conditions results only in an increase of the pulsed breakdown strength of the vacuum. There is no reason to believe that any variation in vacuum conditions should lead to a change of the breakdown mechanism. For electrodes made of copper, aluminum, molybdenum and graphite in all the above cases the

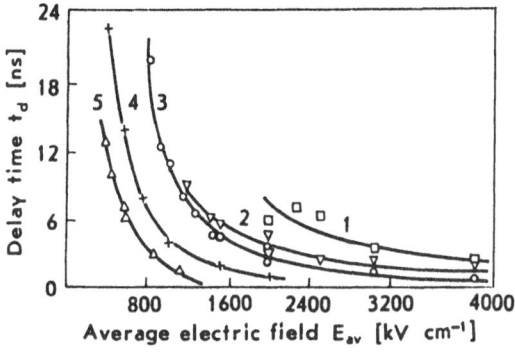

Fig.4.2. Breakdown delay time as a function of the average electric field for electrodes made of molybdenum (*1*), copper (*2*), aluminum (*3*), lead (*4*), and graphite (*5*)

ratio d/t_d is (2 to 2.5)·10^6 cm/s, and for lead electrodes 1·10^4 cm/s. d/t_d is weakly dependent on the average electric field E_{av}. Moreover, it was established [4.6] that the increase of the discharge peak current by an order of magnitude results in an insignificant change in both the shape of current waveforms and the dependence of the time t_c on the gap voltage and length. For ultra-high vacuum conditions the breakdown delay time was discovered to depend only on the average field E_{av} for any electrode material (Fig.4.2) [4.4,5]

In [4.7] a study of the time characteristics for gaps with d = 1-5 mm has been described. Experiments were carried out in "commercial" vacuum (10^{-3} Pa) with a generator that produced single rectangular pulses of voltage up to 300 kV, duration 300 ns, and rise time about 5 ns. The experimental results are presented in Fig. 4.3. The ratio d/t_c for gaps of the order of a millimeter is a constant and equal to about 2.7·10^6 for electrodes made of copper and molybdenum.

The experimental results for the time characteristics of vacuum breakdown of gaps of the order of a centimeter have been described in

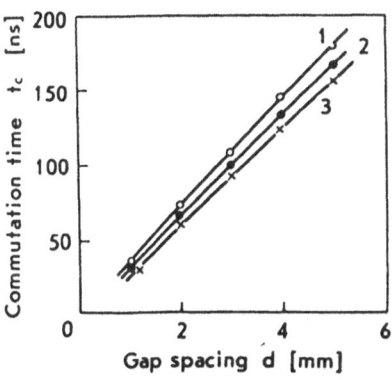

Fig.4.3. Commutation time as a function of the gap spacing for electrodes made of aluminum (*1*), copper (*2*), and molybdenum (*3*)

[4.8]. An Arkad'yev-Marx generator with an operational voltage up to 2000 kV was used. It was observed that the time t_d decreased approximately from 100 to 20 s as the pulse overvoltage coefficient was increased from 1 to 1.5. Under these conditions the rate of gap shortening was about $3 \cdot 10^6$ cm/s.

Kärner and *Bender* [4.9] investigated the breakdown of 1 to 3 cm vacuum gaps with electrodes made of aluminum, stainless steel and copper using a 4-stage $1.2/50\,\mu s$ 600 kV Marx generator. Despite the fact that they employed tail voltage pulses, breakdowns with $t_d = 10$ to 60 μs were recorded. The breakdown delay time was observed to increase with gap spacing.

4.2 Study of Light Emission at Pulsed Breakdown

Knowledge of the time characteristics of vacuum breakdown is still insufficient to provide us with an unambiguous model of how breakdown occurs. Valuable information can be gained by comparing these data with observations of light phenomena in a vacuum gap which accompany the development of breakdown.

4.2.1 Single-Shot Investigations

The results of single-shot electro-optical recording of the glow at the pulsed breakdown of short vacuum gaps have been described in [4.4, 5, 10]. Typical shots of the glow, corresponding to different stages of the breakdown development, are shown in Fig. 4.4. The dc breakdown voltage was 20 kV. In the 4-6 ns after a voltage pulse had arrived at the gap (which approximately corresponds to the time t_d), one or two weakly luminous localized regions appeared on the cathode surface. Such regions usually do not appear simultaneously but during several nanoseconds and not only in the sites of a maximum electric field strength, but also in some peripheral regions. For a given current waveform the instant of the appearance of the cathode glow is associated with the beginning of a rapid current rise with a rate of the order of 10^{10} A/s. A few nanoseconds after the appearance of the first luminous regions their number can reach five. These localized luminous regions at the cathode are named "cathode flares" (CFs). The sites of CF appearance change their location from discharge to discharge. During the period of current rise CFs expand into the gap with increasing brightness.

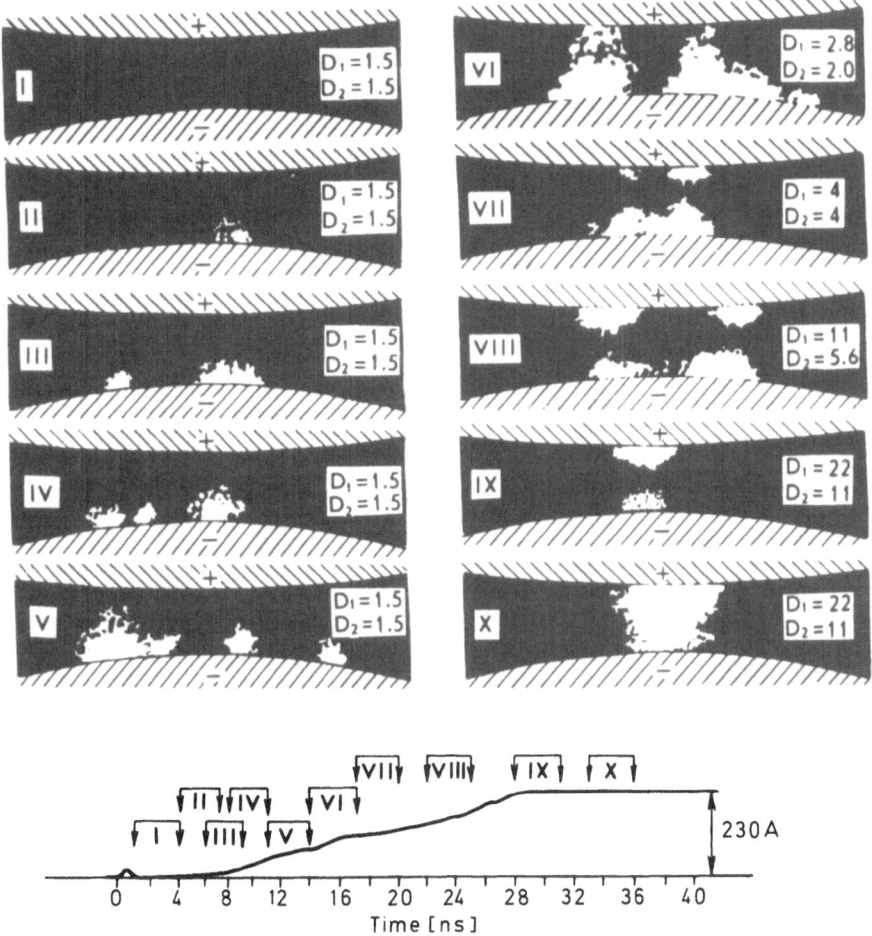

Fig.4.4. Typical photographs of the light emission in a gap, corresponding to different stages of breakdown development (d = 0.35mm, V_0 = 35kV, copper electrodes; D_1 and D_2: apertures of lenses 5 and 6 in Fig.3.5, respectively)

About 15-16 ns after the application of voltage across the gap a glow appears at the anode surface at the sites located opposite to the CFs. By that time, the current reaches about a half of its peak value (100 to 120 A). The glow brightness increases and becomes comparable with the CF brightness only 23-25 ns after the voltage pulse application. The luminous regions occurring at the anode are named "anode flares" (AFs). For a period of 28-30 ns, whilst the current reaches its maximum value limited by the discharge circuit resistance ($\simeq$230 A), AFs occupy about one-third of the interelectrode space ($\simeq$0.1 mm). After 33-36 ns AFs have usually expanded 0.2 mm into the gap. As the current rise terminates, the glow still remains at the cathode, although its brightness becomes somewhat less than that of AFs.

The expansion velocity of cathode and anode flares was identified with the velocity of propagation of the boundary of the corresponding glow. However, it is necessary to take into account that in going from the first to the tenth shot the brightness of glow increases by not less than four orders of magnitude (see the lens stop values in Fig.4.4). Owing to the stopping of the lens it seems as if the CF expansion has stopped after the sixth shot. The velocity of CF expansion was determiend only for the initial stage of their motion, the lens stops remained practically unchanged. For every exposure an average position of the "dense" glow boundary was found from thirty shots. The initial velocity of CF expansion was evaluated to be approximately constant and equal to about $1.7 \cdot 10^6$ cm/s. The velocity of AF expansion was determined from the ninth and tenth shots obtained with the same lens stop. It appeared to be about $2 \cdot 10^6$ cm/s.

By a statistical evaluation of the pictures of the glow front it was established that up to five separately visible CFs usually appeared at the cathode, but only one discharge channel could be seen by the time the current rise terminated. The apparent disappearance of several discharge channels can be explained as follows. Since CFs do not appear simultaneously, one or two leading flares occur during the period of a current rise. The velocity of propagation of the glow boundary was determined using only these leading flares. As the flare brightness increases exponentially, the leading flare becomes much brighter than the others at the time of current rise termination. A similar study of light phenomena was carried out for a 0.3 mm gap between tungsten electrodes at 35 kV. The results obtained were in principle the same as those described above.

4.2.2 The Continuous-Operation Regime

Some additional data which confirm the described mechanism of the breakdown development were obtained at the Institute of High Current Electronics using EOIC in a continuous operation [4.11]. In order to obtain a better spatial resolution, scan pictures in the direction parallel to the electrode axis were taken separately for the cathode and the anode glows (Fig.4.5). A glow that appeared at either of the electrodes was blocked by an opaque screen placed in front of the EOIC photocathode. It can be seen that the initiation of the breakdown starts from the occurrence of several CFs. The sites of occurrence of CFs change from one breakdown to another and are localized at the near-axial

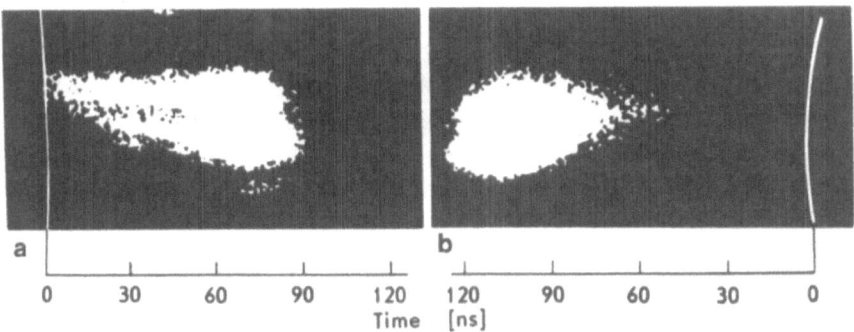

a   b

| 0 | 30 | 60 | 90 | 120 | 120 | 90 | 60 | 30 | 0 |

Time [ns]

Fig.4.5a,b. Scan pictures of the light emission from cathode (a) and anode (b) taken in the direction parallel to the electrode axis. (Indium electrodes; d = 1.25mm; V_0 = 40 kV, t_c = 100ns)

electrode regions. In several photographs, pulsations of the CF brightness can be seen. This seems to indicate that the CF lifetime is limited to several nanoseconds.

The anode glow appears after the beginning of the current rise and expands along the electrode more rapidly than CFs. In the anode glow some separate, brighter sections can be seen, which indicates the existence of several AFs.

The velocity of the plasma expansion into the gap cannot be reliably measured using EOIC recordings obtained in continuous operation. This is particularly the case with the cathode plasma, since its luminosity decreases rapidly in the direction toward the CF periphery because of a significant decrease of its density. Moreover, when the EOIC operates in the swept regime there is no "accumulation effect" intrinsic to the single-shot regime.

The appearance of light in different parts of the vacuum gap was detected by using a much more sensitive photoelectric technique. This study (Chap.9) has completely confirmed the above-described picture of the development of the glow at breakdown and has shown that the velocity of CF expansion is about $2 \cdot 10^6$ cm/s.

When comparing the results described above with those known earlier [4.12-14], a principal difference can be easily revealed. In our experiments, in contrast to those carried out earlier, the glow always appeared at first from the cathode side with the beginning of a rapid rise of the gap current, while the glow from the anode side occurred somewhat later. This discrepancy can be attributed to the fact that the recording equipment used had a much higher time resolution and a greater magnification of the image brightness at the initial steps of breakdown.

4.2.3 Comparison with Other Data

It makes sense to compare our results with the data of other investigations carried out in subsequent years. In 1968 *Maitland* and *Hawley* [4.15] published their results of the EOIC study of the development of the pulsed breakdown of 1 to 17.8 mm gaps with plane electrodes made of stainless steel and copper. The experiment was carried out in "commercial" vacuum (10^{-3} Pa) with the use of two generators. The first one produced voltage pulses up to 100 kV with a pulse duration of 3 μs and a rise time of 0.5 μs with the limiting resistance R_{lim} = $2 \cdot 10^4$ Ω, while the second one generated voltage pulses up to 250 kV with a rise time of 2 μs at R_{lim} = 500 Ω. The electro-optical image converter could operate either in the swept regime (at recording speeds up to 100 mm/μs), or in the single-shot regime (3 shots with a 50 ns exposure and intervals from 0.5 to 5 μs). It was initiated during the voltage pulse rise time. The main conclusion of *Maitland* and *Hawley* [4.15] is that in all cases of breakdown the glow appeared initially at the cathode; at the anode it occurred after a certain time delay depending on the experimental conditions (for instance, with d = 1.25 mm, U = 80 kV the delay time was about 100 ns). When an overvoltage pulse was applied to the cathode, several luminous regions could occur simultaneously, but in the terminating stage the discharge usually passed through a single channel. When the luminous cathode cloud arrived at the anode, an anode cloud started moving in the opposite direction, the anode cloud providing the main contribution to the filling of the gap with the conducting medium for d $\geq$ 5 mm. Thus, the data of [4.15] not only confirmed, in principle, our results but also complemented them by having widened the range of experimental conditions in which the cathode mechanism of breakdown initiation could be realized. The fact that in [4.15] EOIC pictures were not related to the breakdown current oscillograms resulted, in our opinion, in a loss of an important additional information about the development of breakdown.

Let us look at some of the results of *Parker* et al. [4.16]. They studied the pulsed breakdown between wide plane-parallel electrodes made of graphite with gaps d = 1 to 10 mm with the aim of choosing the optimum design for the vacuum diode of a high-power high-current electron accelerator. Using streak-camera photography they established that the onset of light emission always appeared earlier from the cathode side than from the anode side of the diode (Fig.4.6). Because of an insufficient aperture of the recorder and the absence of a light amplifier the cathode glow was detected much later than the beginning of the growth of the gap current. It is possible that the cease

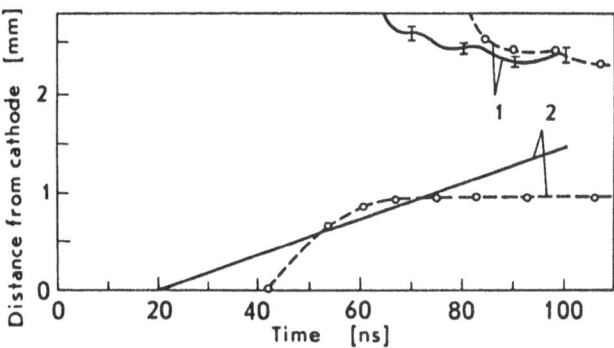

Fig.4.6. Plasma expansion from anode (*1*) and cathode (*2*) at vacuum breakdown. (Graphite electrodes; d = 2.8 mm; V_0 = 200 kV). Dashed lines: recorded light emission front; solid lines: supposed plasma front [4.16]

in the motion of both the cathode and the anode clouds observed in the experiment had the same cause.

Chalmers and *Phukan* [4.17] reported on the EOIC observations of pulsed breakdown of vacuum gaps with d = 0.2 to 1.0 mm, which were formed by a plane anode and a conical cathode with a tip curvature radius of 0.5 mm. The experiments were carried out with copper, aluminum and stainless-steel electrodes under high-vacuum conditions. EOIC operated both in the continuoous-operatiom (maximum recording speed of 1 mm/ns) and in the single-shot regime (frame exposure time 10 ns). *Chalmers* and *Phukan* established that in all cases the glow initially appeared at the cathode. The luminous region did not increase in size and could be seen only near the cathode surface. The anode glow appeared with a delay of 10-30 ns, which increased with gap spacing. The expansion velocity of the anode glow depended on the anode material and accurately agreed with the measurements of *Chiles* [4.13]. Having compared the glow pictures with the gap voltage oscillographic waveform *Chalmers* and *Phukan* drew the conclusion that the transition to arc discharge was due to anode vapor filling the gap. In our opinion, one should not draw the conclusion that no expansion of the cathode glow has occurred basing it on qualitative patterns only. We would also question the validity of the oscillographic measurements of the gap voltage.

Cross and *Mazurek* [4.18] observed the development of a vacuum discharge between a point cathode and a plane anode using the Imacon 600 high-speed camera and associated image intensifier. A standard $1/50 \mu s$ pulse generator was employed. It was established that the cathode light emission has a pulsatory character. The repetition time between pulses increased from 4 to 5 ns after about 100 shots. *Cross*

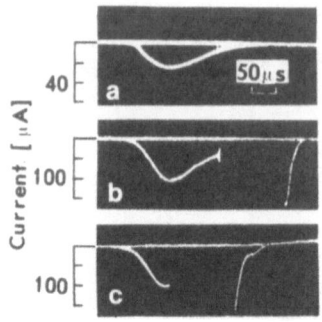

Fig.4.7a–c. Typical current waveforms recorded for a 6 cm vacuum gap: (a) no breakdown; (b) transition to breakdown at a low overvoltage; (c) transition to breakdown at a moderate overvoltage [4.21]

and *Mazurek* attributed this to a decrease in the density of adsorbates on the electrode surface due to discharge conditioning.

Studies of light phenomena at pulsed breakdown of vacuum gaps with spacings up to 15 cm have been carried out in a "commercial" vacuum (10^{-3} Pa) with spherical electrodes up to 20 cm in diameter, made of copper, aluminum, and graphite [4.19-21]. Voltage pulses with amplitudes up to 1 MV and a rise time of 0.5 μs were applied across the gap. Several interesting phenomena in the prebreakdown stage and during the development of breakdown were discovered. Using a thin foil anode transparent to electrons these researchers were able to establish that at voltages lower than the breakdown voltage the prebreakdown electron current is always a single pulse (Fig.4.7a) appearing with some delay after applying the voltage. As the voltage is increased, the time needed to reach the current maximum decreases, while the amplitude increases. When the minimum breakdown field is obtained on the "tail" of the prebreakdown current pulse, the current increases sharply by several orders of magnitude, i.e., breakdown occurs irreversibly (Fig.4.7b). As the voltage is further increased, the time preceding the breakdown shortens (Fig.4.7c).[1]

The integrated glow in the gap was photographed at a voltage lower than the breakdown voltage (for 250 pulses) by a camera with the shutter opened. It was possible to establish that at the surface of both electrodes a diffusion luminous layer ($\approx$5mm thick) is formed and is brighter at the cathode. The luminosity is higher near the axis, i.e., in the region of the stronger electric field. A trace record of this glow using PMT has shown that the light pulse waveform at each pulse application follows the prebreakdown electron current and is synchronized with the latter. Spectral analysis of the radiation has shown that the brighter lines are those of hydrocarbon radicals and carbon monoxide. No spectral lines of metal ions are indicated in this stage of discharge; these lines appear only in the breakdown stage.

---

[1] Similar results have recently been obtained by Kärner and Bender [4.9].

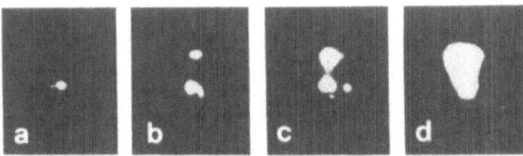

Fig.4.8a-d. Photographs of the light emission at a breakdown of a 10 cm vacuum gap: (a) discharge is cut off 0.5 μs after the onset of breakdown; (b) and (c) the discharge is extinguished in later stages; (d) discharge completed [4.20]

The glow in the initial steps of the breakdown was investigated by the "cut-off discharge" method similar to the one employed in [4.4, 5] in combination with open-shutter photography of the gap. The cut-off in the discharge was accomplished by using a controlled air gap. The photographs relating to a 10 cm gap formed by aluminum electrodes are presented in Fig.4.8. The first photograph shows a discharge cut off 0.5 μs after its beginning. In this period the breakdown current has time to rise up to about one tenth of its peak value of about 500A. A luminous spot near the cathode surface (several spots are often observed) can clearly be seen, while in the rest of the gap there is no glow. In later stages, a luminous cloud propagating from the anode appears. After the completion of discharge, a large anode cloud is observed, its light intensity decreasing with the distance from the cathode surface. These results also indicate that the conducting medium originating from the anode is present only during the process of breakdown development, a fact that does not exclude a significant role of the anode cloud in the last stage of the transition to a low-voltage discharge. *Kassirov* and *Sekisov* [4.20] considered that the current passage during the first (prebreakdown) stage of a pulsed breakdown in "commercial" vacuum is provided by the formation of gas-discharge plasma in the medium of gases desorbed from the electrodes. It is supposed that high-pressure gas layers propagate into the gap from the electrode surface. The gas-discharge plasma, via some secondary processes, "prepares" the cathode for the explosive-like appearance of CF plasma arising from the cathode surface and for the rapid current rise in the discharge breakdown stage.

4.3 Electrode Erosion Studies

4.3.1 Cathode Erosion

Investigation into cathode and anode erosion provides important information for understanding the processes occurring at pulsed breakdown of vacuum gaps. Early studies of cathode and anode erosion [4.22] at breakdown of a 0.35 mm gap with a 35 kV amplitude voltage showed convincingly that during a time of the order of 10^{-8} s after the discharge initiation, extensive melting of the anode (Mo) took place followed by transfer of anode material onto the cathode (Cu), whereas erosion on the cathode was insignificant. Thus, when the current was cut off after 6 ns, no change in the electrode surface was indicated. At t_p = 12 ns some protrusions 1-1.5 μm in height and 10-15 μm in base diameter could be observed on the anode surface. Small molybdenum balls 1-2 μm in size transferred from the anode could be seen on the cathode. At t_p = 18 ns, the boundaries of crystals with craters about 5 μm in diameter underneath them were noticable on the anode, while traces of transferred molybdenum balls 1-2 μm in size and a molybdenum film were visible on the cathode. When the current was cut off after 30 ns, circular craters up to 10 μm in diameter were seen on the anode; their number was of the order of 10^4, which corresponds to the observations by *Maitland* [4.23]. Besides the molybdenum balls and the molybdenum film, a number of micron-sized craters were observed. Their formation was attributed to the occurrence of flares on the cathode. When the total discharge duration was 300 ns, an erosion spot 0.3-0.6 mm in size with a molten central region was observed on the anode, while on the cathode some molten irregularities and craters several micrometers in size could be seen.

4.3.2 The Tracer Method

The mass of the metal transferred from the electrodes during the process was measured by the tracer method [4.22]. The ^{64}Cu isotope of copper was used. It was obtained by activating a section of the hemispheric electrode surface (200 μm thick) by irradiating it with deuterons at 13 MeV. Since the ^{64}Cu isotope has a half-life of 12.8 hour, it was possible to carry out a reliable experimental series at a high activity of the sample. The sensitivity of the mass-measuring method was 10^{-11} g. It turned out that with $t_p \simeq t_c$ = 30 ns no material transfer from cathode to anode was observed, i.e., the specific mass transferred was not more than 10^{-5} g/C. Under these conditions $2 \cdot 10^{-8}$ g of metal were transfered from anode to cathode, which corresponds to a specific mass transfer of $5 \cdot 10^{-3}$ g/C. These data are in agreement with measurements of the electrode material transfer in a spark discharge [4.24]. Note that it is difficult to measure the metal mass transfer from the cathode exactly, because the metal mass removed from the anode is considerably greater.

4.3.3 Anode Erosion

Further research work was carried out with the aim of comparing the results of the studies of the erosion processes with those obtained from investigations of the breakdown light phenomena [4.4,25]. Copper, aluminum and molybdenum electrodes were used. A cut-off spark gap was employed to cut off the current at the right time. Figure 4.9 illustrates typical erosion traces on a copper anode. It can be seen that, as a result of a single incomplete breakdown, several erosion spots form on the anode. The disposal of spots is independent of the structure of the anode. If, however, a crystal boundary occurs within the region of an erosion spot, this region is subjected to destruction to a greater degree than the area of the crystal faces. The appearance of several erosion spots on the anode suggests that the breakdown develops simultaneously through several channels. This finding is in good agreement with the results of electro-optical studies.

Another interesting result is that two types of erosion traces can be observed on the anode. They consist of isolated erosion spots which look like those described in [4.26] and are mostly circular in form. If

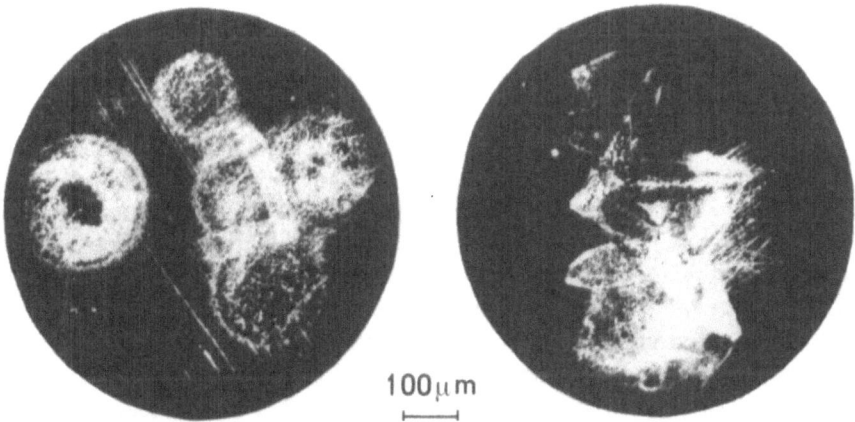

100μ m

Fig.4.9. Typical erosion traces on a copper anode after a single incomplete breakdown

several spots are grouped in one place, stretched erosion traces appear between them, which we have named "touches" [4.25]. The "touches" represent regions with more significant erosion. The "touch" boundaries are subjected to the greatest amount of destruction. We have established experimentally [4.25] that "touches" occur when the discharge channels are situated sufficiently close to each other. One can count the number of discharge channels by the number of "touches" and circular erosion marks much more exactly than by the electro-optical method. By varying the time before the discharge cut-off, we have been able to count two to six CFs formed initially, and during the period of the current rise their number increased up to twelve to sixteen. It is important to emphasize the fact that the regions neighbouring the existing discharge channels are just the places where the conditions most favorable to the formation of new channels are created.

All the results described show that in spite of the initiation of breakdown at the cathode, dominant erosion processes take place mainly at the anode. A pulse of duration of the order of 10^{-8} s is sufficient for the anode erosion to start. All these findings would suggest that the anode is affected by an extremely heavy energy flux which is due to the electron flow emitted by CFs and accelerated in the vacuum part of the gap.

4.4 Nature of the Discharge Current at Breakdown

From the results of the studies described above the conclusion is drawn that the discharge current, after the appearance of a cathode plasma, is dominated by free electrons emitted by the plasma and accelerated in the vacuum part of the gap. Two observations served as background for this conclusion: 1) The absence of light emission between the cathode plasma and the anode when the current has already reached some tens or hundreds of amperes, suggesting an insignificant contribution of the ion component to the discharge current; 2) the AF appearance with a delay of the order of 10^{-8} s from the beginning of the current rise and the material transfer from anode to cathode, that indicates a high density of the energy flux onto the anode. In order to gain some additional arguments in favor of the purely electronic nature of the discharge current, the behavior of the current in a magnetic field [4.27] and the X-radiation from the anode [4.28,29] were studied.

In [4.27] it was supposed that if the discharge current is really a purely electronic one, then on application of a magnetic field trans-

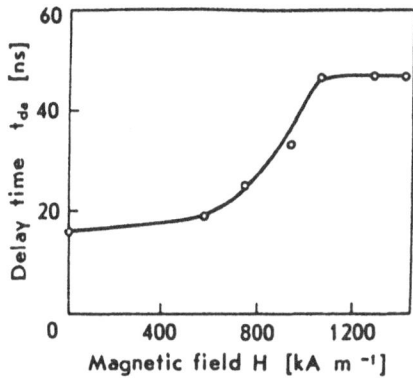

Fig.4.10. The delay time of the occurrence of a current pulse in the anode circuit as a function of magnetic field

verse to the cathode, electrons should be deflected from electric field lines. For non-relativistic electrons, the maximum distance between them and the emitting surface can be evaluated from the relation $y = 2 m_e E/(eH^2)$, where m_e and e are the electronic mass and charge, respectively; E and H are the electric and the magnetic fields, respectively. Therefore, if the vacuum gap length $d > y$, electrons will not reach the anode in the beginning. Experiments were carried out on a system with a point cathode and a plane anode with an electrode separation of 0.5 to 1.2 mm. Rectangular voltage pulses had a rise time of 1 ns, a duration of up to 150 ns, and an amplitude of 20 to 50 kV. The constant magnetic field reached $1.6 \cdot 10^6$ A/m. It was recognized that a part of the electron flow would be directed to the walls of the discharge chamber rather than onto the anode surface. Therefore, the current measurements were carried out separately in the anode and in the cathode circuits.

The delay time for the appearance of a current pulse in the anode circuit as a function of the magnetic field is depicted in Fig. 4.10. Estimates have shown that when electrons emitted by CF reach the anode at the very beginning ($y/d>1$), the delay time t_d remains unchanged. When the magnetic field is sufficiently high (i.e., $y/d<1$), the time t_d is observed to increase with the magnetic field H. This means that the current in the anode circuit starts to flow not at the onset of explosive emission at the cathode, but only when the cathode plasma has approached the anode within a distance y. This shows convincingly that the discharge current is due to free electron flow.

In connection with the results reported above two remarks should be made. First, practically in any kind of discharge the electron component of current in the region far away from the cathode is many times greater than the ion component, because the mobility of electrons in plasma is much greater than that of ions. But, if the vacuum

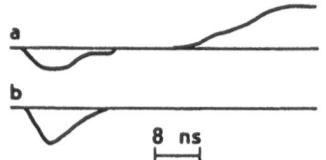

Fig.4.11a,b. An X-radiation pulse waveform (left) and the corresponding breakdown current waveform (right) (a). The calculated time behavior of the X-radiation pulse intensity (in relative units) (b)

discharge current in the spark stage were due to plasma electrons only, i.e., due to bound electrons, there should be no magnetic field effect. Second, the lengthening of the delay time of the onset of vacuum breakdown with the magnetic field, which we established in 1969 [4.27], was the first experimental evidence supporting the increase of the pulsed hold-off voltage in vacuum in the presence of a magnetic field. Subsequently, this effect has found wide application in vacuum lines, in which a strong magnetic field is created by the current of the pulse passed through the line. This phenomenon is named the magnetic insulation effect [4.30].

Another important argument in favor of the statement that the discharge current is free-electron current was gained as a result of studies of X-radiation from the anode in a discharge [4.28,29]. It was important not only to record this radiation but also to measure its duration and elucidate how it depends on the gap length and the discharge parameters, and how its intensity is related to a given breakdown phase.

Experiments [4.28,29] were carried out in "commercial" vacuum with copper electrodes. Typical oscilloscope records of the X-radiation pulse and breakdown current for a gap of d = 0.5 mm are given in Fig.4.11. In the lower part of the picture the calculated time dependence of the intensity of the X-radiation pulse (in relative units) is plotted using the current waveform i(t) and proceeding from the assumption that this current is provided by the electrons extracted from CF and that the intensity of X-radiation is proportional to $i(t) \cdot V^2(t)$, where V is the gap voltage. It is clear that there is a pronounced correlation between the X-radiation waveform and the calculated intensity curve. X-radiation occurs simultaneously with CF and ceases at the instant of transition to the arc discharge regime.

A study of the duration of X-radiation at breakdown of gaps up to 5 mm long using the same techniques has been described in [4.29]. The X-radiation pulse waveform at breakdown of gaps of the order of a millimeter is the same as for gaps of fractions of a millimeter and its duration is proportional to the gap spacing and comparable with the commutation time (Fig.4.12). The results of the studies described indicate the existence of a high-power electron flow that acts on the anode in the period of breakdown current rise.

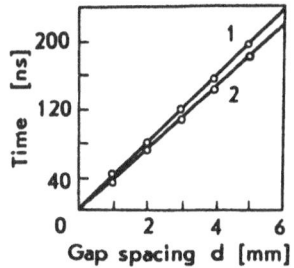

Fig.4.12. Commutation time (*1*) and X-radiation pulse duration (*2*) as a function of gap spacing

4.5 Mechanism of Pulsed Breakdown of Vacuum Gaps

4.5.1 The Role of the Cathode

The experimental results described in [4.4, 10, 31] enabled us to suggest a model for the mechanism of the initiation and development of a pulsed breakdown in vacuum. The experiments unambiguously show the leading role of the cathode in the process of breakdown initiation. Cathode flares appear with a certain delay after the voltage application corresponding to the discharge delay time t_d and afterwards the current starts to increase monotonically. The above discussion proves that cathode flares, which represent plasmoids, are the sites of intense electron emission. The discovery of this emission and the explanation of its existence [4.10, 31-33] is the most important result of the studies of pulsed electrical breakdown in vacuum on the nanosecond time scale. As will be shown below, this emission is due to the explosive phase transition of cathode metal into dense plasma, therefore it is called explosive electron emission (EEE) [4.34]. What is this explosion due to? One of the possible reasons is the heating of cathode micropoints by field electron emission (FEE) current. The explosion can be facilitated by the presence of dielectric films and dielectric inclusions or adsorbed gases at the cathode surface.

As a voltage pulse with an amplitude exceeding the dc breakdown voltage arrives at the gap, FEE starts from cathode micropoints characterized by a high current density. As a result of heating by the FEE current the points explode. The interval from the moment of voltage application across the gap till the explosion of cathode micropoints is the breakdown delay time. The failure of the gap electrical insulation which occurs with the explosion is accompanied by a discharge with a steep negative current-voltage characteristic. In view of the sequence of processes at breakdown, the point explosion accompanied by the formation of cathode plasma and the appearance of intense electron

75

emission from the cathode should be considered as a breakdown initiating event.

4.5.2 The Cathode Plasma and the Electron Current

With the appearance of plasmoids, effective electron sources appear on the cathode and the gap conductance starts to grow rapidly, which is due to the expansion of CF plasma. The interaction of the plasma with the cathode surface persistently creates the conditions which maintain an intense electron emission from the cathode. As is stated in [4.10], plasma which was created explosively stays in contact with the cathode surface and provides an electron source for the current passing through the gap whilst the conductance of the gap is increasing.

Thus, in the studies [4.5, 10] the reason for the appearance of cathode plasma has been established and it has been shown that the current emitted by this plasma is purely electronic. A breakdown event occurs when the electric field at the cathode $E_{av} \geq 10^6$ V/cm. With the field enhancement factor $\beta \geq 100$ the local voltage will be in excess of 10^8 V/cm and the corresponding field emission current density will be $j > 10^8$ A/cm^2. At such values of j fast heating and explosion of the emitter should occur. It should be noted that there had been no direct experimental studies of the process of explosion of field-electron emitters on the nanosecond time scale carried out at the time the above-described results were obtained. The results made it necessary to introduce the assumption that such explosions occur on a time scale of 10^{-9} s. Later on we have confirmed the validity of this assumption.

Independently of our work an article by *Fursey* and *Vorontsov-Vel'yaminov* [4.35] appeared. They observed qualitatively the development of a vacuum arc on the basis of numerous experimental data on the behavior of field-electron emitters at the threshold of the FEE current. An important merit of [4.35], when compared to other efforts to explain the cathode mechanism of the discharge initiation, is not only the attempt to find evidence for the occurrence of explosions at the cathode, but also to represent qualitatively the pattern of the discharge development after the explosion. In particular, an important idea of the role played by the plasma formed at the cathode as a result of an explosion of a micropoint in the appearance of new explosions on the cathode was suggested in [4.35]. However, no attention was paid to the possibility of the existence of a free-electron current in a vacuum discharge after the explosion of micropoints on the cathode. *Fursey* and *Vorontsov-Vel'yaminov* [4.35] limited themselves to the statement that the vacuum discharge further develops in the cathode metal vapor.

4.5.3 Anode Phenomena

The above-described results of studies into electron emission from the plasma formed during the explosion of the cathode micropoints provide an understanding of the anode erosion, the appearance of anode vapor, and the X-radiation pulse. Electrons emitted by CF, after being accelerated in the vacuum section of the gap, transfer their energy to the anode surface. The power flux on the anode ($d = 0.35$ mm, $V_0 = 35$ kV, $R = 150$ Ω) is estimated to be 10^9 W/cm^2 during about 14 ns, the diameter of the erosion trace on the anode surface being $2 \cdot 10^{-3}$ cm [4.31-33]. The electron free path in the body of the anode is of the order of 10^{-4} cm. If one neglects all heat losses, it turns out that the specific energy input to the anode is significantly in excess of the specific sublimation energy for copper. It is obvious that these are estimates of "upper" limits, but they show that intense and even "explosive" evaporation of anode material and AF formation may occur. According to estimates the mass of metal evaporated from the anode is of the order of 10^{-8} g, i.e., several orders greater than the mass of metal evaporating in a CF. Thus, in the discharge final stage, the anode can be the main supplier of the conducting medium to the vacuum gap. The AF brightness will be much greater than that of the CF. In the experiments described in [4.13,36], the movement of such a bright AF has been revealed. It is difficult to record the CF light emission using the method of a rotating mirror, since its brightness in the initial period of the current rise is several orders of magnitude lower than the AF brightness. This is also the reason why the use of an image intensifier in the continuous-operation regime [4.16] or in the single-shot regime [4.14,37], without special measures for lenghtening the light-flow path of picture transmission, leads to incorrect results. Data subsequently obtained with an electro-optical apparatus substantiated our statement of the primary nature of the CF glow [4.18,38].

Estimates obtained for gaps of the order of a millimeter show that the power density can be greater than 10^8 W/cm^2 and the specific energy introduced into the anode surface layer during the commutation time can be in excess of the specific sublimation heat. Therefore, the dominant effect of the anode material on the vacuum breakdown characteristics, as established in [4.39,40], becomes understandable, since during the conditioning process the cathode is still covered with the deposit of the anode material.

Our experiments contradict the hypotheses of *Maitland* and *Howley* [4.23, 26] about the role played by the anode craters. The shortcoming of their method is that they examined electrode erosion only after a complete breakdown. As is shown above, the anode craters similar to those observed in [4.24, 26, 41] appeared only in the stage of the current rise and were absent in the period of the breakdown initiation.

Thus, during the growth of the discharge current the electron flow transfers a high-power pulse to the anode surface which gives rise to the anode flares, anode erosion, and X-radiation. The opposing motion of the anode and the cathode flares causes the gap to fill up completely with the conducting medium which leads to the termination of the current rise and to a decrease of the voltage to a level which is characteristic for an arc discharge. During the time period between the micropoint explosion and the termination of the current rise there exists a spark stage of the breakdown. This is followed by an arc stage of the discharge in which the current is limited by the external circuit resistance only. From a physical point of view, its duration is determined only by the value of current. Immediately after the spark the arc discharge operates in the vapor of the cathode and the anode material. Later on, provided that no anode spots form on the anode, only the cathode spot supplies the conducting medium.

5. Cathode Processes in a Pulsed Vacuum Discharge

The above-described studies have established that the solid-to-plasma transition of the cathode material occurs due to a high energy concentration in the cathode microvolume. One of the reasons for such a high energy concentration is the Joule heating of the cathode microprotrusions by the FEE current. However, no studies were carried out of the explosion of a field-electron emitter on a nanosecond time scale. The mechanism of the cathode processes after the discharge initiation remained unexplained.

Since, as will be clear from what is said below, the EEE phenomenon is inseparably linked with a loss of cathode material, which accounts for the formation of cathode plasma, the investigation of the mechanism of metal loss should provide essential information about the phenomena occurring in the emission region on the cathode. In connection with this we carried out a detailed study of cathode erosion. The observation of erosion traces using a scanning electron microscope (SEM) enabled us to obtain a number of important data on the processes of crater formation, and the regeneration and displacement of emission centers (ECs). By investigating the erosion mechanism we were able to determine the current density on the cathode and, consequently, the energy balance of its operation. These data, obtained on a nanosecond time scale, have led us to the conclusion that the cathode phenomena occurring in a vacuum discharge are quite nonstationary and represent a series of microexplosions accompanied by EEE.

5.1 EEE Initiation by High-Density FEE Current

In Chap.2 we considered our present knowledge on FEE at threshold current densities and on the transition of FEE to the "vacuum arc". The results of the work reviewed suggest that if some critical current density uniquely related to the electric field is achieved, the initially stable FEE changes over to vacuum breakdown owing to the emitter explosion. However, there has been a lack of reliable data on the development of this process because apparatus with a resolving power high enough to record such a fast process did not exist. Moreover, there has been no information found on the limiting values of the FEE current density and the electric field at a pulse duration $t_p < 10^{-6}$ s. In this respect it has been of interest to study the transition of FEE to vacuum breakdown in the range of pulse durations from 10^{-9} to 10^{-6}s.

5.1.1 Experimental Conditions

The study was carried out for monocrystal tungsten point emitters with a tip angle $\theta < 15°$ and a tip radius $r_e = (1 \text{ to } 5) \cdot 10^{-5}$ cm. In the earlier experiments [5.1,2] sealed-off glass tubes were used. A grounded tungsten grid with a transmittance of 80% was mounted in the cathode-electron collector space to eliminate the displacement current which could well be in excess of the FEE current at a pulse rise time of $\sim 10^{-9}$ s. Using a molybdenum getter cooled with liquid nitrogen the tube pressure was maintained at a level of 10^{-7} Pa. Later on, a metal-glass tube was designed [5.3], in which after heat treatment the same pressure could be maintained by continuous pumping with the use of a device of the Varian type. In contrast to the sealed-off tubes used formerly, we were able to obtain much better matching to the pulse generator by wave impedance in this metal-glass tube. This allowed us to carry out experiments with pulses of a duration less than 1 ns and a rise time shorter than 0.5 ns.

DC current-voltage characteristics (CVCs) were measured for every emitter and plotted as $\log(i/V^2)$ versus $1/V$. The experimental set-up [5.3], using an amplifier with a bandwidth of 150 MHz in the collector circuit, allowed us to measure CVCs accurately in the nanosecond range and so compare them to the dc characteristics. Shown in Fig.5.1 are the Fowler-Nordheim curves for two tungsten emitters. The lower and the upper sections of these curves were measured in the dc and the pulsed regime, respectively. The departure of the characteristic shape from the straight lines associated with the space charge effect is observed at the current density $j = 5 \cdot 10^6 - 10^7$ A/cm^2, thus showing good agreement with the data of [5.4]. The emitter radii were determined from the Drechsler-Henkel formula [5.5] $r_e = 0.0074 \, V_{-5}^{3/2}$, where V_{-5} [V] is the voltage corresponding to the emitter current of 10^{-5} A; r_e being measured in Angströms.

In order to obtain the most complete information it was necessary to measure with a sufficient amplitude and time resolution the current per pulse for three stages of the process: the field emission, the explo-

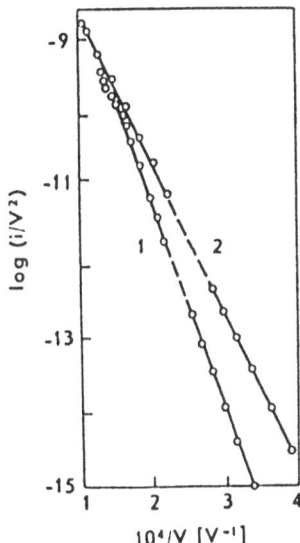

Fig.5.1. Current-voltage characteristics of two monocrystal tungsten point field emitters

sion, and the initial stage of breakdown. Since these stages may be significantly different in the current and the duration, the current from the collector was successively fed - with various attenuations - to a series of oscilloscopes with different time bases. In the earlier experiment [5.1,2] the maximum current sensitivity was $7 \cdot 10^{-2}$ A/mm, and so we were unable to record the field emission current together with the explosion phase current at relatively long times before the explosion ($>10^{-8}$s). Later on, due to the use of an amplifier, we were able to increase the current sensitivity up to 10^{-4} A/mm [5.3].

The investigation was carried out in several regimes: 1) With $t_p = 4 \cdot 10^{-6}$ s, the voltage was stepped by 1% up to the value at which an explosion took place at the pulse plateau. As the explosion occurred, the inrush of current triggered the oscilloscopes operating on standby.

2) With $t_p = 4 \cdot 10^{-6}$ s, an overvoltage was applied across the gap to increase the prebreakdown current density. The overvoltage factor was set as V_0/V_{cr}. In this case V_{cr} was determined from the dc current-voltage characteristics as the voltage at which the FEE current density reached some critical value ($5 \cdot 10^7$ A/cm^2) [5.4,6,7] on a microsecond time scale. The V_0/V_{cr} value was varied from 1 to 10.

3) With $t_p = 4 \cdot 10^{-8}$ and $3 \cdot 10^{-7}$ s, the voltage was stepped up to the value at which the point explosion occurred at the end of the pulse. Current waveforms were recorded that showed the transition of FEE to the initial stage of breakdown.

4) For the study of explosion at the limiting value of the high current density, pulses were used with a duration of 0.7–5 ns.

5.1.2 Description of EEE Current

Typical waveforms of the collector current for $V_0/V_{cr} > 1$ obtained with various oscilloscope time-bases are presented in Figs.5.2b to d. Their analysis enabled us to distinguish four characteristic phases of the process (Fig.5.2a). The first one (I) is the prebreakdown (field emission) phase investigated in detail on the microsecond time scale. The second one (II), the transition phase is associated with an explosive destruction of the emitter. The current in this phase rapidly increases at a rate $di/dt = 5 \cdot 10^7$ to 10^8 A/s for $(1$ to $5) \cdot 10^{-8}$ s. After this current increase period the third (III) phase is revealed, in which the rate of current rise is an order of magnitude less (e.g., $5 \cdot 10^6$ to 10^7 A/s). Then the fourth (IV) phase follows with the rate of current rise somewhat higher than in Phase III. In the overvoltage regime the character of current rise is essentially different (Fig.5.2e). It is clear that one is unable to make a precise demarcation into several phases for this regime like in the above-described cases. One can suppose that the duration of Phases I and III is shortened, and Phase II merges into Phase IV. Another characteristic feature of this regime is the appearance of current bursts against the background of the monotonically increasing current some time after the emitter explosion has taken place. The amplitude of the bursts increases with the overvoltage coefficient. It is found that the rate of current rise di/dt steeply increases as the

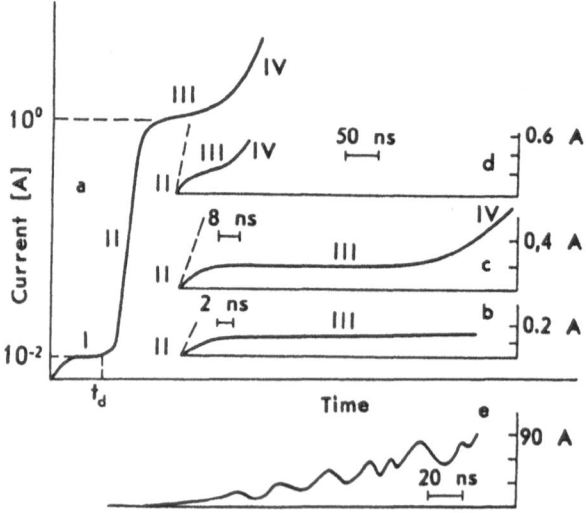

Fig.5.2a–e. Current behavior at a transition from FEE (phase I) to EEE (phases *II–IV*) (a). Collector current waveforms at an explosion of a field emitter without overvoltage (b–d), and with an overvoltage (e)

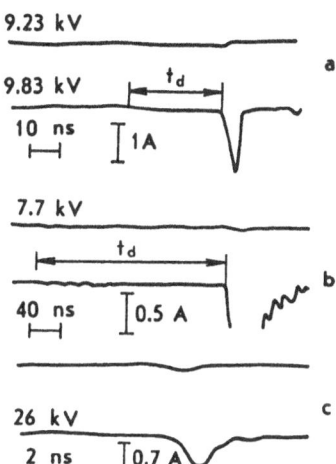

9.23 kV

9.83 kV $\quad \longmapsto t_d \longrightarrow$

10 ns $\quad \mathbf{I}\, 1A$

a

7.7 kV

$\longmapsto \quad t_d \quad \longrightarrow$

40 ns $\quad \mathbf{I}\, 0.5\, A$

b

26 kV

2 ns $\quad \mathbf{I}\, 0.7\, A$

c

Fig.5.3. Collector current waveforms illustrating the transition from FEE to EEE

overvoltage coefficient is increased. At $V_0/V_{cr} \simeq 1.5$ di/dt $\simeq 5 \cdot 10^8$ A/s.

In order to observe the current behavior in the pre-explosion stage for $V_0/V_{cr} > 1$, measurements were carried out with $t_p = 4 \cdot 10^{-8}$ and $3 \cdot 10^{-7}$ s [5.3]. Oscillograms illustrating the behavior of two field emitters under such conditions are shown in Fig.5.3. For emitter *1* (Fig.5.1) the current waveforms are stable up to V = 9.23 kV and follow the voltage waveforms. This suggests that the current is a purely field-electron one. With a small increase in the voltage ($\simeq 600$V), an explosion occurred (see the second trace in Fig.5.3a). Here two phases are prominent: the field emission one with a current of $\simeq 0.2$A and the transient (explosive) one in which the current has reached $\simeq 1.5$ A for 5 ns. The explosion delay time was 35 ns. These data correspond to the last point on the Fowler-Nordheim curve for this emitter. The same effect for emitter *2* (Fig.5.1) is illustrated by the current waveforms in Fig.5.3b. Stable field-electron emission was observed up to V = 7.7 kV. As the voltage was increased by approximately 3%, an explosion occurred with a delay time of $2.4 \cdot 10^{-7}$ s. The field emission current waveform shows a spontaneous increase not long before the explosion, analogous to that observed on the microsecond time scale [5.4,6,7]. The existence of this phenomenon enable us to define the physical explosion delay time as the period before the beginning of the spontaneous current rise. Shown in Fig.5.3c are the current waveforms for $t_p < 2$ ns. Both current traces were obtained at almost the same voltage. The upper one corresponds to a purely field-electron current. On the lower one the transition to a breakdown with a delay time of less than 1 ns can be identified. The Fowler-Nordheim plot recorded after the action of this pulse has shown that the destruction of the emitter tip took place. The rate of the current rise at the moment of explosion was 10^9 A/s.

5.1.3 The Point Explosion Delay Time

By using emitters with various tip radii and by varying the voltage pulse amplitude we were able to find the relationship between the point explosion delay time t_d, and the pre-explosion emission current density j. In calculating the current density it was conventionally assumed that the emitting area was equal to r_e^2. A plot of t_d versus j is given in Fig.5.4. It can be seen that all the experimental points fall on a straight line with a slope equal to -2. It follows that the product of the square of the current density and the emitter explosion delay time is a constant over a wide range of delay times (approximately eight orders of magnitude) and current densities j (about three orders of magnitude). We find from this plot that $j^2 t_d = 4 \cdot 10^9$ A^2s/cm^4. Also shown in Fig.5.4 is a plot of the delay time t_d versus the electric field at the emitter tip E_0 for the same experimental points. As the electric field E_0 is increased from $7 \cdot 10^7$ to $1.3 \cdot 10^8$ V/cm, the crictical current density increases from $4.5 \cdot 10^7$ to $2.2 \cdot 10^9$ A/cm^2 resulting in shortening

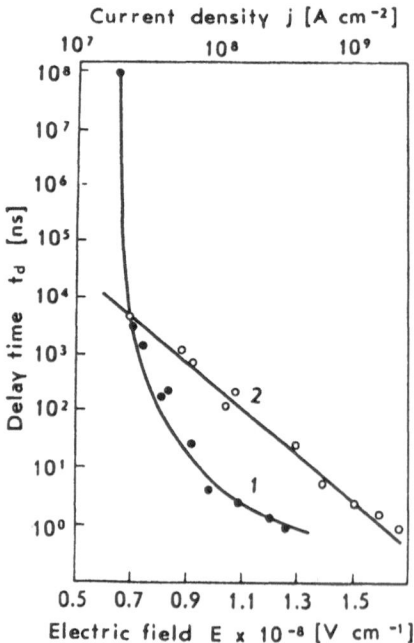

Fig.5.4. Explosion delay time of a tungsten field emitter as a function of the electric field (*1*) and the current density (*2*)

of the emitter explosion delay time from $4 \cdot 10^{-6}$ down to $1 \cdot 10^{-9}$ s. Decreasing the field E_0 only from $7 \cdot 10^7$ to $6.5 \cdot 10^7$ V/cm results in a lengthening of the time t_d from $4 \cdot 10^{-6}$ to 10^{-1} s. Thus, the experimental results indicate an extremely strong dependence of the explosion delay time on the electric field at the point tip.

5.1.4 Calculation of the Emitter Heating

Now we consider the heating of a point emitter by the self electron emission current. In spherical coordinates, the equation of heat balance for the cathode can be written in the form

$$\rho c \frac{T}{t} = \left(\frac{2 \partial T}{r \partial r} + \frac{\partial^2 T}{\partial r^2} \right) + \frac{i^2(t) \kappa(T)}{\Omega^2 r^4} , \tag{5.1}$$

where ρ, c, and κ are, respectively, the density, the specific heat, and the resistivity of the cathode material; $\Omega = 4\pi \cdot \sin^2(\theta/4)$ is the solid angle of the point cone. Since the emitters used in the experiment had a small angle θ and $h \gg (\lambda t_d / \rho c)^{1/2}$, the emitter may be assumed to be cylindrical in order to simplify the calculation, and heat conduction losses can be neglected. Moreover, we shall ignore, in the first approximation, a surface heat source. After simplifying (5.1) we obtain

$$\int_0^{t_d} j^2(t) dt = \rho \int_{T_0}^{T_{cr}} \frac{c(T)}{\kappa(T)} dT . \tag{5.2}$$

If we assume $j(t) = \text{const.}$, $\kappa = \kappa_0 T$, $c(T) = \text{const.}$, and $T = T_0$ at $t = 0$, then we can find from (5.2) the time t_d, after which some critical temperature T_{cr} will be reached at the tip, i.e.,

$$j^2 t_d \simeq \frac{\rho c}{\kappa_0} \ln \frac{T_{cr}}{T_0} , \tag{5.3}$$

$$t_d \simeq \frac{\rho c}{j^2 \kappa_0} \ln \frac{T_{cr}}{T_0} . \tag{5.4}$$

It is clear from (5.3) that the value of the product $j^2 t_d$ is independent of the current density and determined by the physical properties of the emitter material. To compare this with experiment it is necessary to take into account that Joule's heating is produced by the current with a density π times less than the conventional emission current density defined as i/r_e^2. Therefore, the right-hand sides of (5.3,4) should be multiplied by π^2. Then, setting $T_{cr} = T_m$ we obtain for tungsten $j^2 t_d = 2 \cdot 10^9$ $A^2 s/cm^4$. If T_{cr} is assumed to be the boiling point at atmospheric pressure, then $j^2 t_d = 2.7 \cdot 10^9$ $A^2 s/cm^4$, i.e., the $j^2 t_d$ value is weakly dependent on the choice of T_{cr}. The simplified calculation predicts a $j^2 t_d$ value which is about a factor of 2 lower than the experimental one.

For a more rigorous analysis of emitter heating one should take into account the surface heat source (the Nottingham effect) as well as the dependence of the field emission current on the temperature which can reach high values just before the explosion. The problem of nonstationary heating of a cylindrical emitter, taking into account the Nottingham effect, the field-assisted thermionic emission, and the temperature dependence of the material resistivity, was solved in [5.8]. The relation between the current density j and the time t_d was obtained in the form

$$j^2 t_d \simeq \frac{2.2\pi^4}{4} \frac{\rho c}{\kappa_0} \simeq 55 \frac{\rho c}{\kappa_0} . \qquad (5.5)$$

From (5.5) we obtain for tungsten $j^2 t_d \simeq 4.5 \cdot 10^9$ $A^2 s/cm^4$, the value being in good agreement with the experimental results. Later, a similar problem was solved for a conical emitter [5.9-11].

It should be noted that a relation $\int i^2(t)dt = $ const. analogous to (5.3) was found experimentally in the studies of the electrical explosion of thin wires with $j = 10^7$ to 10^8 A/cm^2 and with the explosion delay time ranging from 10^{-8} to 10^{-6} s [5.12]. This relation named an "integral of action" defines the energy accumulated in the wire during the pre-explosion period. The "integral of action" for every conductor material remains constant until the moment of the explosion and is independent of the wire cross-sectional area, its length and the current density [5.12]. For most of the metals its value ranges from 10^8 to 10^9 $A^2 s/cm^4$ [5.12]. This is a direct indication of the identical nature of the phenomena that are responsible for the explosion of thin wires and field-electron emitters.

5.1.5 The Vacuum Discharge Delay Time

The results of the investigation into the explosion of the field-electron emitters allow one to analyze more carefully the breakdown delay time as a function of the average electric field for electrodes of large area (Fig.4.2). Before using (5.5), let us pay attention to the fact that for field emission current densities $j > 5 \cdot 10^7$ A/cm^2 the electronic space charge at the emitter tip strongly affects the path of the current flow in a diode with a field-emission cathode. To describe the path of the current flow one should apply the "3/2-power" law [5.4]

$$j = \frac{1}{9} \sqrt{\frac{2e}{m_e}} \frac{V^{3/2}}{d_{eff}^2} , \qquad (5.6)$$

where d_{eff} is a certain effective gap length ($V \simeq E_0 d_{eff}$). For a crude estimate $d_{eff} \simeq r_e$ can be assumed. Then we obtain from (5.5,6)

$$t_d = \frac{2.2 \cdot 10^4 \; m_e c \rho r_e}{2 e \kappa_0 \beta^3 E_{av}^3} . \qquad (5.7)$$

It has to be recalled that the measured time t_d is a sum of the true delay time t_d' and the time of the current rise to a level of $0.1 i_{max}$. The values of t_d' were computed for the two possible values in the case when several cathode flares occur simultaneously (n = 1 and n = 4) (Sect.7.6). Figure 5.5 shows an example of the dependence of t_d' on E_{av} drawn on a logarithmic scale for aluminum electrodes. This plot is

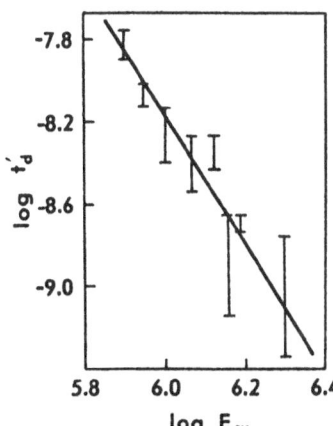

Fig.5.5. True breakdown delay time as a function of the average electric field for aluminum electrodes

a straight line with a slope of -3. From the plot one can easily determine the constant in (5.7) and further the value of r_e/β^3. For aluminum $r_e/\beta^3 \simeq 2 \cdot 10^{-12}$ cm. For real values $10^{-6} < r_e < 10^{-5}$ [cm], quite acceptable values of $80 < \beta < 170$ are obtained. Thus, we have shown that for electrodes of large area the breakdown initiation can also be satisfactorily explained in terms of EEE which occurs at cathode micropoints. The analogy established enabled us to suggest a simple method for the determination of the micropoint field enhancement factor. Introducing the value t_d = const. into the expressions for $t_d(E_0)$ and $t_d(E_{av})$ we obtain

$$\beta = E_0/E_{av} . \tag{5.8}$$

The calculation showed that for such materials as molybdenum, copper, and aluminum the value of β is in the range of 50 to 120, which corresponds to the values usually measured at vacuum breakdowns. It should be noted that for the real micropoint heights ($10^{-4} < h < 10^{-3}$ [cm]) such a comparison of $t_d(E)$ between the point and the large-area electrodes can be made for delay times $t_d < h^2 \rho c/\lambda = 10^{-8}$ to 10^{-6} s.

In conclusion, we note that the majority of the pulsed breakdown studies have been carried out under "commercial" vacuum conditions with the use of electrodes made of commercially pure materials. Therefore, a detailed comparison of the experimental data with the results predicted by (5.5,7) is unsatisfactory. It can only be said that - qualitatively - the character of the function $t_d(E_{av})$ remains the same as that for the "pure" conditions, but the breakdown electric field decreases. There may be several reasons for this:

1) Forming of carbon-containing films on the electrodes from products of the decomposition of vacuum oil in a discharge. Actually, from Fig.4.2 it is clear that for graphite electrodes the function $t_d(E_{av})$ has the same form as for metals, but the breakdown electric field is 1.5 to 2.5 times lower.

2) The presence of non-metallic films and inclusions on the cathode surface. As shown in Chap.2, the cathode regions having non-metallic inclusions and films are often more efficiently operating electron sources than the micropoints.

3) The presence of adsorbed gases and vapors on the cathode surface. Both the experiments [5.13-15] and estimates [5.14-16] show that the pulsed gas desorption in a strong electric field can result in a rapid formation of the gas-discharge plasma at the cathode, thus stimulating the initiation of the breakdown. The contribution of the adsorbed gas in applications of a pulsed electric field of $E \simeq 6 \cdot 10^7$ V/cm is equiva-

lent to a 2- to 3-fold enhancement of the electric field at the cathode [5.16].

5.2 Erosion of Point Cathodes

If the voltage pulse duration is longer than the explosion delay time of a field-electron emitter, the emitter tip is destroyed. The question whether the destruction has indeed an explosive nature can be answered upon having observed the history of the erosion process. The simplest method for studying erosion is by varying the duration of the EEE current pulses. In this case the amount of metal removed from an emitter can be determined by photographing the emitter prior to and after the EEE current pulse [5.17].

When we began our work on emitter erosion, only the publications by *Dyke* et al. [5.18, 19] were known to describe some rare measurements of the emitter deformation due to melting at the FEE transition to a vacuum arc. On measurement of the current in the "arc stage", *Dyke* et al. concluded that the emitter deformation could, in principle, be attributed to the Joule heating of the emitter tip by the arc current. However, these data appeared to be insufficient for inferring the character and mechanism of the cathode erosion, particularly in the high-current-density regime. Therefore, we have carried out a systematic study of these mechanisms for the cathodes having a typical geometry of the field-electron emitters in both the nanosecond and the microsecond ranges of pulse duration [5.20, 21].

5.2.1 The Fast Current Rise

In operating on the nanosecond time scale, single voltage pulses of duration t_p = 5, 20, 40, and 80 ns were used. From the EEE current waveforms given in Fig. 5.6a-d it can be seen that $t_d < 10^{-9}$ s, after which the current rises with a rate of the order of 10^9 A/s. On the traces corresponding to t_p = 80 ns a peak of current is observed about 40 ns after the voltage application. Shown in Fig. 5.7 are some typical profiles of molybdenum points before and after the action of the EEE current pulse. In the case of t_p = 5, 20, and 40 ns the metal was as a rule removed only from the point tip, while its lateral surface remained untouched. The tip region contiguous with the site of evaporation was melted. With t_p = 80 ns, a noticeable erosion of the lateral surface de-

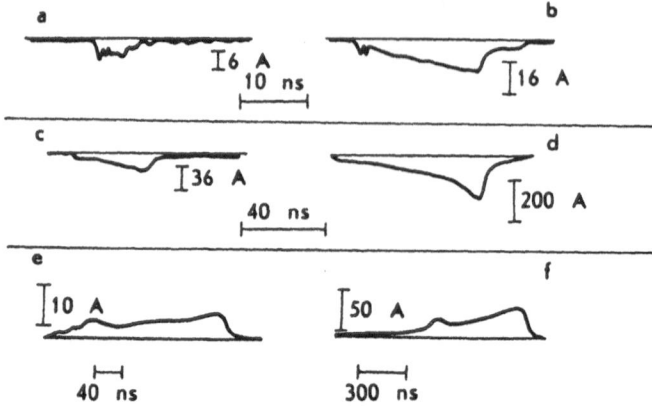

Fig.5.6. Typical EEE current waveforms obtained for molybdenum point cathodes

creasing with the distance from the point tip was observed for cathodes with $\theta > 6° - 8°$. Sometimes the initial phase of the process was also indicated with t_p = 40 ns. In the case of t_p = 40 ns the molten tip could be seen in the scanning electron microscope, while there were no visilbe

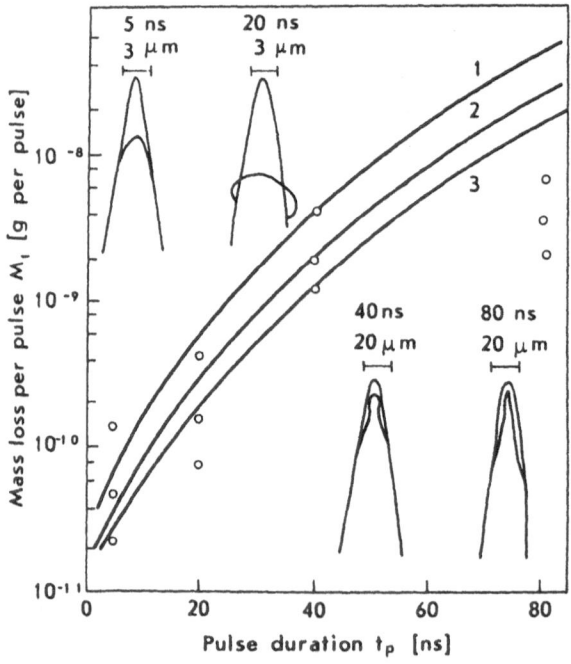

Fig.5.7. The mass of metal removed from the tip of a molybdenum point for the first current pulse, M_1, as a function of pulse duration. The point taper angle θ = 8° (1), 16° (2), and 24° (3). (Dots: experiment; solid lines: calculation)

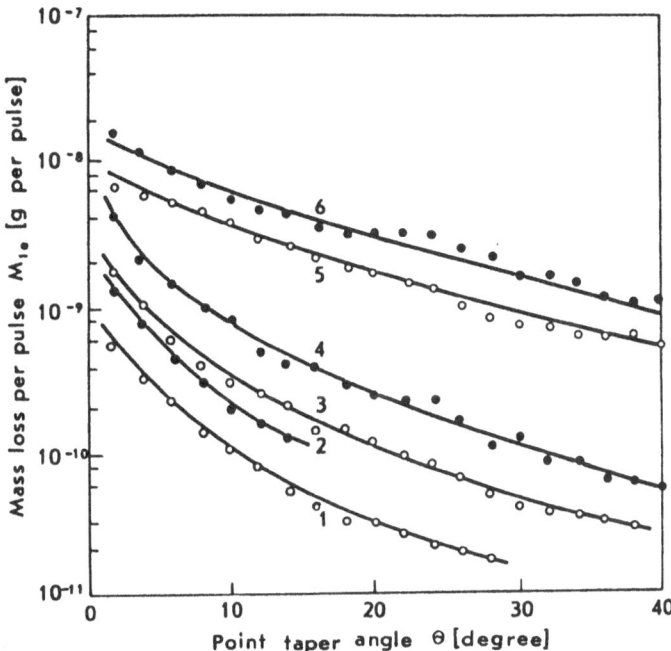

Fig.5.8. Experimentally obtained M_1 as a function of θ for field emitters made of molybdenum ($1,3,5,6$) and tungsten ($2,4$). t_p = 5 ($1,2$), 20 ($3,4$), 40 (5) and 80 ns (6).

traces of erosion on the rest of the point surface; with t_p = 80 ns microcraters appeared on the lateral surface, which indicated the occurrence of new emission centers.

Chosen as a basic experimental parameter was the relationship between the mass of metal removed from the point tip M_{1e} and the taper angle θ. Between 150 and 200 measurements were made for each value of pulse duration. When processing the results, all the point cathodes were divided into groups according to the angle θ within the group being 2°. The data were averaged over every group. The relationship obtained is presented graphically in Fig.5.8. It can be seen that as the angle is increased from 2° to 40°, the mass M_{1e} decreases by approximately one or two orders of magnitude. The decrease of the M_{1e} values with increasing θ was subsequently confirmed in the study of the erosion of tungsten points [5.22]. No effect of the point tip radius r_e on the mass M_{1e} was found. Practically no effect due to the spread in the M_{1e} values or the change of cathode material was seen on the electron current waveforms. Using the data of Fig.5.8, one can plot the mass removed from the point tip as a function of the pulse duration. This relationship is given in Fig.5.7 for three values of the taper angle θ. As is clear from this figure, the erosion of the point tip does not end with the initiation event. The metal mass M_{1e} sharply increases with the pulse duration. When some new emission centers (ECs) appear on the lateral surface region adjoining the point tip, not only does the increase of M_{1e} become slower with time, but the character of the damage to the tip itself changes.

5.2.2 The Slow Current Rise

In order to obtain the microsecond pulses of the EEE current without a transition to the discharge arc regime, it is necessary to widen the electrode separation d considerably. In our experiment the gap spacing was 3 cm. Figures 5.6e,f show some typical current pulse waveforms corresponding to this case. The average rate of the current rise was about $2.5 \cdot 10^7$ A/s for molybdenum points and about $1.7 \cdot 10^7$ A/s for copper ones ($V_0 = 30$kV). It has been established that in the case of the molybdenum emitters with $\theta \leq 10°$ both "strong" and "weak" erosion of points can be observed. Listed in Table 5.1 are the data for several points obtained for the cases when a significant tip erosion could be observed. For points with $\theta > 10°$ only weak erosion was always indicated.

Table 5.1. Calculated and experimentally obtained values of the mass removed from molybdenum points in a single voltage pulse

Point condition	Point number	θ [deg]	$M_{1e} \cdot 10^8$ [g]	$I \cdot 10^4$ [A²s]	$M_{1c} \cdot 10^8$ [g]
Annealed Mo	1	8	2.8	3.9	3.0
	2	8	2.6	3.9	3.0
	3	8	4.0	7.0	4.7
	4	7	3.3	2.7	2.8
	5	7	1.8	2.7	2.8
	6	7	5.4	5.3	4.4
Unannealed Mo	1	8	3.2	7.6	5.0
	2	8	2.4	2.2	2.0
	3	10	2.0	3.9	2.4
	4	7	2.5	2.2	2.3
	5	6	4.2	5.3	5.0
	6	8	4.9	5.3	3.8

Note: Here M_{1c} is the calculated mass loss per pulse; $I = \int_0^{t_p} i^2(t)dt$.

By studying the erosion of copper emitters it has been established that the erosion was significant for all points with a taper angle of $3°$ < θ < $7°$. The average (over 50 points) value of the mass removed from the cathode was about $2.5 \cdot 10^{-8}$ g. With angles θ = 10 to $20°$, both strong and weak erosion was observed. Increasing θ beyond $20°$ resulted in a strong decrease of the erosion of the point tips and in the formation of craters on their lateral surfaces.

It should also be noted that the character of the point erosion and its rate had no noticeable effect on the electron current waveforms. This is apparently due to the presence of a large fraction of material in the non-plasma phase in the matter removed from the point.

5.2.3 The Point Erosion Rate

The experimental results enable the point erosion rate and the current density in the destruction zone to be evaluated. As can be deduced from Fig.5.7, the average velocity of the propagation of the boundary of the destruction zone v_b = h'/t_p (h' is the height of the evaporated part of the point) can reach 10^5 cm/s for a few nanoseconds after the initiation of EEE current, a value comparable with the velocity of sound in the metal. This suggests that the erosion processes possibly have an explosive nature. As t_p is increased, the velocity v_b decreases and with t_p = 1.3 μs it is of the order of 10^4 cm/s ($\theta < 10°$).

For the cases where the current flows only through the eroding emitter tip, one can estimate the average current density in the phase transition region by the end of the pulse. Estimates for molybdenum and tungsten have shown that j decreases with time from $2.7 \cdot 10^8$ A/cm^2 (t_p = 5 ns) down to $1.1 \cdot 10^8$ A/cm^2 (t_p = 40 ns). With t_p = 1.3 μs and significant erosion of the point j $\simeq$ $5 \cdot 10^7$ A/cm^2 (for molybdenum and copper). The j values obtained indicate that in the initial stage of EC formation the current density in the emission center is not in excess of $3 \cdot 10^8$ A/cm^2. This is in good agreement with the result that for t_d $\simeq$ 10^{-9} s the prebreakdown current density is of the order of 10^9 A/cm^2.

5.2.4 Erosion Due to Joule Heating

Let us find an analytic relation between the mass removed from a cathode and the cathode geometry, the current flowing through it, and the cathode material constants, proceeding from the assumption that the main cause for the destruction of the cathode is the Joule heat release due to the flow of the discharge current. Using (5.1) and neglecting the term $\lambda \nabla^2 T$ we obtain the solution of (5.1) in the form

$$T = T_0 \exp\left[\frac{\kappa_0 \int_0^t i^2(t)dt}{\Omega^2 c\rho r^4}\right]. \tag{5.9}$$

From (5.9) one can determine r_{cr} at which some critical temperature T_{cr} is reached at the point where cathode destruction sets in

$$r_{cr} = \left[\frac{\kappa_0 \int_0^t i^2(t)dt}{\Omega^2 \rho c \, \ln(T_{cr}/T_0)}\right]^{1/4}. \tag{5.10}$$

From (5.10) the propagation velocity of the boundary of the destruction zone $v_b = dr_{cr}/dt$ can be found. If the current increases as a linear function of time $(i = kt)$, then

$$v_b = \frac{3}{4} \frac{k^{1/2}}{t^{1/4}} \left[\frac{\kappa_0}{3\Omega^2 \rho c \, \ln(T_{cr}/T_0)}\right]^{1/4}. \tag{5.11a}$$

For a constant current $(i = i_0)$

$$v_b = \frac{1}{4} \frac{i_0^{1/2}}{t^{3/4}} \left[\frac{\kappa_0}{\Omega^2 \rho c \, \ln(T_{cr}/T_0)}\right]^{1/4}. \tag{5.11b}$$

Shown in Fig.5.9 are some typical plots of the time dependence of v_b. In the same place the time function of the heat wave velocity $v_h = [\lambda/(\rho c)]^{1/2} t^{-1/2}/2$ is plotted. It is clear from the plots, that there are certain restrictions on the applicability of (5.9), because $v_b > v_h$ is valid not for all values of Ω, k, and i_0 so that one cannot neglect the influence of the heat conduction. Let us evaluate the role of the heat conduction with the use of (5.9), considering the ratio of the heat conduction losses to the Joule heat release

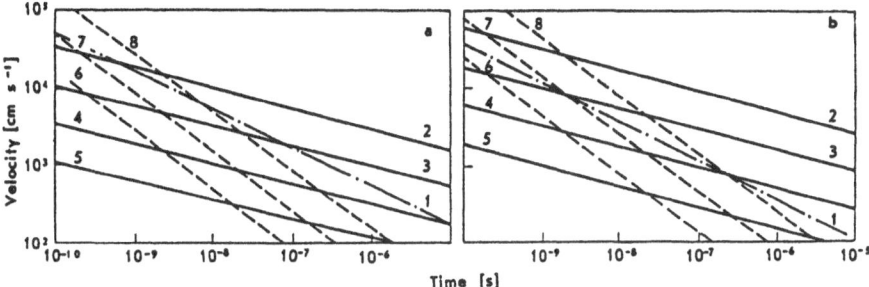

Fig.5.9a,b. The time dependence of the heat wave velocity (*1*) and the destruction boundary velocity (*2 - 8*) for plane copper (a) and molybdenum (b) cathodes. $di/dt = 10^{10}$ (*2*), 10^9 (*3*), 10^8 (*4*), and 10^7 A/s (*5*); $i_0 = 1$ (*6*), 10 (*7*), and 100 A (*8*)

$$\xi = \lambda \Delta T / \left[\frac{i^2(t)\kappa_0 T}{\Omega^2 r^4} \right] \simeq \frac{4\lambda \int_0^t i^2(t)dt}{\rho c \, i^2(t) r^2} \left[4\ln \frac{T_{cr}}{T_0} + 3 \right]_{r=r_{cr}}, \qquad (5.12)$$

from which it follows that

$$\xi \simeq \alpha_T \Omega / (kt^{1/2}) \qquad \text{for} \quad i = kt ,$$

$$\xi \simeq \sqrt{3} \, \alpha_T \Omega t^{1/2} / i_0 \qquad \text{for} \quad i = i_0 , \qquad (5.13)$$

where $\alpha_T = 2.2 \cdot 10^5$ A/$\sqrt{s}$ for molybdenum and $\alpha_T = 1.5 \cdot 10^6$ A/$\sqrt{s}$ for copper at $T_{cr} = 6 \cdot 10^3$ K. This indicates that the contribution of the heat conduction increases with time at a constant current, and that heat conduction decreases with an increasing current. In the latter case the greatest effect occurs within the shortest period. For example, with $k = 10^9$ A/s and $t = 2 \cdot 10^{-9}$ s, the factor $\xi \leq 1$ for $\theta < \theta_{lim} \simeq 30°$ and $\simeq 12°$ for molybdenum and copper, respectively. Thus, the above experimental results can be described in terms of the nonstationary Joule regime for $\theta < \theta_{lim}$. Assuming $r_{cr} \gg r_e$, we can find the mass of metal removed from the cathode by the end of the pulse as follows:

$$M_{1c} = \frac{1}{3} \rho \Omega r_{cr}^3 = \frac{1}{6\sqrt{\pi} \, \sin(\theta/4)} \left[\frac{\rho^{1/3} \, \kappa_0 \int_0^t i^2(t)dt}{c \, \ln(T_{cr}/T_0)} \right]^{3/4} . \qquad (5.14)$$

95

5.2.5 Comparison with Experiment

Figure 5.7 illustrates, together with experimental data, some plots of M_{1c} versus t_p. For all pulse durations except $t_p = 80$ ns the calculated and the measured M_1 values differ by no more than a factor of two. Therefore, one can ascertain that the Joule heating within the current constriction area could now account for the cathode erosion up to a time $t_p = 40$ ns. In the case of $t_p = 80$ ns the mass of metal removed from the tip turned out to be an order of magnitude less than that predicted by the model suggested. This is due to the appearance of some new ECs on the lateral surface of the point (the velocity v_b becomes less than the velocity of EC propagation along the lateral surface of the point v_{EC}).

Let us now turn to the data obtained in the microsecond range of pulse duration (Table 5.1). Both the experimental and the calculated values of the mass of metal removed are listed. Comparison of these values indicates that the suggested model satisfactorily describes the erosion for the cases when $v_{EC} < v_b$. The same conclusion could be drawn upon analyzing the data on the erosion of copper points. Thus, for $\theta = 3°$ to $7°$ it follows from (5.14) that $M_{1c} = 1.7 \cdot 10^{-8}$ g, while $M_{1e} = 2.5 \cdot 10^{-8}$ g.

5.3 EEE Current Density Measurements

The cathode current density at vacuum discharge is most important, since it is essential in understanding the mechanism of the electron emission. The results of a study on cathode erosion enabled us to suggest a rather simple method for determining the current density. For a point cathode, this problem could be solved in a simpler way than for a plane cathode.

5.3.1 Current Density of a Point Cathode

When deriving (5.14), we assumed that all the electron current flows through the disintegrating point tip, although a significantly greater part of its lateral surface is covered with plasma. The reasonable agreement between the calculated and the measured values of the mass M_1 for $\theta < \theta_{lim}$ suggests that in this case practically all the current goes into the flare plasma through the phase transition region, i.e., the emitting area on the cathode smaller than the cathode surface area covered with plasma. The average density of the current flowing through this area is defined here as $\bar{j} = i(t)/\Omega r_{cr}^2(t)$, or

$$\bar{j} = \sqrt{\frac{3\rho c \ \ln(T_{cr}/T_0)}{t\kappa_0}} \quad \text{for } i = kt , \tag{5.15a}$$

$$\bar{j} = \sqrt{\frac{\rho c \ \ln(T_{cr}/T_0)}{t\kappa_0}} \quad \text{for } i = i_0 . \tag{5.15b}$$

As follows from (5.15), the current density $\bar{j}$ decreases as $t^{-1/2}$ and is weakly dependent on the cathode material constants. Values of $\bar{j}$ predicted by (5.15a) show good agreement with those obtained experimentally. For copper, with t_p ranging from 5 to 100 ns we have $\bar{j} = 10^9$ to 10^8 A/cm^2.

5.3.2 Current Density from a Massive Cathode

Measuring the current densities on a massive cathode is a far more complicated task than on a point one, since in the former case one should not neglect heat conduction. We shall point out in Chap.10 that the well-known methods for the determination of the cathode spot current density give values which are too small. The study described below has enabled us to suggest a somewhat different method for measuring the current density on massive cathodes, which has been called an "erosion method".

This method was developed while investigating the re-usability of point cathodes operating in the EEE regime [5.17, 23]. Shown in Fig.5.10 are several pictures of profiles of the same point after a certain number of shots. It can be seen that, as a result of the current passage, not only metal is removed from the point tip, but there are

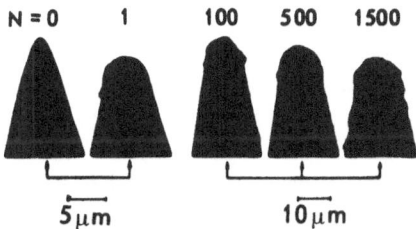

Fig.5.10. The variation of the profile of the molybdenum emitter tip varying with the number of current pulses, N

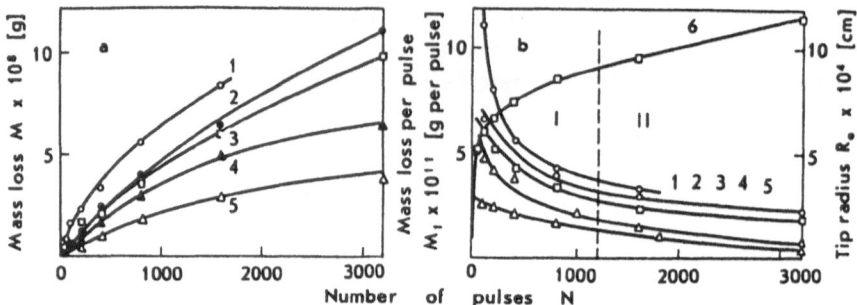

Fig.5.11a,b. Plots of the functions M(N) (a) and M_1(N) (b) for nickel (*1*), molybdenum (*2,3*), aluminum (*4*), and copper (*5*) points and a plot of the function R_e(N) for a molybdenum point (*6*). d = 1mm, V_0 = 30kV, t_p = 5 (*1,2,4,6*) and 10 ns (*3,5*); θ = 10° (*1,2,4*), 12° (*6*), and 24° (*3,5*)

considerable changes in the structure of its surface. The point surface at the tip and near it is covered with a great number of microprotrusions. An increase in the current pulse duration is accompanied by an increased probability of the appearance of microprotrusions of a relatively large size. The electric field at the microprotrusion tips is enhanced approximately by an order of magnitude, therefore, even at a point tip radius as small as R_e = 5-10 μm ($E_{av} \simeq 10^7$ V/cm), such a field enhancement is quite sufficient for the explosion of micropoints to occur during the pulse rise time. Thus, owing to the effect of the regeneration of micropoints, it is possible to use point cathodes with a tip radius of some tens of micrometers [5.17,23].

Presented in Fig.5.11a are some plots of the mass M removed from a point versus the number of the EEE current pulses, N. The various curves correspond to point cathodes made of different metals [5.23]. These curves allow the removed mass M_1 per pulse to be determined by graphical differentiation. Figure 5.11b shows the relationship between M_1 and N using the data of Fig.5.11a. In this figure a typical tip radius of a molybdenum point R_e as a function of the number of shots, N, is plotted. On the M_1(N) curves, two characteristic regions can be distinguished. In Region I, for small numbers N, a strong relationship between M_1 and N can be seen, while in Region II this dependence is significantly less pronounced. The transition between Regions I and II occurs at $8 \le R_e \le 12$ [μm] and $N \ge 10^3$. When $R_e >$ 12 μm, the mass M_1 is weakly dependent on the taper angle and becomes a more unambiguous characteristic of the specific mass removed for given cathode material. One can see from Fig.5.11b that the least loss of material takes place in the case of the copper cathode, although the variation in the M_1 values for the materials used is small. From the M_1 values in Region II, determined for the various pulse

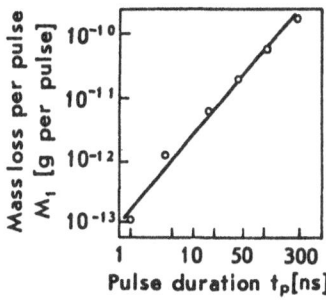

Fig.5.12. A plot of the function $M_1(t_p)$ for a copper point. $N = 10^5$

durations at N = const., one can ascertain how the rate of mass removal changes with t_p. In Fig.5.12 the dependence $M_1(t_p)$ for copper is given, from which it is clear that the cathode material is removed not only at the moment of the explosion of the primary microprotrusions, but also at subsequent times, i.e., the plasma generation by EEE never ceases while the current is flowing. Estimates have shown that under the experimental conditions described the cathode erosion rate amounts to 10^{-5} to 10^{-4} g/C.

In Region II (Fig.5.11) for all the metals examined the volume V_1 of the material removed per current pulse is 10^{-13} to 10^{-12} cm^3, i.e., its linear size can be estimated as $\sqrt[3]{V_1}$ = 0.4 to 1.0 μm $\ll R_e$. This means that Region II corresponds to a quasi-plane massive cathode. Thus, for a duration of the EEE current pulses of the order of 10^{-8} - 10^{-7} s and $R_e \geq 10^{-3}$ cm we really have dealings with massive cathodes.

5.3.3 Measurements Based on Erosion

The "erosion method" is suggested for just such massive cathodes. It is based on the experimentally established fact that on application of the next voltage pulse emission centers usually appear on the hemispherical top of the cathode leading to its erosion. If a point cathode is used, the radius of its tip increases during its operation. In the case of a cylindrical emitter this would mean a decrease of its height H_e. Let us represent a volume element V_1 of the mass removed as a cone-shaped hole of solid angle Ω at the cathode surface. We will also assume that the removal of metal always occurs from that cathode site where electron emission with a current density j takes place. Since it is unknown whether all emission centers appear at the same time or their number increases with the current from cathode, we shall consider the two limiting cases: 1) the current per EC is not in excess of a certain value

i_1 and the increase of current is entirely provided by EC multiplication, i.e., $n = i(t)/i_1$, where n is the number of emission centers; 2) on the operating (hemispherical) surface of the cathode n emission centers appear at the same time and the current rise is ensured by an increase of current in every center: $i_1 = i(t)/n$. In the first case, the total volume of metal removed from the cathode per pulse is determined by

$$V_{tot} = nV_1 = \frac{1}{3\Omega^{1/2}n^{1/2}} (i/\bar{j})^{3/2} = \frac{i \cdot i_1^{1/2}}{3\Omega^{1/2}\bar{j}^{3/2}} . \tag{5.16}$$

On the other hand, this volume may be represented as follows: for a point with a small angle

$$V_{tot} \simeq 2\pi R_e^2 \, dR_e/dN , \tag{5.17a}$$

for a cylinder

$$V_{tot} = 2\pi R_e^2 \, dH_e/dN . \tag{5.17b}$$

Setting the right-hand side of (5.16) equal to that of (5.17) we obtain: for a point

$$R_e = \left[\frac{i \cdot i_1^{1/2} N}{2\pi\Omega^{1/2}\bar{j}^{3/2}} \right]^{1/3} , \tag{5.18a}$$

for a cylinder

$$G_e = \frac{dH_e}{dN} = \frac{i_1^{1/2}}{6\pi\Omega^{1/2}\bar{j}^{3/2}} \left[\frac{j^3}{R_e^6} \right]^{1/3} . \tag{5.18b}$$

For the second case we supposed that at the moment when the voltage pulse is applied n emission centers appear on the cathode at the same time, the n value being proportional to the emitter tip area:

$$n = 2\pi R_e^2 \Pi , \tag{5.19}$$

where Π is the number of emission centers per unit area (cm^{-2}). Subsequent EC operation results in the formation of n craters. In this case the expression for V_{tot} takes the form

$$V_{tot} = nV_1 = \frac{i^{3/2}}{3(2\pi)^{1/2}\Omega^{1/2}\Pi^{1/2}\bar{j}^{3/2}R_e} , \tag{5.20}$$

and, allowing for (5.17), one can write:
for a point

$$R_e = \left[\frac{2i^{3/2}}{3\pi(2\pi)^{1/2}\Omega^{1/2}\Pi^{1/2}\bar{j}^{3/2}}\right]^{1/4} N^{1/4} , \tag{5.21a}$$

for a cylinder

$$G_e = \frac{dH_e}{dN} = \sqrt{\frac{1}{6\pi(2\pi)^{1/2}\Omega^{1/2}\Pi^{1/2}\bar{j}^{3/2}} \frac{i^3}{R_e^6}} . \tag{5.21b}$$

The volume element of the material removed from the cathode can be represented as a cylindrical hole in the cathode surface, through the base of which the emission current flows, the depth of which increases as metal is removed. It is easy to show that both in the case of simultaneously operating emission centers and in the case of their multiplication with increasing current we shall have $R_e \propto N^{1/3}$ and $G_e \propto (i^3/R_e^6)^{1/3}$. The object of the experiment was to find out the relationship between the tip radius R_e and the number of EEE current pulses N for a point cathode and the dependence of the erosion rate G_e on the parameter i^3/R_e^6 for a cylindrical one.

Experiments were carried out with copper emitters in the "commercial" vacuum (10^{-2} Pa). Voltage pulses of duration 1.5 ($V_0 = 20$kV), 5, 20, 50, 100, and 300 (30kV) ns were applied to the electrode gap at a repetition rate of 10 to 30 Hz. The point taper angle θ was 5 to 7°. The gap spacing d was adjusted in the range of 3 to 18 mm, allowing the current amplitude to be varied from 1.5 to 260 A. For each value of d and t_p between five and seven points were tested. When plotting the function $R_e(N)$ it was found that for a given gap, the same pulse duration and an identical number of shots, the spread of R_e was not in excess of 10% after 10^3 shots for all t_p. This suggests that there is apparently no essential correlation between the change in the point tip radius and the gap spacing. Some plots of $R_e(N)$ for all pulse durations are given in Fig.5.13. This relationship can be approximated by

$$\bar{R}_e = \bar{B} \cdot N^\alpha . \tag{5.22}$$

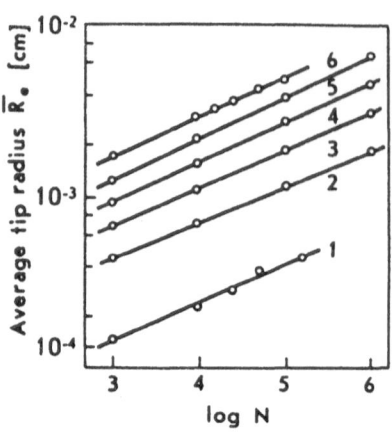

Fig.5.13. Plots of the function R_e (N) for copper emitters with t_p = 1.5 (1), 5 (2), 20 (3), 50 (4), 100 (5), and 300 ns (6)

Values of $\overline{B}$ and $\overline{\alpha}$ are tabulated in Table 5.2. Comparing (5.18a, 21a) with (5.22) and with the data of Table 5.2 indicates that the values of parameter α for all the pulse durations are close to 0.25.

5.3.4 Experimental Data

The operation of single cylindrical emitters was investigated with wires of radius 15, 25, and 40 μm, which protruded 1.5 cm above the cathode holder. Voltage pulses of duration 50 ns were applied to the anode at a repetition rate of 25 Hz. The current amplitude was varied from 20 to 120 A by varying the gap spacing (3 to 18 mm) and the voltage (V_0 = 15 to 33 kV). The conditions were adjusted so that EEE would be initiated during the voltage pulse rise time and EC would appear only on the top part of the emitter. The shortenting of the emitter height per pulse was measured as an average for $5 \cdot 10^4$ to $2 \cdot 10^5$ current pulses. Given in Fig.5.14 are the experimental results plotted as a function of G_e versus i^3/R_e^6.

Table 5.2. The values of $\overline{B}$ and $\overline{\alpha}$ corresponding to different values of the pulse duration

t_p [ns]	$\overline{B} \cdot 10^4$ [cm]	$\overline{\alpha}$
1.5	0.3	0.225
5	1	0.21
20	1.47	0.22
50	1.93	0.23
110	2.4	0.24
300	3.16	0.235

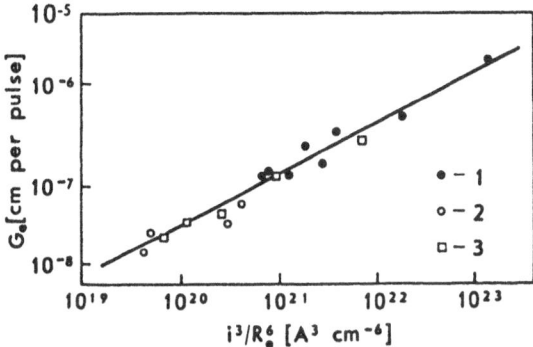

Fig.5.14. Plots of the function $G_e(i^3/R_e^6)$ with t_p = 50ns. R_e = 15 (*1*), 25 (*2*), and 40 μm (*3*)

It can be seen that all the experimental points fit rather well on a straight line with a slope of 0.5, thus the experimental relationship $G_e(i^3/R_e^6)$ can be approximated by the expression [5.20, 24]

$$G_e = 4 \cdot 10^{-18} \, (i^3/R_e^6)^{0.5} \, . \tag{5.23}$$

Comparison of the experimental relations (5.22, 23) with (5.21) enables one to conclude that an EC is a region of constricted current with a spherical geometry and that (5.19) is valid. Moreover, such a comparison allows the conclusion drawn that with t_p = const. and, with the condition of the EEE initiation during the voltage pulse rise time being satisfied, the product $\Omega\Pi\bar{j}^3$ is a constant within the experimental error. In view of this fact, the very slight dependence of the factor $\bar{B}$ on the gap spacing d found in the experiments becomes understandable. Indeed, if $\Omega\Pi\bar{j}^3$ = const., then $B \propto i^{3/8} \propto d^{-3/8}$, see (7.11).

Using (5.21b, 23) we find that with t_p = 50 ns the product $\Omega\Pi\bar{j}^3$ = $1.12 \cdot 10^{32}$ $A^3 \cdot sr/cm^8$. We can now estimate the parameters of the emis-

Table 5.3. Ω and $\Omega\Pi$ as a function of the cathode current density

$j \cdot 10^{-8}$	$\Omega\Pi \cdot 10^{-4}$	Ω [sr]		
[A/cm²]	[cm⁻²]	$\Pi = 10^5 \, cm^{-2}$	$\Pi = 10^6 \, cm^{-2}$	$\Pi = 10^7 \, cm^{-2}$
3	415	41.5	4.15	0.45
5	90	9.0	0.9	0.09
7	32.7	3.27	0.327	0.0327
10	11.2	1.12	0.112	0.0112
20	1.4	0.14	0.014	0.0014

sion area. The results are listed in Table 5.3. It is obvious that the angle Ω cannot be in excess of $2°$. As to the EC density, the value of 10^6 cm^{-2} can be considered to be fairy accurate. These values are estimated from SEM photographs of the cathode (one EC falls on an area of $10 \cdot 10$ μm^2). As follows from Table 5.3, by the end of a pulse of duration 50 ns the mean current density on the cathode is not less than $(3 \text{to} 5) \cdot 10^8$ A/cm^2.

5.4 Microstructure of the Cathode Surface

It has been mentioned above that the passage of the EEE current pulses is accompanied by erosion of the cathode leading to changes on the cathode surface. To examine the cathode surfaces at different values of the EEE current pulse parameters, two independent techniques were used: 1) SEM observations; and 2) measuring the effective electric field enhancement factor at micropoints [5.25].

5.4.1 Erosion Traces in SEM

The examination of erosion traces in SEM was carried out both for smooth electrically polished emitters after the action of a single current pulse and for emitters exposed to multiple pulsing. Most of the tests were conducted with copper and molybdenum cathodes. In the tests with electrically polished surfaces needles with taper angle $20°$ - $30°$ and tip radius 10-20 μm were used as the cathode. Every such cathode could be thought of as being a macrotip suitable for SEM observations. Using commercially pure materials for emitters in combination with oil pumping we were able, with $V_0 = 30$ kV and $0.08 \leq d \leq 0.8$ [cm] ($E_{av} \simeq 10^7$ V/cm), to ensure $t_d \simeq 10^{-9}$ s owing to the fact that the EEE initiation was facilitated by the availability of dielectric inclusions and films. The current pulse duration was controlled in the range $1.5 \cdot 10^{-9}$ to $5 \cdot 10^{-6}$ s. With $t_p \ll t_c$, we managed to observe the formation of the microstructure in the stage of current rise (di/dt = 10^9 to 10^{10} A/s). With $t_p \gg t_c$, it was possible to compare the erosion pattern in the spark and the arc discharge stages.

It was established that the elements of damage to the cathode surface are microcraters. Cathode microcraters formed in a time of the order of 10^{-8} s were first observed in the experiment described in [5.26]. Every microcrater forms as a result of the displacement of the liquid metal surrounding the EC under the action of the pressure developed in the emission region and of the subsequent solidification of the melt. It turned out that from the appearance, shape and distribution of the microcraters on the cathode, it was possible to deduce information on both the surface structure and the processes concerning emission centers (their origin, function, destruction and movement).

The appearance of microcraters, the distribution density, and the "qualitative" structure alter depending on the experimental conditions. The simplest form of microcrater is the "germ crater". As a rule, the germ crater occurs where there are defects in the cathode surface. Microcraters on a smooth surface are first seen when t_p = 5 ns. As t_p is increased up to about 100 ns, the crater size increases to around 3 to 5 μm. A further increase in the pulse duration results mostly in the appearance of the crater substructure (Fig.5.15). The crater substructure manifests itself in that new ECs appear on the rims of the existing craters. As is clear from Fig.5.15, the liquid metal is displaced in an irregular manner. The melt layers adjoining the cold cathode surface cool rapidly and form rolls which pile up on top of each other. This could already be seen at t_p = 20 ns. For example, in Fig.5.15a three or four such layers can be seen. This shows that the process of the liquid metal displacement has a sporadic nature and indicates that several cycles of EC appearance and disappearance took place. From Fig.5.15 it also follows that the "hotter" sections of liquid metal stretch, acquiring the form of micropoles, and solidify while elongating. The direction of stretching is close to the tangent to the cathode surface. The longest of the micropoles appear to lie on the cathode surface owing to the plasma pressure. On the micropole tips tiny drops solidify. The detachment of such a drop from the micropole results in the formation of a micropoint and a microparticle, both of which are less than 1 μm in size.

The results described question the dominant role of the stretching of micropoints from the liquid metal under the action of the space charge field [5.22, 27]. Firstly, the size of the stretched points ($h \simeq 10^{-4}$ cm) is several orders of magnitude greater than the thickness of the space charge layer ($\leq 10^{-6}$cm). Secondly, if the micropoints were stretched only by the space charge field, they would be normal to the cathode surface. These results explain in a simple way a well-known experimental result [5.28, 29] that in an arc discharge microdrops leave the cathode spot in a direction close to the tangential one.

The analysis of the photographs shows that new ECs arise not only on crater rims, but also at a certain distance from them. As a result of the appearance of new ECs the distribution density of craters on the tip increases and the zone in which craters are formed broad-

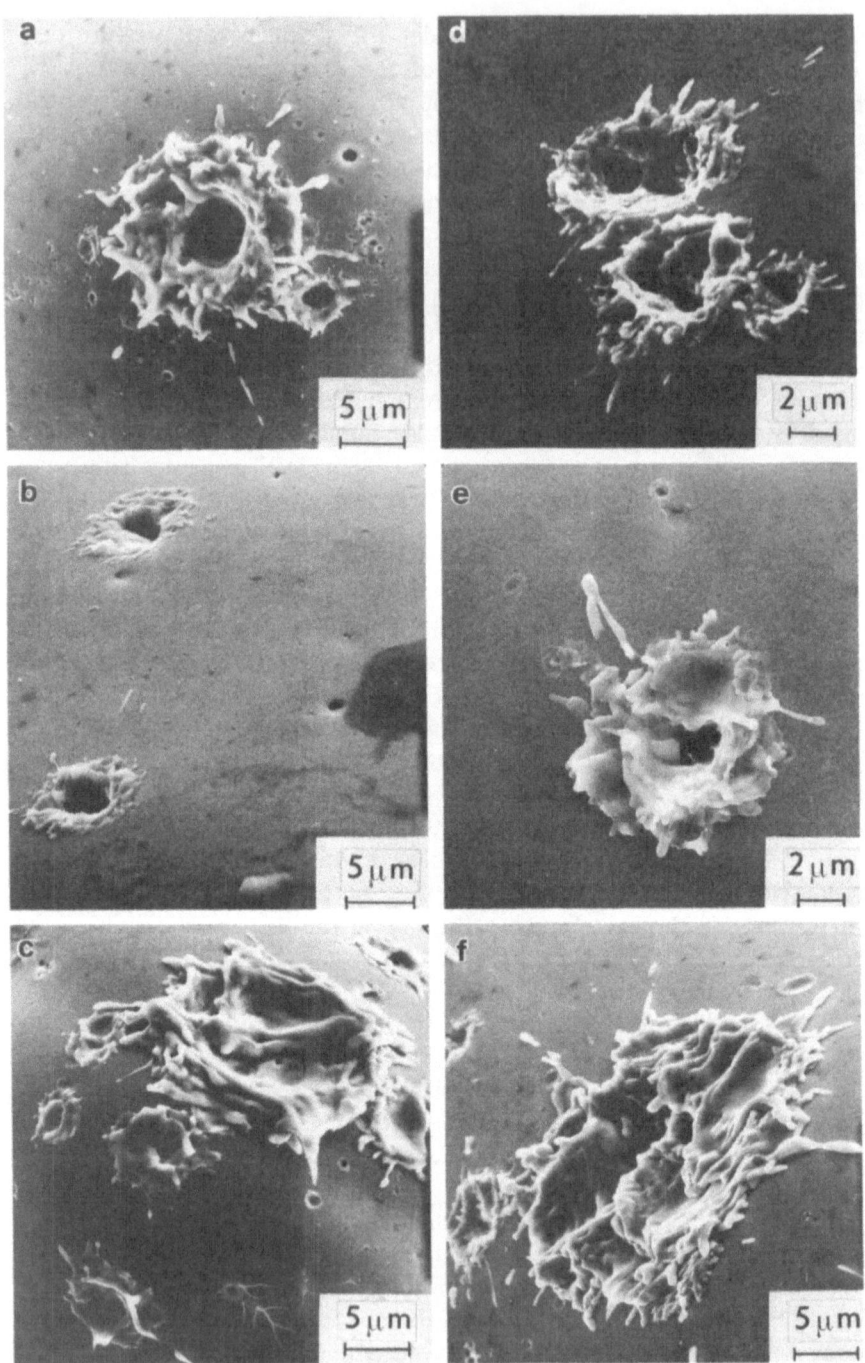

Fig.5.15a–f. Micrographs of the copper cathode surface after a single current pulse with t_p = 20 (a), 50 (b), 100 (c), 300 (d), 1300 (e), and 5000 ns (f). d = 3 mm, V_0 = 30 kV

106

ens. At t_p = 5 to 20 ns erosion traces are observed only on the top of the cathode. Increasing t_p results in the propagation of crater formation boundary onto the lateral surface of the point. At t_p = 5 μs ($t_p \gg t_c$) this boundary is more than 500 μm away from the top, i.e., its propagation velocity $v_{EC} \geq 10^4$ cm/s is in agreement with the calculated velocity of propagation of the glow boundary [5.20]. According to estimates obtained by analyzing photographs of the cathodes (di/dt = 10^9 A/s), the density of the crater distribution is of the order of 10^6 per square centimeter.

Photographs of the top part of copper and molybdenum cathodes after a great number of shots are shown in Fig.5.16. Here, it can also be seen that the microcraters are the elements of the cathode surface damage. The cathode microrstructure forms as a result of successive superposition of a great number of microcraters. As t_p is decreased, the mean crater diameter also decreases. For instance, at t_p = 1.5 ns the copper cathode surface is already wavy with the characteristic size of the irregularities being 0.1 to 0.2 μm. No microprotrusions with a large height-to-base ratio could be observed.

The examination of cathodes made of titanium, nickel, niobium, aluminum, copper, and lead leads one to the conclusion that the surface damage of these metal cathodes has common features.

5.4.2 The Field Enhancement Factor

The influence of current pulse parameters on the formation of microstructure of the cathode surface can be judged by the change of the field enhancement factor at micropoints which are formed from the liquid phase in the EC operation. Since our experiments were carried out in a "commercial" vacuum, it turned out that the application of the commonly used technique for measuring the factor β from the Fowler-Nordheim plots was practically impossible. Therefore we used the method described in Sect.5.3 in combination with the results of tests on the durability of cathodes [5.30]. In these tests it was established that as the cathode operates and its tip radius R_e increases, the delay of the current rise with respect to the moment of arrival of the voltage pulse at the gap starts to show up in the waveforms. Since the time t_d is uniquely related to the local field at the micropoints E_0, the factor β can be determined from (5.7b) assuming that the electric field at the top of an emitter of radius R_e takes the value E_{av}. Experimentally, it turned out to be most convenient to determine from the EEE current waveforms the "limiting" moment when $t_d \simeq$ 1-2 ns and $E_0 \simeq 10^8$ V/cm (Fig.5.4).

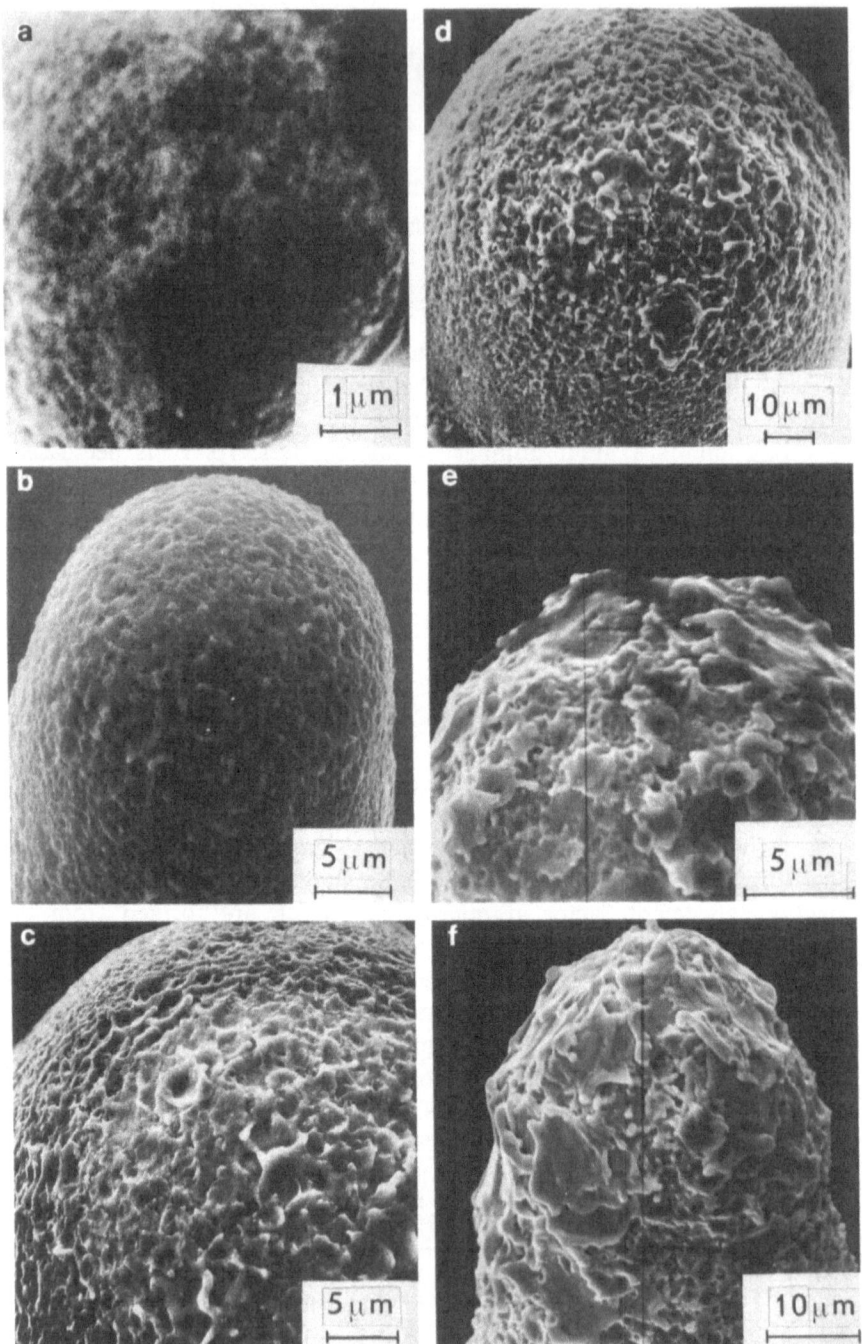

Fig.5.16a-f. Micrographs of the top section of copper ($N = 10^5$) (a-d) and molybdenum emitters ($N = 10^4$) (e,f). $t_p = 1.5$ (a), 5 (b,e), 20 (c), 50 (f), and 100 ns (d); d = 3mm, $V_0 = 30$kV

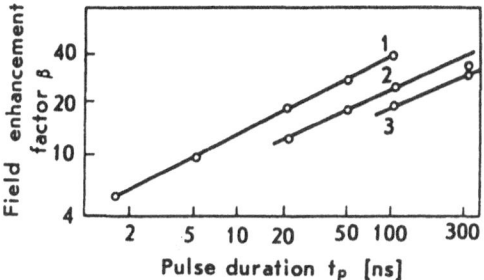

Fig.5.17. Field enhancement factor at micropoints as a function f of pulse duration for d = 3 (*1*), 12 (*2*), and 18 mm (*3*)

Figure 5.17 represents the dependence of β on t_p for three values of d. The weak effect of β on the gap spacing indicates that the EEE current amplitude has a small influence on the geometry of the micropoints as they are formed. The factor β is significantly dependent on the pulse duration: $\beta \propto t_p^{0.4-0.5}$. With t_p = 1.5 ns its value can be estimated as $\beta \simeq 5$, i.e., the micropoint tip radius is commensurable with its height. This agrees well with the SEM photographs of the cathode surface.

The above-described results are qualitatively similar to the data [5.31-33] obtained under UHV conditions for wire cathodes (Mo, Cu). *Jüttner* and co-workers [5.31-33] used the Fowler-Nordheim characteristics to investigate the parameters of the micropoints formed on the cathode as a function of the discharge pulse duration after a thorough conditioning of the electrodes by high-voltage nanosecond breakdowns. Histograms of the field enhancement factor [5.31] for several EEE current pulse durations are given in Fig.5.18. It can be seen that the histograms are shifted towards smaller values of β for the shorter pulse

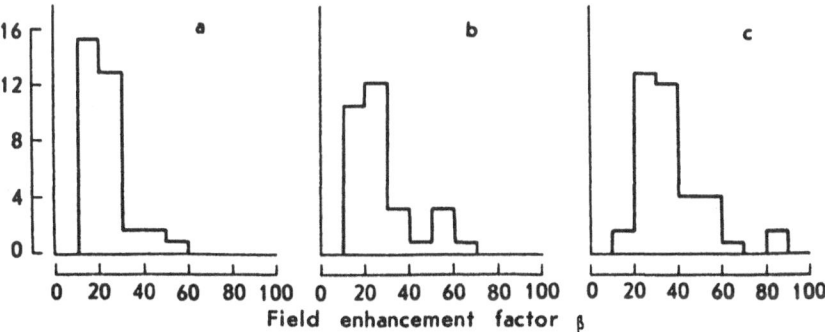

Fig.5.18a-c. Histograms of β values after breakdown conditioning with 3 (a), 5 (b), and 10 ns (c) pulses [5.31]

durations. Average β values are about 30, 25, and 18 for $t_p = 10, 5,$ and 3 ns, respectively. With $t_p = 10$ ns, the most probable values of the emitting area are in the range of $10^{-12} \leq S_e \leq 10^{-10}$ [cm^{-2}].

In an earlier investigation [5.34] it was established that a decrease of t_p to values of about 1 ns resulted in the failure of the regeneration of micropoints at subsequent discharges. This process was investigated in more detail with the use of SEM by *Jüttner* and co-workers [5.31-33], who were able to carry out their investigations under good vacuum conditions. They ascertained that there exists a minimum time of passage of the discharge current (3 to 5 ns) required for the formation of microcraters. They considered that the "polishing effect" which they observed at short durations of the EEE current pulses is due to the time lag in the motion of liquid metal from the crater. It should be noted that it is necessary to allow for the time it takes to form a liquid layer capable of movement.

5.5 The Contribution of Droplet Ejection to Cathode Erosion

From the SEM photographs of the cathode surfaces as well as from direct and photographic observations of cathodes at nanosecond discharges it is clear that part of the metal leaves the cathode in the form of molten microparticles. A special consideration of the emission of microparticles is required for several reasons. Firstly, the microparticles ejected may provide information about the action of the pressures in the EC region, which caused their acceleration. Secondly, from the pattern of the microparticle ejection one can draw some conclusions about the dynamics of the EC processes. Thirdly, a well-founded analysis of erosion characteristics requires data on the fraction of droplets in the total mass removed from the cathode.

In a study described in [5.35] the probe method was employed. Glass plates with copper grids, on which a graphite film of a thickness of the order of 10^{-6} cm was deposited, served as probes. The probes were placed 2 to 3 cm from the top of the cathode parallel to the electrode axis. The copper-graphite meshes were examined for deposited particles in a transmission electron microscope with a magnification of up to 25000X. A cathode in the form of a single emitter made of a thin wire 30 to 100 μm in diameter protruded 1.5 cm from the center of the cathode holder. The anode, a plane disc 1.8 cm in diameter, was 6 mm away from the cathode top. In order to stop the anode erosion products from being deposited on the probe, the latter was surrounded with a ring of 1.8 cm inner diameter which protruded 4 mm above the anode plane. Voltage pulses of 30 kV were applied across the gap at a repetition rate of 10 to 100 Hz. "Commercial" vacuum was maintained in the discharge chamber. The pulse duration was varied from 10 to 300 ns.

Experiments with cathodes made of various materials carried out at t_p = 50 ns showed that emission of microparticles always took place. The microparticle size was, as a rule, not in excess of 1 μm. The emission of microparticles from a copper cathode was studied in more detail in view of the fact that most other erosion experiments were carried out with copper cathodes. Probes were introduced into the discharge chamber only after application of about 10^4 conditioning pulses to the gap, by which time the cathode top had acquired a hemispherical shape, thus ensuring an equal probability for the appearance of emission centers anywhere on its surface. Therefore, one may suppose that the microparticles scatter from the cathode top isotropically. Special experiments with probes placed at an angle of 45° to the cathode axis, in which anode erosion was absent ($t_p \leq$ 50ns), confirmed this supposition. The total amount of material removed from the cathode was determined by the decrease of height of the cathode. In order to obtain the sediment sufficient for analysis (300 to 600 particles) on the probes, $3 \cdot 10^5$ to 10^7 current pulses were applied to the gap. The spherical shape of practically all particles indicated that their formation took place in the liquid phase (Fig.5.19a). The minimum drop size discernible in the electron microscope was 0.025 μm. Apparently, drops of a smaller size were also present, but they could not be distinguished from the microparticles formed as a result of the migration of vapor atoms to the energetically unbalanced sections of the graphite film. For all the pulse durations the diameters of the largest drops were less than the crater radius corresponding to a given pulse duration.

Figure 5.20a shows the distribution of the diameters of the drops. In all cases, with the exception of t_p = 10 ns, the distribution curves have maxima which are typically shifted to the region of high particle dispersion. The position of the maximum remains almost unchanged as the pulse duration is increased, the maximum itself corresponding to a microdrop diameter of about 0.1-0.2 μm. This result can be explained

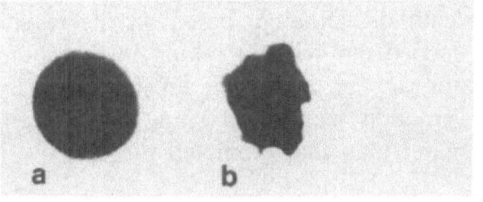

Fig.5.19a,b. Photographs of particles leaving copper (a) and graphite (b) cathodes

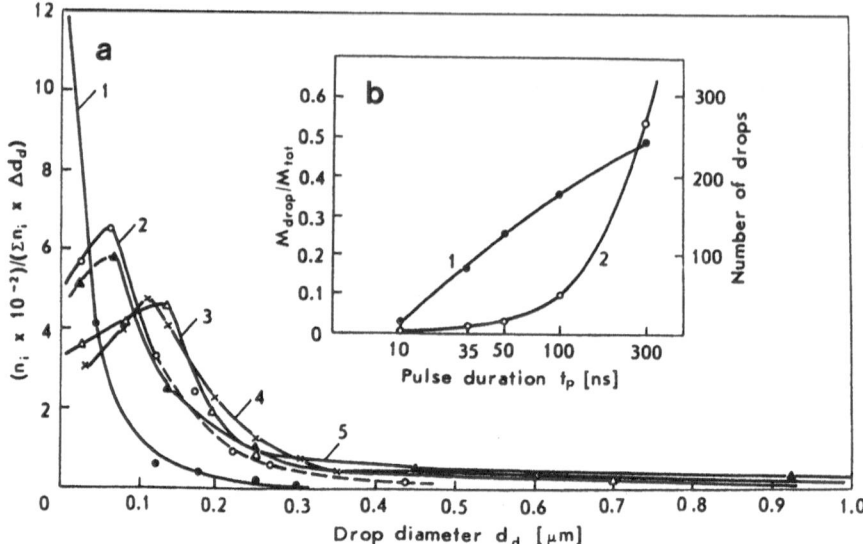

Fig.5.20a,b. Distribution of the drop diameter at t_p = 10 (*1*), 35 (*2*), 50 (*3*), 100 (*4*), and 300 ns (*5*) (a); the drop fraction (*1*) and the number of particles (*2*) as a function of t_p (b)

by noting that with a pulse duration of hundreds of nanoseconds, ECs appear and disappear many times. By analogy with arc discharge studies [5.36], the number of drops per unit charge carried across the gap was calculated. It turned out to be about $(1 \text{ to } 3) \cdot 10^7 \text{ C}^{-1}$. The mean number of particles leaving the cathode per pulse increased with t_p (Fig.5.20b). Proceeding on the assumption that all the drops had a spherical form and scattered isotropically, estimates were made of both the liquid phase volume for each group of particles and the total volume of the drop fraction. It turned out, for instance, that in spite of the fact that the fraction of particles more than 0.5 μm in diameter was not in excess of 10% (t_p = 50ns), they carried about 80% of the total drop fraction volume. It is also interesting that no liquid particles were ejected from craters during the first five nanoseconds (Fig.5.20b). As the pulse duration was further increased, the drop fraction increased and asymptotically approached the value that had been determined in the study of the quasistationary vacuum arcs ($\simeq$ 55%) [5.29].

Thus, the above-described measurements of the current density and the energy balance on the cathode, although based on the results of cathode erosion studies and carried out without taking into account the mass loss in the form of drops, appear to be well substantiated. By taking into account the drop fraction due to erosion, the current density $\bar{j}$ should be somewhat greater than that predicted by (5.21).

5.6 Pressure in the Emission Zone

The formation of the cathode microstructure and the ejection of molten microdrops are due to the action of the pressure developed in the emission zone. It is of interest to estimate this pressure.

The existence of high pressures in the emission zone was found experimentally [5.37]: when a graphite cathode is operated in the EEE regime, the liquid phase traces can always be seen on its surface. Shown in Fig.5.21a is a photograph of a fragment of the graphite cathode top surface after $4 \cdot 10^5$ current pulses. Macrostructures tens of micrometers in size can be clearly seen. Their rounded shape is an indication of the fact that part of the cathode material was in the molten state during the process of emission. Figure 5.21b illustrates the appearance of the cathode surface between these macrostructures. The microstructure of these cathode sections has been formed by a great number of microcraters superimposed on each other. Also seen here are traces of graphite melting, however the craters are not so pronounced as those on metal cathodes are metallic. Some additional arguments in favor of the existence of the liquid phase on the cathode are obtained in examining the sediment deposited on the probes. It is shown that along with particles of irregular shape with sharp edges, which seemed to be formed due to the thermoelastic stresses [5.38], there exist spherical particles (Fig.5.19b). The results obtained indicate that the conditions for graphite melting are created not only in the emission zone, but also at a distance of tens of micrometers from this zone.

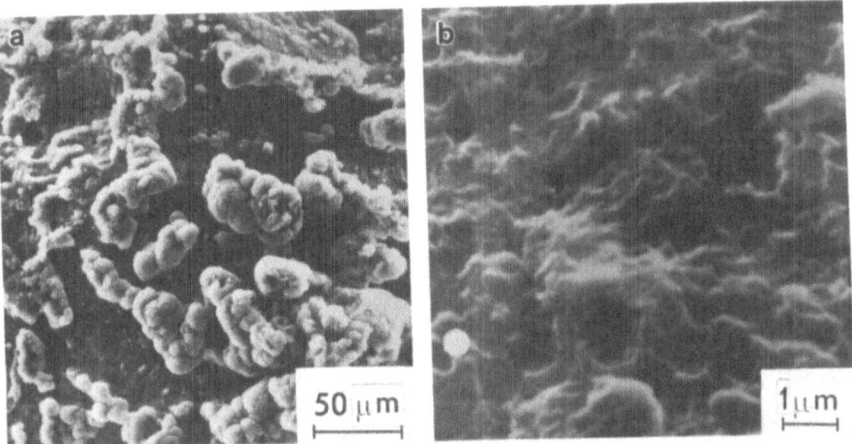

Fig.5.21. Micrographs of two fragments of a graphite cathode surface exposed to $4 \cdot 10^5$ current pulses

The minimum temperature and pressure that ensure the transition of graphite into its liquid state are determined by the triple point in the phase diagram. According to the data of [5.39], the triple point of graphite corresponds to a pressure $p_{tr} \simeq 10^7$ Pa and a temperature of $4100 \leq T_{tr} \leq 5000$ [K]. As follows from the phase diagram, the formation of the liquid phase is possible as a result of either vapor condensation or direct transition of solid graphite into liquid. The realization of the former way is not likely, as the vapor is practically completely ionized (Sect.6.2). The direct transition of solid graphite into liquid is the most probable reason for the graphite cathode melting.

We shall now estimate the pressure p_e developed in the emission zone from the distribution of the plasma density along the CF radius, see (6.2), as

$$p_e \simeq p_{tr}(r_m/r_{cr})^2 . \qquad (5.24)$$

Here r_m is the characteristic radius of the molten objects on the cathode. Putting $r_m = 2 \cdot 10^{-3}$ cm and $r_{cr} = 10^{-4}$ cm, we obtain $p_e \simeq 4 \cdot 10^9$ Pa.

The same order of magnitude for the pressure in the emission zone is predicted by the MHD calculation of the initial stage of the explosion of a point cathode (Sect.6.4). Indeed, as is clear from the data of Fig.6.8, when the maximum energy input ($\epsilon_{np} \simeq 2.5 \cdot 10^4$ J/g) is reached in the cathode material, which is in the state of nonideal plasma ($\rho_{np} \simeq 10^{-1}$ g/cm^3), the pressure in this material is $p_e = \epsilon_{np} \rho_{np} \simeq 2.5 \cdot 10^9$ Pa. If at the same time the current density in the emission zone $j = 5 \cdot 10^8$ A/cm^2, then the specific pressure force acting on the cathode will be $(12 \text{ to } 20) \cdot 10^{-4}$ N/A.

5.7 Formation of Cathode Microstructure

The findings discussed in the present chapter lead to the conclusion that the explosive emission processes really occur at the cathode and that they have a cyclic nature. This is suggested by the observation that an emission center appears in an explosive manner, exists for a limited time ($t_c \leq 10^{-9}$ to 10^{-8} s), disappears, and then a new emission center occurs near the former one. The mechanism of this process is schematically illustrated in Fig.5.22.

It is assumed that the cathode surface is free from adsorbed gases, contaminations, and foreign inclusions. As the next voltage pulse is applied across the vacuum gap, the original micropoints explode. Due to the formation of CF plasma the current flowing through the gap starts rising. Correspondingly, the current flowing through each exploding point also starts to rise. In a time $h/v_b = 10^{-4}$ cm/10^5 cm/s $\simeq 10^{-9}$ s the original micropoints are destroyed at their bases. Thus, the primary ECs must cease because of a sharp increase in the heat conduction energy loss. Hence the time of the first cycle will be $t_c \simeq 10^{-9}$ s.

In accordance with the model by *Rakhovsky* [5.40], we suppose that due to a rapid increase of the cathode drop, favorable conditions are created for the remaining original micropoints and, hence, for the maintenance of the emission current. Then the current flow would be accompanied by a "smoothing" of the cathode surface for $t_p \gg 3$ to 5 ns, which contradicts the experimental data. Thus, the maintenance of

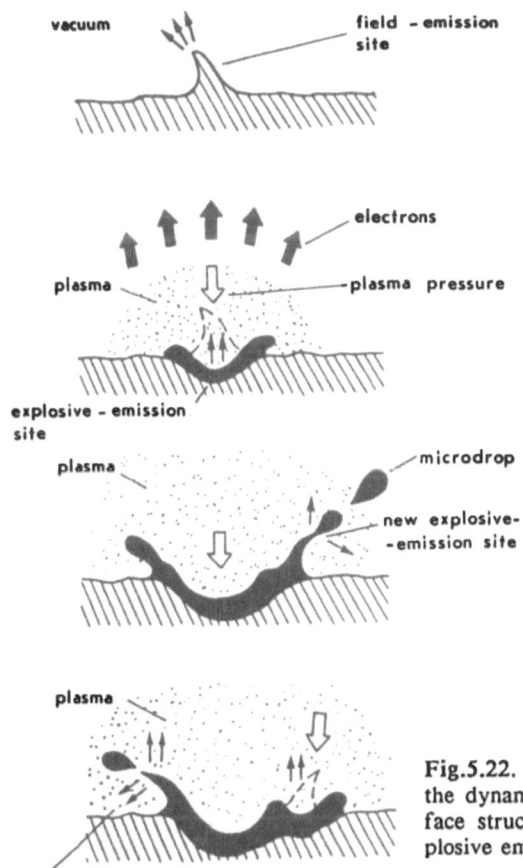

Fig.5.22. A schematic representation of the dynamics of the formation of the surface structure on a cathode revealing explosive emission

the emission from the cathode after the end of the first cycle should be accounted for by some other processes. We believe that, because of the redistribution of the current between a smaller number of most efficiently operating ECs, the conditions for the continuation of emission from them are ensured. Numerical simulation of the emission and erosion processes in a single EC will be discussed in Chap.11. Here, we only mention that from the calculated data the following result can be derived: If EC operates for several nanoseconds, a layer of liquid metal of about 1 μm thickness forms beneath it. The same result was obtained by *Hantzsche* [5.41].

The liquid metal will be affected by the plasma pressure tending to displace it to the periphery of the crater. However, the liquid-metal layer is unable to start moving instantly because of the inertial and the viscous forces affecting it. *Hantzsche's* estimates have shown that a time of 2 to 5 ns is required in order that a liquid layer of thickness 0.1 to 1.0 μm to occur at a distance of 2 μm [5.41]. Thus, several nanoseconds should pass before the liquid metal will be noticeably displaced under the action of the pressure. If the pulse duration is shorter than this time, the disappearance of the microprotrusions and the smoothing of the cathode surface will mainly take place and no new micropoints will appear.

At longer pulse durations a noticeable displacement of the liquid metal will be observed. At first, the liquid layer is displaced to the periphery, thus forming a ring-shaped ridge. Subsequently, this ridge, due to the action of surface tension, pressure fluctuations, and hydrodynamical phenomena, disintegrates into separate "stems", which, in their turn, decompose into micropoints and microdrops. The maximum velocity of the displacement of the liquid metal can be found from the relation [5.42]

$$v_{1_{max}} = \sqrt{2p_e/\rho} , \qquad (5.25)$$

provided that the plasma pressure force is essentially greater than the surface tension and viscous forces. For example, for copper from (5.25) it follows that $v_{1_{max}} \leq 10^5$ cm/s. A more detailed analysis of the motion of the liquid layer from the crater and of the escape of microdrops has been given by *Hantzsche* [5.43]. For copper drops from 0.1 to 1.0 μm in diameter the escape velocity is estimated as $(1 \text{ to } 7) \cdot 10^4$ cm/s. *Utsumi* and *English* [5.36] carried out experiments with low-current vacuum arcs and found drops about 1 μm in size leaving the cathode with a velocity of up to $5 \cdot 10^4$ cm/s.

The use of nanosecond pulses comparable in duration with the period of a single explosive emission cycle made it possible to suggest two other possible reasons for the spontaneous jumping of emission centers to the "colder" regions of the cathode. They are directly related to the drop formation. One of these has been discussed by *Hantzsche* [5.44]. He considered that just after the break-away of a drop a strong electric field occurs between the drop and the tip of the liquid point, and an intense field-assisted thermionic emission starts which is capable of leading to explosion. We have considered another reason [5.16,45,46]. Namely, that before the drop breaks away, when a very thin neck remains between the drop and the point, there are favorable conditions for the drop to explode. Further details are deferred to Chap.8, we only note here that such explosions may determine the duration of the cycle.

After the end of the current pulse the plasma rapidly disintegrates leaving a trace in the form of a crater with uneven edges on the cathose surface. The final state of both the crater and the edge irregularities is dominated by the competing processes of cooling and solidification of molten objects, mainly due to the heat conduction, and by smoothing of small protrusions under the action of the surface tension forces. The characteristic times of these processes are estimated in [5.16,41,47]. The characteristic time of disintegration of a liquid microprotrusion can be estimated from (5.25), assuming that the pressure $p_\sigma = \sigma_t / r_e$ [5.48]

$$t_\sigma = h/v_\sigma = \sqrt{h^3 \rho/(\sigma_t \beta)} \,, \tag{5.26}$$

where σ_t is the surface tension factor; v_σ is the rate of deformation of the point tip. The time of solidification of the point can be estimated, assuming that it is directly related to the removal of the phase transition heat $\lambda \Delta T = \rho \epsilon_m v_{cf}$, where ϵ_m is the specific heat of melting and v_{cf} is the velocity of the crystallization front [5.48]

$$t_\lambda \simeq \rho \epsilon_m h^2/(\lambda T_m) \,. \tag{5.27}$$

The condition for the micropoints to be preserved from destruction is the inequality $t_\lambda < t_\sigma$ or [5.48]

$$h < \lambda^2 T_m^2/(\rho \epsilon_m^2 \sigma_t \, \beta) \,. \tag{5.28}$$

Setting $\beta = 10$ to 100 we obtain for copper $h < 10^{-3}$ to 10^{-4} cm, which is in good agreement with the experimental data described above.

6. Cathode Flare Plasma

The above-described studies of the vacuum pulsed breakdown phenomenon provide convincing evidence that the explosive-like occurrence of the cathode flare plasma plays a principal role in the kinetics of the discharge development. It is shown that the efficient interaction of the CF plasma with the cathode surface, which is characterized by a high emitting power of the plasma, provides a high discharge current flow through the gap. Therefore, a further insight into the physics of the vacuum breakdown and discharge is impossible without a detailed knowledge of the parameters and the dynamics of the CF plasma.

However, such studies anticipate complications because of some peculiarities. In contrast to the so-called "main body" of the plasma, the CF plasma is a current-carrying plasma ball with an extension of 10^{-1} to 1 cm and a time scale of 10^{-8} to 10^{-7} s. Typical for this plasma ball are high gradients of particle concentration and internal electric field as well as the essentially directed velocities of the components. These specific features of the CF plasma make it a difficult topic for research. Nevertheless, a great deal of necessary information has been obtained as a result of studies carried out at our Institute as well as at other laboratories. The present chapter gives the data on the CF plasma expansion velocity and other parameters and on the dynamics of the radiation emitted by the plasma. It concludes with some ideas on the kinetics of the processes occurring in the plasma and provides a model of the plasma expansion.

6.1 Velocity of CF Plasma Expansion

CF plasma expansion at vacuum breakdown gives rise to various effects. It is therefore possible to evaluate the initial expansion velocity by using several different techniques. One of these has already been mentioned. It consists of the determination of the expansion velocity of the glow boundary using a single-frame-operating EOIC [6.1]. Later

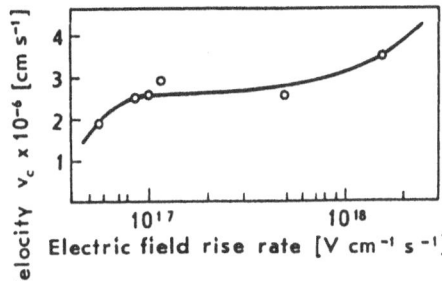

Fig.6.1. Expansion velocity of CF plasma as a function of the rise rate of the electric field at the tips of tungsten field emitters

on, at the Institute of High Current Electronics a number of other methods were employed, too.

6.1.1 The Grounded Grid and Collector Method

The grounded grid and collector method was first employed in the study of the explosion of field-electron emitters [6.2]. The device in which these experiments were carried out turned out to be well suited for measuring the plasma expansion velocity. Actually, when the CF plasma formed at the explosion of a field-electron emitter reaches the grounded grid, an arc discharge occurs between the cathode and the grid, and the collector current sharply decreases. By measuring the time of the current passage in the collector circuit t_{col} with the known emitter-grid separation d, one can find the velocity of the cathode plasma expansion $v_c = d/t_{col}$. The experiments were carried out under high-vacuum conditions with the use of a high-transparence grid at relatively low voltages, thus the effect of the anode plasma on the time t_{col} was practically eliminated.

Given in Fig.6.1 is the expansion velocity of the CF plasma produced by the explosion of a tungsten point plotted against the rise rate of the electric field at the point tip.

The results show that for low overvoltage coefficients (the points on the left-hand side of the plot) the velocity of the plasma expansion is (2 to 2.5)$\cdot 10^6$ cm/s. On the whole, the velocity v_c is seen to increase with the overvoltage coefficient. This may be due to increased energy delivered to the cathode material prior to the explosion. Later, the measurements of the velocity v_c under "commercial" vacuum conditions for cathodes made of various metals as needles with a tip radius of 10^{-2} cm were carried out at our Institute. One of the curves obtained is given in Fig.6.2. It is clear that the amplitude of the applied voltage pulse does not significantly affect the plasma expansion velocity. For Al, Mo, and Cu cathodes $v_c = 2.4 \cdot 10^6$, $2.2 \cdot 10^6$, and $2.4 \cdot 10^6$ cm/s were obtained, respectively.

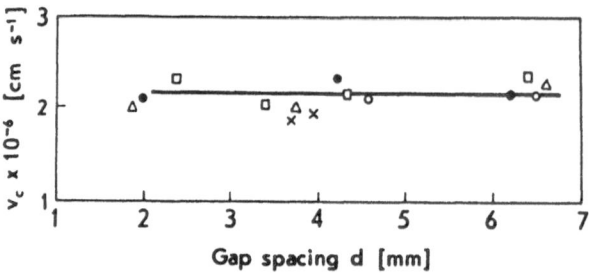

Fig.6.2. Expansion velocity of CF plasma as a function of gap spacing at various values of the applied voltage (Mo electrodes): ○ - 105 kV, × - 110 kV, △ - 120 kV, ▢ - 125 kV, ● - 175 kV

6.1.2 The Photoelectric Method

The photoelectric method was first applied to establish the times of the appearance of light in a vacuum gap at dc and pulsed breakdowns [6.3] (Chap.10). The propagation velocity of the cathode plasma front for copper and molybdenum electrodes amounted to about $2 \cdot 10^6$ cm/s. In the experiments described in [6.4] this velocity was evaluated from an increase of the plasma glow intensity at different distances from the top of the point cathode. It was determined by the "cut-and-try" method, i.e., in every case a value of the velocity was found at which the calculated coordinate-and-time dependence of the plasma radiation intensity was in accordance with the experimental observations. Using this method the plasma front velocity was determined for a number of materials (Cu: $2.6 \cdot 10^6$; Mo: $2 \cdot 10^6$, Pb: $1.1 \cdot 10^6$ [cm/s]).

6.1.3 The Transverse Magnetic Field Method

The main feature of the transverse magnetic field method has been described in Sect.4.4 [6.5]. When $y = 2m_e E/(eH^2) < d$, i.e., the electrons emitted by CF reach the anode after a certain delay time t_d. The velocity of propagation of the CF emission boundary across the gap can be determined from the relation $v_c = (d-y)/t_d$. It was observed that v_c decreased with H increasing approximately from $2 \cdot 10^6$ to $1.6 \cdot 10^6$ cm/s. This is associated with the plasma drift in perpendicular electric and magnetic fields [6.6].

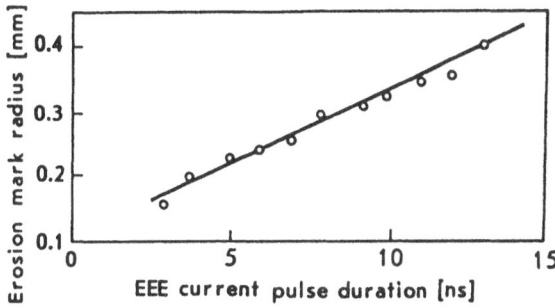

Fig.6.3. Radius of an erosion spot on a copper anode as a function of EEE current pulse duration

6.1.4 The Method of the Anode Erosion Mark

In order to determine the transverse velocity of the cathode plasma expansion we employed a method based on the fact that in the discharge spark stage the anode erosion is caused by a heavy electron flow. For a point cathode the erosion trace has the shape of a circle. From the rate of increase of the erosion spot radius the transverse velocity of the CF expansion can be estimated [6.7]. Experiments were carried out with a molybdenum point cathode and a copper plane anode with $d = 0.35$ mm and $V_0 = 35$ kV. A plot of the dependence of the anode erosion spot radius on the pulse duration is given in Fig.6.3. The erosion spot radius, as estimated from this plot, increases at a rate of $(2.2$ to $2.3) \cdot 10^6$ cm/s. The velocity of the transverse expansion seems to be of the same order of magnitude.

In this section we have discussed some methods used to determine the velocity of the plasma expansion during the initial stage, i.e., for a time of the order of 10^{-8} to 10^{-7} s. Subsequently, this velocity may become lower. This will be considered in more detail in Sect.7.2.

6.2 CF Plasma Parameters

6.2.1 CF Plasma Density

First estimations and measurements of CF plasma parameters were made at our Institute [6.8-14]. The plasma density distribution in CF was evaluated from the experimental data on the mass of the cathode material removed per unit time and from the plasma expansion velocity.

Let us assume that in a local region near the cathode surface a completely ionized plasma is being generated from cathode erosion products. We suppose that the plasma cloud is formed and expand in a spherically symmetrical manner with the velocity v_c. The heavy-particle density n_i can be evaluated from the mass conservation law [6.11, 12]

$$n_i(r,t) 4\pi r^2 dr = \frac{\dot{M}(t - r/v_c)}{m_i} dt , \qquad (6.1)$$

where $\dot{M}$ is the mass of the cathode material removed per unit time; (m_i is the ion mass). From (6.1) we obtain

$$n_i(r,t) = \frac{\dot{M}(t - r/v_c)}{4\pi m_i v_c r^2} . \qquad (6.2)$$

Plots of the function $n_i(r)$ calculated for three possible values of the mass removed per second are given in Fig.6.4. It follows that the plasma density near the cathode surface can be as high as 10^{19} cm^{-3}.

The existence of a dense plasma near the cathode was later confirmed experimentally with the use of optical interferometry [6.15]. For this a ruby laser with a pulse duration of 30 ns was employed. A vacuum spark was initiated between copper electrodes (needle cathode and hemispherical anode) with a gap spacing d = 0.7 mm. A voltage was applied across the gap through a controlled spark gap from a pulse-forming line with wave resistance of 5 Ω, which had been charged up to 20 kV, with a maximum discharge current of 4 kA. The pulse duration was 100 ns. The electron density n_e was measured at t = 10, 30, and 70 ns. It turned out that with t = 10 and 30 ns the electron density near the cathode was $n_e \simeq 2 \cdot 10^{18}$ cm^{-3}. With t = 70 ns the electron density was $n_e \simeq 10^{19}$ cm^{-3} at r $\simeq$ 100 μm. *Bugaev* et al. [6.15]

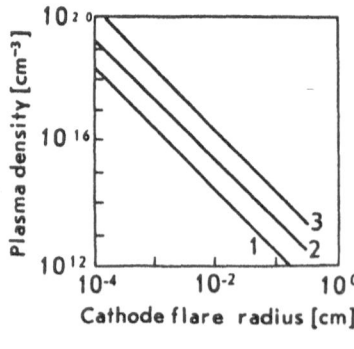

Fig.6.4. Radial distribution of the CF plasma density for the mass of cathode material removed per second $\dot{M} = 10^{-4}$ (*1*), 10^{-3} (*2*), and 10^{-2} g/s (*3*)

estimated the erosion rate from the measured cathode plasma density distributions and discharge current. It was about $5 \cdot 10^{-5}$ g/C, which is in rather good agreement with the results of some other researchers. This may be considered a substantial argument for the correctness of the above estimates.

The optical interferometry technique only enables one to determine the dense-plasma density ($n_e \geq 10^{17} cm^{-3}$ can be measured using a ruby laser). This is why laser interferometry in combination with electro-optical photography is widely utilized in studying the plasma generated in high-current vacuum diodes and self-magnetically-insulated vacuum lines used for ICF problems. For instance, an experiment carried out with a vacuum diode (current: 100 kA, voltage: 300 kV, pulse duration: 70 ns, gap spacing: d = 3 mm) has been reported [6.16]. The distribution of electron density was measured using interference holography. Analysis showed that the electron density was $n_e \simeq 10^{20}$ cm^{-3} at a distance of 100 μm away from the cathode. In another investigation [6.17], using the laser Schlieren method and electro-optical photography, spatial distributions of the electron density and temperature in a magnetically insulated vacuum line were determined. The experimentally obtained electron density near the cathode (r $\simeq$ 100 μm) was $n_e \simeq 10^{17}$ cm^{-3} and dropped to a half of its former value about 400 μm away from the cathode.

Thomson scattering was used to measure the parameters of the plasma edge layers [6.18]. A a ruby laser was used with a pulse duration of 25 ns and an energy of 2 J. The spectrum was scanned in steps of 1.0 nm. The electrode separation was 5 to 17 mm, the applied voltage 10 to 15 kV, and the current amplitude 5 to 7 kA. The diagnostic laser beam intersected the diode axis at a distance of 2.5 mm from the cathode 200 ns after the spark initiation of the breakdown at the cathode. It was established that as the discharge current was increased within those limits, the electron density rose from $5 \cdot 10^{13}$ to $2 \cdot 10^{14}$ cm^{-3}.

Jüttner with co-workers [6.19, 20] used Langmuir probes to obtain the density distribution of plasma produced by a pulsed vacuum arc cathode spot (i = 30 to 100A). They established that the electron density n_e falls as r^{-2}. According to their estimates $n_e \simeq 10^{19}$ cm^{-3} at a distance r $\simeq 10^{-3}$ cm from the cathode. A mass-spectroscopic study of the parameters of CF plasma produced by an explosion of micropoints on liquid-metal cathodes [6.21] has revealed that such electron densities really exist near the surface of explosively emitting cathodes.

The experimental data indicate that the plasma density near the cathode can actually reach a value of the order of 10^{19} cm^{-3} and then it falls away from the cathode. However, the true nature of the space-time distribution of particles in cathode flares (CFs) still remains unclear.

6.2.2 CF Plasma Composition and Temperature

A spectral study of the plasma composition and the temperature in CFs were first described in [6.13]. Aluminum electrodes were used (a needle cathode and a plane disc anode). A voltage pulse with parameters $V_0 = 30$ kV, $t_p = 7$ ns was applied across the gap with a separation $d = 0.5$ mm, the current amplitude reaching 100 A. The CF plasma mainly consisted of singly- and doubly-ionized aluminum atoms. The radiation of Al^{+2} ions appeared much earlier than that of Al^{+1} ions. The intensity of the spectral lines dropped with distance from the cathode (Fig.6.5). From the character of the Al^{+2} afterglow, which resembled a recombination afterglow, the existence of triply-ionized aluminum atoms in the CF plasma was deduced. Assuming a local thermodynamic equilibrium *Baksht* and co-workers [6.13, 14] determined the electron temperature T_e from the ratio of the intensities of Al^{+2} lines at the wavelengths $\lambda = 452.9$ and 447.9 nm, yielding $T_e \simeq 4.5 \pm 0.8$ eV. *Hinshelwood* [6.22] who used the same method with Al^{+2} and other lines found that the electron temperature is 3 to 6 eV.

In some later studies the above-reported values of CF plasma parameters were confirmed. The above-mentioned study [6.21] has shown that ions with charge numbers $z = 1$ to 5 were present in the CF plasma. It was observed that for $z = 1$ to 3 the higher the charge

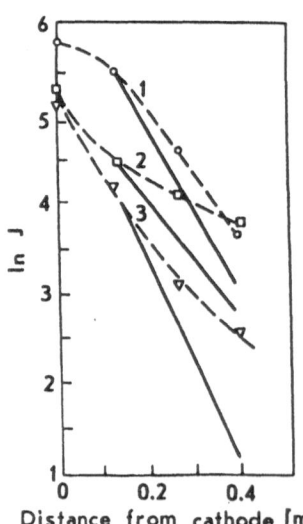

Fig.6.5. Al III spectral line intensity as a function of the distance from the cathode with $\lambda = 451.2$ (*1*), 572.2 (*2*) and 415 nm (*3*). Solid lines show calculated results

number, the higher the corresponding ion fraction was, while for z = 3 to 5 the reverse was true. The plasma electron temperature was about 5.5 eV. According to [6.17], the cathode plasma electron temperature in a magnetically insulated vacuum line was $T_e = 2$ eV at a distance of (3 to 4)$\cdot 10^{-2}$ cm from the cathode. From Thomson's scattering experiments it follows that $T_e = 1.4 \pm 0.2$ eV at 2.5 mm from the cathode [6.18].

Elementary processes occurring in the CF plasma have been analyzed by *Litvinov* [6.14], who considered the inelastic processes of the kind a+b → a′+b′. To estimate the collisions numerically it is necessary to know the relaxation time τ_{ab}. The reaction will take place if the corresponding relaxation time τ_{ab} is small ($\tau_{ab} < t$, t is the time from the beginning of CF expansion) and the free path is less than the flare size ($v_{Ta}\tau_{ab} < v_c t$, v_{Ta} is the thermal velocity of the component a). In determining the relaxation time τ_{ab} the time and coordinate dependence of the plasma density is taken into account. From this, the time, for which the given reaction is essential, can be found. For an electron-impact ionization ($a_i + e = a_{i+1} + 2e$) the reaction times $t_{a_i e}$ are determined and listed in Table 6.1. In order to find out about the types of ion which constitute the flare, it is necessary to compare $t_{a_i e}$ with the time $t'_{a_i e}$ for which the local ionization equilibrium can be established ($n^i \simeq n^{i+1}$). The inequality $t_{a_i e} > t'_{a_i e}$ is the condition that ions with a charge number i are present in the flare. As can be seen from Table 6.1, for an aluminum cathode, ions of all orders of ionization except the fourth should be present in the CF plasma. However, since the radiation from the triply-charged ion Al^{+3} is attributed to the UV region, it is impossible to record it with the instruments used. At the same time, the existence of the ion recombination afterglow indicates

Table 6.1. Reaction times for an electron-impact ionization in CF plasma

Reaction	Ionization pot. [eV]	$t_{a_i e}$ [s]	$t'_{a_i e}$ [s]	n^i/n [%]
$Al^0 + e \rightarrow Al^{+1} + e$	6	$1.3\cdot 10^{-8}$	$4.9\cdot 10^{-12}$	1
$Al^{+1} + e \rightarrow Al^{+2} + e$	18.82	$6\cdot 10^{-10}$	10^{-11}	16
$Al^{+2} + e \rightarrow Al^{+3} + e$	28.44	$5\cdot 10^{-11}$	$3\cdot 10^{-11}$	84
$Al^{+3} + e \rightarrow Al^{+4} + e$	119.96	$2.2\cdot 10^{-20}$	$3.2\cdot 10^{-6}$	-

the presence of Al^{+3} ions in the CF plasma. The data of Table 6.1 confirm the experimentally established fact that the intensity of Al^{+2} lines is higher than that of Al^{+1} lines.

Solving a set of differential equations describing the population of higher Al^{+2} levels, *Litvinov* [6.14] found the line strength of these ions as a function of the distance from the cathode (solid lines in Fig.6.5). *Litvinov* attributed the deviation of the calculated curves from those obtained experimentally to the fact that the recombination of Al^{+3} ions which undergo transition preferentially to the higher Al^{+2} levels is not accounted for in the calculation. The recombination is generally observed after the pulse terminates, i.e., when the plasma has covered a distance from the cathode $r \geq v_c t_p \simeq 0.2$ mm.

6.3 EEE Current Effect on the Dynamics of the Plasma Light Emission

By analyzing carefully the results of [6.13] one notices two significant features. Firstly, the recombination pointed out by *Baksht* et al. [6.13] seems to occur throughout the plasma only after termination of the current pulse. Secondly, the electron temperature T_e was measured by *Baksht* et al. [6.13] 20-25 ns after the current pulse, i.e., when the plasma had expanded and consequently cooled down. Therefore, it is reasonable to suppose that the electron temperature could be significantly higher at earlier times. These observations triggered us to investigate in more detail the space-time characteristics of the CF plasma glow, depending on the rate of current rise and its duration. The availability of such information allows one to decide whether there exists any noticeable energy flux from the internal regions of the cathode flare toward its periphery. The respective investigation was carried out at our Institute [6.23,24] under the following experimental conditions:

(i)	$V_0 = 48$ kV	$t_p = 75$ ns	$d = 1$ and 2 mm ,
(ii)	$V_0 = 24$ kV	$t_p = 15$ ns	$d = 1$ mm ,
(iii)	$V_0 = 24$ kV	$t_p = 75$ ns	$d = 2.8$ mm ,
(iv)	$V_0 = 24$ kV	$t_p = 200$ ns	$d = 4$ mm .

This set of experimental conditions allowed current rise rates di/dt to be obtained in the range of $7 \cdot 10^8$ to 10^{10} A/s. The electrodes were copper or aluminum (a needle cathode and a plane-disc anode). The vacuum system was pumped by a diffusion pump. It has been established [6.24] that the radiation spectrum has two components appearing at different times - the continuous and the line spectra.

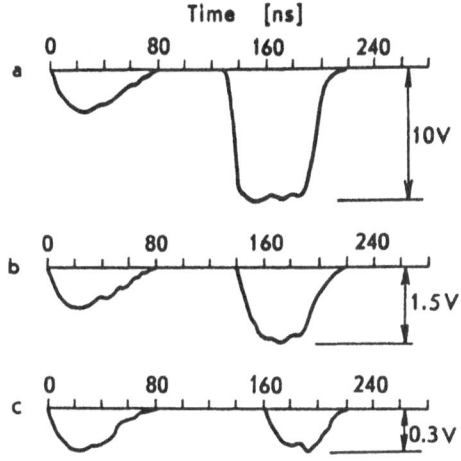

Time [ns]

Fig.6.6a-c. Time and space evolution of the radiation intensity of CF plasma (right) and a reference spark in air (left) under conditions (i) at a distance from the cathode l = 0 (a), 0.2 (b), and 0.6 mm (c)

The continuous-spectrum radiation ceases 10 to 15 ns after the termination of the current pulse. Its intensity is directly related to the rate of current rise, di/dt. This is indicated by the fact that under the experimental conditions (ii) there was no continuous radiation signal, while under the conditions (i) this signal was much stronger than under the conditions (iii). Figure 6.6 shows the space-time evolution of the continuous radiation under the conditions (i). It is typical that the signal amplitude sharply decreases with the distance from the cathode. The velocity of propagation of the glow boundary estimated from the radiation pulse waveforms was not lower than $2 \cdot 10^6$ cm/s. The velocity was measured both along and across the discharge axis. No appreciable difference between the two velocities was observed, suggesting that the CF plasma expansion is spherically symmetrical. The information obtained in [6.23,24] allows one to identify the kind of radiation emitted by the plasma. This is of technical importance. The bremsstrahlung and the recombination radiation intensities are known to be proportional to the product of the ion and electron densities [6.25], while the ion line intensity is proportional to the ion density. Hence, the intensity of continuous radiation falls with decreasing plasma density much stronger than the line radiation intensity. This appears to account for the halt of the cathode plasma glow boundary observed in a number of experiments using recorders of integrated visible radiation [6.26,27].

We now consider the results of a study on the time dependence of the line intensity recorded at different distances from cathode. The following lines corresponding to the visible range were observed: $Al^{+2}(\lambda = 589.6, 452.9, 470.1,$ and $572.2nm)$, $H_\alpha(\lambda=656.2nm)$, $H_\beta(\lambda= 486.1nm)$, $C^{+1}(\lambda=657.8nm)$, $Cu^0(\lambda=521.8$ and $515.3nm)$. The common feature of all these lines is the time lag for the appearance of light

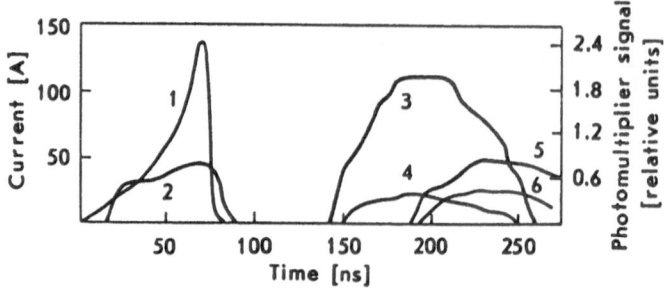

Fig.6.7. Combined oscillographic records of the EEE current (*1*) and the PMT signals due to the emission of continuum (*2*) and line radiation Al III (452.9nm) (*3*), Al III (572.2nm) (*4*), H_α(*5*), CuI (521.8nm) (*6*) obtained for experimental conditions (iii) at a distance 0.2 nm from the cathode

with respect to the onset of current rise. Figure 6.7 illustrates the combined waveforms of the diode line radiation and current. It is clear that the line radiation appears after the termination of the current pulse. Shown in Fig.6.7 are only the waveforms of a few of the lines observed; no principal difference is observed for other lines. Analysis of H, Cu, and Al line radiation waveforms recorded at various distances from the cathode has shown that the line radiation practically appears at the same time throughout the CF plasma volume.

Thus, the spectral investigation [6.12, 24] reveals that the characteristics of the radiation emitted from the CF plasma are dependent on the current flowing through it. This unambiguously indicates that a significant amount of energy is delivered to the CF plasma by the explosive emission current.

6.4 A Model for CF Plasma Expansion

6.4.1 The Adiabatic Model

The fact that the CF plasma expands with a constant velocity has led us to the conclusion that this expansion is likely adiabatic [6.11, 12]. By analogy with [6.28], we assume that energy is rapidly and efficiently delivered into a certain initial microvolume of the cathode material with radius r_0, up to a specific value ϵ_0. Further, during plasma expansion, the accumulated energy changes into kinetic energy of radial motion of the plasma particles. When the CF radius becomes much greater than the initial radius r_0, the role of the parameter r_0 becomes less and less significant. The expansion becomes inertial in nature and the mass velocity of the particles tends asymptotically to the limiting value $v_\infty = (2\epsilon_0)^{1/2}$. As a result of total energy conservation in the particle volume it can be shown [6.27] that the velocity of motion of the leading layers, v_{max}, is

$$v_{max} = \sqrt{\frac{4\gamma}{\gamma - 1}\, \epsilon_0} = \sqrt{\frac{2\gamma}{\gamma - 1}}\, v_\infty \,. \qquad (6.3)$$

Let us assume that for the case of a cathode flare with $v_{max} = v_c$ and adiabatic parameter γ is 1.67 and 1.24 for atomic gas and plasma, respectively [6.28]. Then, using (6.3), one can estimate the specific energy ϵ_0 accumulated in the initial volume of the cathode material. According to estimates, ϵ_0 is $(3 \text{ to } 8)\epsilon_s$, where ϵ_s is the sublimation energy. Thus, the objective is to find at what stage (heating, explosion or expansion of the cathode material) the specific energy is released.

While being heated, the cathode material undergoes several phase changes; i.e. the solid phase ($\rho = 3 \text{ to } 20 \text{ g/cm}^3$), then the nonideal plasma ($\rho_{np} \simeq 1 \text{ to } 10^{-2} \text{ g/cm}^3$) and, in the end, the classical ideal plasma ($\rho_{ip} \leq 10^{-2} \text{ g/cm}^3$).

The energy released in the solid phase ϵ_c can be estimated from the considerations of Sect.5.1. Using (5.3) one can write

$$\epsilon_c = \frac{j^2 t_d \kappa}{\pi^2 \rho} \,, \qquad (6.4)$$

where κ is the resistivity.

The presence of π^2 in the denominator of (6.4) is due to the fact that the real field emission current density is smaller by a factor of π than the conventional one (Sect.5.1). Substituting the experimentally found value of $j^2 t_d = 4 \cdot 10^9 \text{ A}^2 \cdot \text{s/cm}^4$ into (6.4) and setting the specific resistance of tungsten equal to a value which is 30 to 100 times greater than that at room temperature we obtain $\epsilon_c \simeq (2 \text{ to } 6) \cdot 10^3 \text{ J/g}$, i.e., it is approximately equal to the specific sublimation energy.

The result obtained is easily understood from a physical point of view. If an energy density of the order ϵ_s is instantly introduced into solid matter, the matter will expand into vacuum with a velocity of the same order of magnitude as that of the velocity of sound v_s. The time for the matter to "off-load" will be of the order of x/v_s, where x characterizes the space scale of the microvolume into which the energy is introduced ($r_e \leq x \leq h$). Hence, the specific energy delivered into the microvolume for the period during which the matter "off-loads" will be

$$\epsilon_c \simeq \frac{j^2 x \kappa}{v_s} \,. \qquad (6.5)$$

Setting $j = 10^9$ A/cm^2, $x = 10^{-5}$ to 10^{-4} cm, $v_s = (4$ to $5) \cdot 10^5$ cm/s and the above given value of κ, we obtain $\epsilon_c = 10^3$ to 10^4 J/g.

6.4.2 MHD Calculation

As to the energy released in the nonideal plasma phase, it is difficult to estimate it numerically because of a strong dependence of the conductivity on material density and specific heat energy. The self-consistent problem involving both expansion dynamics and heating can be solved only by using a computer. The first successful attempt was demonstrated in [6.29], where a MHD calculation of the initial stage of the explosion of a field-electron emitter has been described. Taken into account is the fact that the cathode material goes through several steps of phase state. The MHD method developed for numerical simulation of the explosion of wires has been employed [6.30]. The resistivity κ was assumed to be a function of the material density and the specific heat energy. The form of this function is established using a combination of theoretical and experimental results. In order to take into account the two-dimensionality (2-D) of the matter expansion, a method is used which describes 2-D processes combining 1-D equations for the motion both along the point radius r and along the axis z. The emitting point is divided into layers along the z-axis. Each layer is divided into meshes (rings) along the radius. Current is assumed to flow along the z-axis.

Calculations have been carried out for copper points with a tip radius $2 \cdot 10^{-5}$ cm and taper angles of 12°, 20°, and 40°. The time function of the current is taken in the form $i(t) = a+kt$, where the parameter a is chosen so that at $t = 0$ the current density at the point tip is about 10^9 A/cm^2 and the rate of current rise k ranges between 10^9 and 10^{10} A/s. The calculation showed that with $k = 10^{10}$ A/s, independently of the value of the point tip angle, the explosion of the point tip occurs at the moment $t = 0.5$ ns. The specific internal energy is distributed in the exploded matter nonuniformly and reaches $(2$ to $5) \cdot 10^4$ J/g. The distributions of the specific energy and the matter density along the z-axis at $r = 0$ for $t_1 = 10^{-9}$ s and $t_2 = 2 \cdot 10^{-9}$ s are shown in Fig.6.8. From Fig.6.8 it is clear that intense heating occurs in a narrow layer ($\sim 10^{-4}$cm) contiguous to the unexploded metal. The high energy release in this layer is associated with a rapid decrease of the resistivity in going from the metal to a plasma state. As the specific energy is increased to $\epsilon_{np} \simeq 2 \cdot 10^4$ J/g, the resistivity sharply rises. Subsequently, its rise becomes slower and the decrease of the current density during plasma expansion results in a decrease of the energy release. The calculation of the plasma heating is limited to a density n

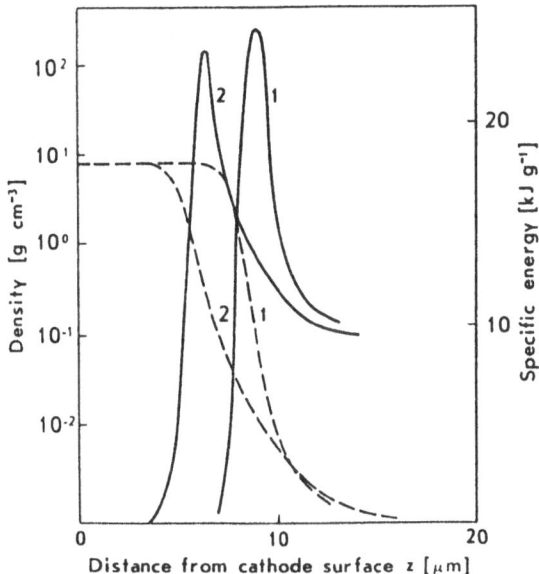

Fig.6.8. Distributions of the density (dash lines) and the specific heat energy introduced into the matter (solid lines) along the z-axis at r = 0 for $t_1 = 10^{-9}$s (*1*) and t = $2 \cdot 10^{-9}$s (*2*)

$\geq 10^{19}$ cm^{-3}, since with n < 10^{19} cm^{-3} the plasma is characterized by two different temperatures.

From the calculated values of the specific heat energy and the material density, using tabulated plasma thermodynamic functions, the temperature and the ionization degree of the matter with n $\leq 10^{21}$ cm^{-3} are found. The values of these quantities lie in the ranges: $3 \leq T \leq 5$ [eV] and $2 \leq n_e/n_i \leq 3$. At denisities of the solid state, the specific energy reaches $8 \cdot 10^3$ J/g, which is in good agreement with the above estimates. According to calculation, the velocity of plasma expansion is (2to 3)$\cdot 10^6$ cm/s. The mass to charge ratio of the exploded matter amounts to $8 \cdot 10^{-4}$, $4 \cdot 10^{-4}$, and $9 \cdot 10^{-5}$ g/C for $\theta = 12°$, $20°$, and $40°$, respectively.

A calculation for k = 10^9 A/s showed that the explosion of the point tip occurs in 1.5 ns. The pattern of the explosive process and the plasma parameters differ insignificantly from those at k = 10^{10} A/s. Thus, MHD calculation makes it possible to estimate the energy releases and the plasma parameters in the initial stage of the explosion and the CF formation.

6.4.3 The Model of an Ideal Plasma

We now consider the energy release and related processes for a classical collisional ideal plasma [6.24]. In a coordinate system which moves with the plasma, the specific energy released due to the Joule heating ϵ_{ip} can be found from the relation [6.31]

$$\frac{j^2}{\sigma} = n_i m_i \frac{d\epsilon_{ip}}{dt} , \tag{6.6}$$

from which, taking into account (6.2), $j(r,t) = i(t)/(4r^2)$, and $dt = dr/v_c$, we obtain

$$\epsilon_{ip} = \int_{r_{ip}}^{r} \frac{j^2}{\sigma m_i n_i} \frac{dr}{v_c} = i^2/(4\pi\sigma\dot{M}) \left[\frac{1}{r_{ip}} - \frac{1}{r} \right] . \tag{6.7}$$

Here, $\sigma[\Omega/cm] = 15\ T_e^{3/2}[eV]/z$ is Spitzer's conductivity of the plasma [6.32], z is the average ion charge, r_{ip} is the characteristic radius, starting from which the plasma can be considered as ideal. Setting $T_e = 4.5$ eV, $z = 2$ to 3, $\dot{M} = 10^{-3}$ to 10^{-4} g/s, $r_{ip} = (3$ to $5)\cdot 10^{-4}$ cm, $r \gg r_{ip}$ and determining the current from (7.1), we obtain $\epsilon_{ip} \simeq (1$ to $5)\cdot 10^5$ J/g. From this it follows that the energy released in the state of an ideal plasma is not less than that for nonideal plasma.

It is necessary, however, to emphasize that the Joule energy determined above is no other than the kinetic energy acquired by electrons in an electric field. Therefore, it is important to investigate the means by which way this energy may "transfer to other degrees of freedom". Let us now consider two elemental processes in the CF plasma - Coulomb's electron-ion collisions and the electron-induced ionization. We first evaluate the electron drift velocity v_d in the electric field. As $j = i/(4\pi r^2) = en_e v_d \simeq ezn_i v_d$, then, taking into account (6.2), we have

$$v_d = \frac{im_i v_c}{ez\dot{M}} . \tag{6.8}$$

For typical values of the quantities given in (6.2) we find $v_d \simeq 10v_c$.

The probability of the electron-to-ion energy transfer can be expressed as

$$P_{ei} = \int_{r_{ip}}^{r} n_i Q_c v_{Te} \frac{2m_e}{m_i} \frac{dr}{v_d} = \frac{\dot{M}}{4\pi m_i v_c} \frac{2Q_c v_{Te} m_e}{v_d m_i r_{ip}} , \tag{6.9}$$

where $Q_c = 3 \cdot 10^{-13} \, z^2/T_e$ is the cross-section of Coulomb's electron-ion collisions. For typical values of the quantities involved in (6.9) we obtain $P_{ei} \sim 10^{-3}$. Taking into account the ratio $v_d/v_c \simeq 10$ and assuming that the plasma is quasi-neutral ($n_e \simeq zn$), it follows that the energy transferred by an electron to an ion in Coulomb's collision is not more than 10% of its energy on the average, i.e., about 1 eV.

The probability of the electron-induced ionization is

$$P_{ea} = \int_{r_{ip}}^{r} n_i Q_i v_{Te} \frac{dr}{v_d} , \qquad (6.10)$$

where Q_i is the ionization cross-section [6.33]. From (6.10), it can be estimated that $0.1 \leq P_{ea} \leq 1$. Consequently, a significant fraction of the electron's kinetic energy is used up for ionization. Moreover, from the ratio $v_d/v_c \simeq 10$ it is clear that in the CF plasma singly- and doubly-ionized atoms should be present.

Let us now consider how the relatively "hot" electrons, which tend to scatter, affect the dynamics of motion of the relatively "cold" ions. If the CF plasma ions move at a velocity of about $2 \cdot 10^6$ cm/s, their kinetic energy is 50 to 100 eV. The problem of the CF plasma expansion is analogous to that of the formation of high-velocity jets observed in the operation of vacuum arcs. *Plyutto* [6.34] suggested that the electron gas pressure plays an essential role in the acceleration of the plasma ions. According to the ideas of *Plyutto* et al. [6.35], the ion acceleration should be due to a high gradient of the electron density in the cathode spot. The energy an ion can acquire in the electric field due to the density gradient can be expressed as

$$e(V_0 - V) = 1.5\epsilon_e \log(n_e^0/n_e) , \qquad (6.11)$$

where ϵ_e is the average electron energy in the cathode spot region; n_e^0 is the maximum electron density in the cathode spot. Assuming $\epsilon_e = 5$ to 8 eV and $\log(n_e^0/n_e) = 3$, *Plyutto* et al. [6.34] have obtained $e(V_0-V) = 20$ to 40 eV. The nature of the acceleration of vacuum arc cathode jets was also discussed in [6.36,37].

A hydrodynamic model for the expanding CF was carefully analyzed by *Litvinov* [6.38-40]. He considered the double-component plasma consisting of electrons and ions with a mean charge z. In the limit, when the flare radius becomes much greater than the radius of the initial volume of the cathode matter into which the main energy is introduced, the following relation is obtained for the flare expansion velocity

$$v_c = \int_{n_i}^{n_i^0} \sqrt{\frac{\partial(p_e + p_i)}{m_i \partial n_i}} \; \frac{\partial n_i}{n_i} \; . \tag{6.12}$$

Assuming that the conditions for flare expansion are close to the adiabatic ones, i.e., $p_i = n_i kT_i^0(n_i/n_i^0)^{\gamma_i}$; $p_e = n_e^0 kT_e^0(n_e/n_e^0)^{\gamma_i}$; $\gamma_i = \gamma_e = \gamma$, from [6.12] it is found that

$$v_c = \frac{2}{\gamma - 1} \sqrt{\gamma \frac{(kt_i^0 + zkT_e^0)}{m_i}} \; . \tag{6.13}$$

For $z = 3$, $kT_e^0 \simeq kT_i^0 \simeq (\gamma - 1)\epsilon_s m_i$ from (6.13) it follows that $v_c = 2 \cdot 10^6$ cm/s (Al, Mo, Cu) and $v_c = 1 \cdot 10^6$ cm/s (Pb), the values being in good agreement with experimental data [6.4, 23]. Also obtained in [6.38] was an expression for the cathode plasma expansion velocity under the isothermal conditions, which predicts, for the same initial conditions, v_c values 20 to 30% greater than those found from (6.10). The non-essential difference between the v_c values obtained theoretically in the adiabatic and the isothermal approximation can be attributed to an abrupt decrease of the pressure gradient with increasing CF radius.

A recent theoretical investigation should be mentioned [6.41] which computes the dynamics of a thin plasma layer near the cathode of a magnetically insulated cylindrical vacuum line. The initial layer parameters are taken as follows - thickness: 10^{-4} cm, material: aluminum, ion density: $2 \cdot 10^{20}$ cm^{-3}, temperature: 5 eV, plasma current density $j \leq 10^5$ A/cm^2. Calculated are the spatial plasma density distribution, the velocity of expansion of the plasma boundary and the electron temperature. It is established that the mean velocity of the plasma boundary is about $2.4 \cdot 10^6$ cm/s, the electron temperature rises to $T_e = 14$ eV, i.e., the Joule heating of the plasma is efficient.

Thus, the above experimental data provide us with the following mechanism of the physical processes in the near-cathode plasma region. The cathode material is rapidly heated by the field-assisted-thermionic emission current and starts to "off-load" into vacuum. The specific energy delivered to the matter in the solid state is about $5 \cdot 10^3$ J/g. Then the matter rapidly ($t < 1$ ns) changes to a state of nonideal plasma and acquires an average specific energy of $(2$ to $5) \cdot 10^4$ J/g. It seems that just in this phase the velocity of motion of the CF external layers reaches the value of about $2 \cdot 10^6$ cm/s. When the matter has changed into a state of ideal plasma, only the electron subsystem proceeds to heat up intensively, because the frequency of the electron-ion collisions abruptly falls and, hence, the electron-to-ion energy transfer becomes less intensive as a result of a decrease in the particle concentration. The processes of excitation and electron-induced ionization, however, are as intense as before. It seems that, in this region the charge composition of plasma ions finally forms. Moreover, herein the main ion mass is finally accelerated up to an energy of the order of 100 eV due to the collective processes of the energy transfer from the electrons being continuously "drawn" through the ion "frame".

7. Current Passage in the Spark Stage of Breakdown

In the spark stage of breakdown the heavy discharge current through the vacuum gap convincingly shows the high emissive power of the CF plasma. Cathode flares, however, represent nonstationary plasma and their emissive properties seem to have some peculiarities. Indeed, as model experiments with field-emission cathodes have shown, the character of the EEE current waveforms is essentially dependent on the overvoltage coefficient. At low overvoltages, a decrease in the EEE current rise is observed, which vanishes at high overvoltages (Fig.5.2). These peculiarities are qualitatively similar to those established for cathode-ignited discharge tubes [7.1]: at low arc currents in the igniting gap, the transition to the high-current stage can be significantly delayed, while at high currents the delay stage practically disappears. The latter is also evident from *Flinn's* experiment [7.2]. All that suggests that the change in the character of EEE current waveform can be accounted for by the rate of generation of the cathode plasma and the variation of its emissive properties during expansion.

Moreover, in studying the breakdown spark stage it was observed more than once that there were spikes on the current waveforms [7.3-6] (Fig.5.2). Similar current spikes were observed in cathode-ignited tubes [7.1,2]. In [7.3,4] it was shown that the current bursts are accompanied by the constriction of the electron beam toward the discharge axis, an increase of the cathode plasma potential, and the appearance of positive ions accelerated toward the anode, the energy of which is an order of magnitude or more above the gap voltage.

From the above it follows that the studies of the CF plasma emissive properties and those of the nonstationary behavior of emission are of independent interest. The present chapter describes the results of these studies and explains, by using these results, their background and the mechanism of the current rise at vacuum breakdown.

7.1 Electron Emission from CF Plasma into Vacuum

In the peripheral layer of a CF plasma, contiguous to vacuum, the condition of quasineutrality may not be fulfilled [7.7]. The characteristic parameter having the dimension of length, which appears in the plasma problems taking account of space charge, is the Debye length [7.8]

$$L_D = \sqrt{\frac{kT_e}{4\pi n_e e^2}} \ . \tag{7.1}$$

Estimates show that for typical values of CF plasma density the inequality $L_D \ll v_c t$ is valid, i.e., charge separation at the emission boundary occurs in a very narrow region adjoining the CF plasma front. It is natural to adopt the field-assisted thermionic emission as the mechanism of electron emission from the CF plasma boundary. In this case the current density is defined by [7.9]

$$j_e = \frac{4\pi m_e e k^2 T_e^2}{h^3} \exp\left[-\frac{\phi - (e^3 E)^{1/2}}{kT_e}\right] , \tag{7.2}$$

where h is Planck's constant. Provided that the plasma electrons obey the Maxwell-Boltzmann statistics, the following relation for the electron work function will be valid [7.9]:

$$\phi = kT_e \ln\left[\frac{2(2\pi m_e kT_e)^{3/2}}{\mathbb{Z} n_e h^3}\right] . \tag{7.3}$$

Substituting (7.3) into (7.2) results in

$$j_e = \mathbb{Z} e n_e \sqrt{\frac{kT_e}{2m_e}} \exp\left[\frac{e^2 E}{kT_e}\right] . \tag{7.4}$$

Expression (7.4) indicates that the CF plasma emits a so-called thermal current, since the electron density n_e is small and hence the work function ϕ is also small. Electrons leave the CF plasma with a thermal velocity $v_{T_e} \simeq 10^8$ cm/s $\gg v_c$, i.e., the electron emission from the plasma may be treated using "stationary" models.

Having left the CF plasma, the electrons move into vacuum. Their self-space charge distorts the applied electric field. In other words, the electron current density emitted by the CF plasma should be defined in this case by the Child-Langmuir law (the "3/2 power" law: $j_{3/2} \propto V^{3/2}$). The electron emission from the CF plasma has, however, some specific features. Firstly, the plasma emitting surface expands at a high rate. Secondly, the plasma density at the emission boundary decreases as the CF expands. Thirdly, because of the cyclic nature of cathode processes (in particular, cathode erosion), the CF plasma density distribution will be non-monotonic. These factors may result in three situations.

If the thermal current density is $j_e > j_{3/2}$, an electric field which decelerates electrons and accelerates ions arises at the emission boundary, and a virtual cathode forms near it. The acceleration of the ions (and the emission boundary) can be estimated from [7.10]

$$m_i \frac{dv_c}{dt} \lesssim e \; H(4\pi n_e T_e) \; . \tag{7.5}$$

Substituting (6.2) for n_e into (7.5) and solving the resulting equation, *Bazhenov* et al. [7.10] found

$$\Delta v_c \lesssim \frac{5}{2} \left[\frac{e^2 \dot{M} T_e}{m_i^3} \right]^{1/5} \left[\ln \frac{v_c t}{r_0} \right]^{2/5} . \tag{7.6}$$

If $j_e = j_{3/2}$, the electric field at the emission boundary vanishes, and (7.4) can be rewritten as

$$j_e = \frac{z e \dot{M}}{4\pi m_i v_c^3 t^2} \sqrt{\frac{k T_e}{2\pi m_e}} . \tag{7.7}$$

If $j_e < j_{3/2}$, the electrons exert no screening effect on the emission boundary, and the ions are decelerated under the action of the applied electric field. In this case the plasma behaves as if it expands into a medium with a pressure $p_E = E^2/(8\pi)$, and the velocity of expansion may be defined as [7.10]

$$v_c' = v_c[1 - (p_E/p_{pl})^{(\gamma-1)/2\gamma}] , \tag{7.8}$$

where p_{pl} is the initial plasma pressure determining the velocity v_c.

Taking into account the peculiarities mentioned above, we shall discuss below some results of experimental studies on electron emission from CF plasma.

7.2 Electron Emission from CF Plasma, Experimental Studies

Experiments devoted specifically to the study of the emissive properties of CF plasma were started independently and practically simultaneously by *Korop* and *Plyutto* [7.3,4], and by scientists of our Institute [7.5,6]. All the experiments were carried out with cathodes having the form of a single needle, in order that the EEE delay time might be reduced to a minimum ($t_d < 10^{-9}$s) and a single CF could form. In the experiment [7.3,4] the needle tip was placed in the center of a hemispherical anode of 1-cm radius. The electron-current density distribution at the anode was investigated with the use of collectors placed at angles of 0°, 45°, and 60° with the needle axis. It was established that at a voltage $V_0 \leq 80$ kV the EEE current increased throughout the pulse ($t_p = 50$ns) with a satisfactory reproducibility from pulse to pulse (Fig.7.1a). The current rise could be satisfactorily explained using the Child-Langmuir law for a spherical electrode configuration, provided that the plasma emission boundary propagated from the cathode isotropicly with a velocity of (2 to 2.5)·10^6 cm/s.

At a voltage $V_0 \geq 80$ kV, 1.5·10^{-8} s after pulse application several short-time spikes ($\leq 10^{-8}$s) appeared in the EEE current waveforms, the amplitude of which was 1.5 to 3 times greater than the predicted current (Fig.7.1b). Simultaneously with the spikes a sharp rise (4-6

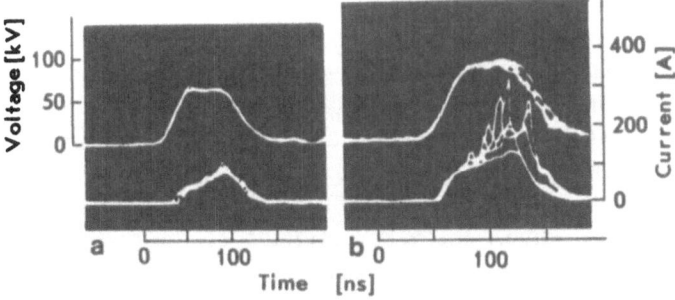

Fig.7.1a,b. Voltage and current waveforms obtained in the regime without current bursts for 20 subsequent pulses (a), and in the regime with current bursts for 10 subsequent pulses (b) [7.4]

139

fold) of the current to the axial collector and a decrease of the current to the other collectors were observed.

The transition from a stable to an unstable regime of current extraction from the CF plasma was investigated in more detail in [7.6, 11]. Most of the experiments were carried out with a needle–plane electrode configuration.

To elucidate the influence of the voltage amplitude on the principal emission characteristics experiments were carried out on two installations which differed in the level of operating voltage (for the first $V_0 = 10$ to 40 kV, $t_p = 5 \cdot 10^{-9}$ to $1.3 \cdot 10^{-6}$ s; for the second $V_0 = 180$ to 200 kV, $t_p = 3 \cdot 10^{-8}$ to 10^{-7} s). Some typical diode current waveforms are shown in Fig.5.6. In the initial stage, the current monotonically increases due to the shortening of the vacuum gap and the increase of the CF emitting area. The rate of the monotonic current rise can be varied in a wide range (from 10^7 to 10^{11} A/s) by varying V_0 and d. In this stage good reproducibility of the oscillographic traces in form and amplitude has been noted for a great number of pulses. As the current reaches a certain value i_s for given V_0 and d, the electron emission from CF becomes unstable. For example, in the case of $V_0 = 30$ kV and d = 2 mm in $t_s \simeq 40$ ns after the CF appearance some current bursts occur with a duration of the order of 10^{-8} s and an amplitude 1.5 to 2 times greater than the current value just before the burst. After the bursts the current drops down to values that correspond to its monotonic increase. The current variations are accompanied by oscillations of the gap voltage. In spite of the random nature of the current bursts the average duration of the stationary stage is established to increase with the gap spacing. For instance, with $V_0 = 30$ kV and d = 3 cm the time of monotonic current increase reaches 400 ns. Shortening the duration of the stationary stage, as well as decreasing the gap spacing, results in a noticeable decrease in both the burst duration and the difference between the burst amplitude and the preceding current value. For example, with $V_0 = 30$ kV and d = 3 cm these bursts are rather weakly pronounced and look like low-amplitude oscillations on the background of a high average rate of current rise ($\simeq 10^{10}$ A/s). A similar pattern for the current instabilities and their nature is observed when a plane anode is exchanged for a hemispherical one. The pattern of current flow remains unchanged when the voltage is increased to a level of about 200 kV. With d $\leq$ 10 mm the current bursts shift to almost the very beginning of the pulse. Their amplitude can be 1.5 to 2 times larger than the current preceding the burst.

Typical waveforms of the anode current density along the discharge axis and at its periphery obtained with $V_0 \leq 30$ kV are shown

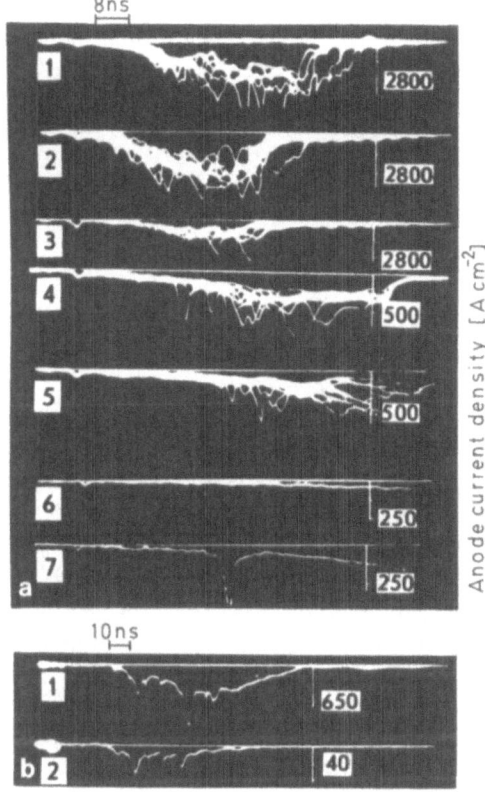

Fig.7.2. Oscillographic records of the anode current density obtained under the following conditions: (a) V_0 = 20 kV, d = 1 mm, r = 0 (*1*), V_0 = 30 kV, d = 1 mm, r = 0 (*2*), V_0 = 30 kV, d = 1 mm, r = 0.8 mm (*3*), V_0 = 30 kV, d = 2 mm, r = 0 (*4*), V_0 = 30 kV, d = 2 mm, r = 2 mm (*5*), V_0 = 20 kV, d = 4 mm, r = 0 (*6*), and V_0 = 30 kV, d = 4 mm, r = 0 (*7*); (b) V_0 = 200 kV, r = 0, d = 7 (*1*) and 17 mm (*2*). r is a distance from the discharge axis

in Fig.7.2. In the initial period of emission the current density monotonically increases following the total current. During this period the anode current density waveforms $j_a(t)$ are well reproducible for a great number of pulses. At a certain time t_s some sharp transient spikes of duration $\simeq 5 \cdot 10^{-9}$ s with an amplitude 2 to 5 times higher than the preceding j_a value appear in the current density waveforms. The bursts along the discharge axis are in synchronism with the spikes of the current waveforms. The behavior remains unchanged as the voltage is increased to 200 kV (Fig.7.2b). The average time t_s until the appearance of spikes lengthens as the gap spacing d is increased and the voltage V_0 is decreased.

141

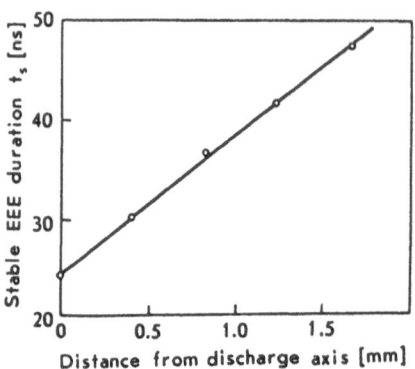

Fig.7.3. Stable EEE stage duration as a function of the distance from the discharge axis

Collector measurements were also made to investigate the behavior of the current density j_a away from the discharge axis. Analysis of current density waveforms offered the following conclusions: (i) the starting moments of the current rise on and away from the axis coincide, but the rate of the current density rise decreases with the distance from the axis; (ii) the time structure of the waveforms, $j_a(t)$, at the periphery is similar in nature to that at the axis, but the average time t_s increases with the distance from the axis (Fig.7.3), while the spike amplitude decreases. From Fig.7.3 it can be concluded that the time between the appearance of the first bursts of electron current on the axis and away from it can be much greater than the duration of the burst itself. This means that the conditions leading to a failure of stable emission are not attained in all directions at the same time, but occur at first in the near-axis region, where the current density is maximum. In our investigation with a hemispherical anode (d=4mm, V_0=20-40 kV), the average values of the current density measured in both the stable and the unstable stages are in good agreement with the values calculated from the relation i(t)/S, where S is the hemispherical anode surface area (i.e., for such a geometry there is no preferential direction in which the transition to the unstable regime would develop first). This result differs from that of [7.3,4] and seems to be due to an inessential effect of the anode shape in the experiment [7.3,4] (R_{CF} = $4 \cdot 10^{-2}$cm << d=1cm). In our experiment $R_{CF}/d \simeq 0.25$.

It was observed that in the range of pressures 10^{-3} to 10^{-1} Pa the above patterns were unchanged. As the pressure was increased up to about 1 Pa, the oscillations in the total current and the current density waveforms disappeared. Their smooth increase was observed until the transition to the arc regime, with the only distinction that di/dt and dj_a/dt were 1.5 times greater. This points to the fact that the residual-gas ions exert a stabilizing effect on the current and partially compensate for the electronic space charge.

142

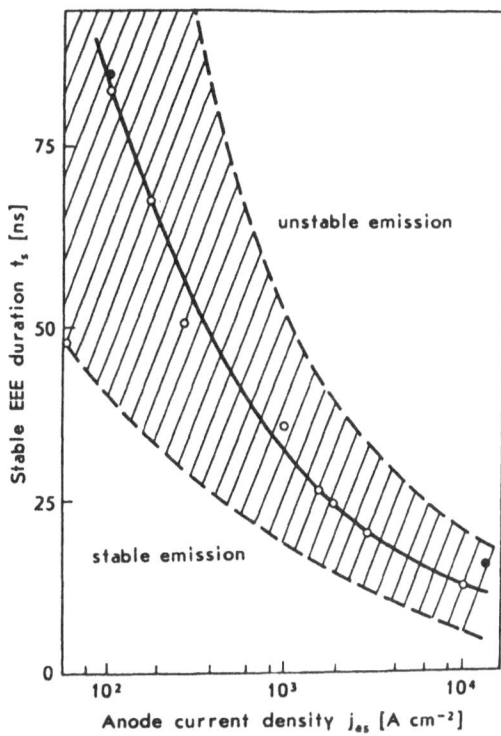

Fig.7.4. Stable EEE stage duration as a function of the anode current density just prior to the burst. Dots denote average values

It can be assumed that the beginning of an unstable current passage is caused by the transition of the plasma emitter (CF) to a saturation regime [7.12]. Therefore it appeared interesting to study the dependence of the stationary stage duration on the current density extracted from the emission boundary. Since direct measurements of the emission current density are impossible, we made an attempt to establish a correlation between the time t_s and the corresponding current density at the beam axis. It turned out that for the conditions under consideration a single-valued dependence of t_s on the axial current density at the moment preceding the burst j_{as} shown in Fig.7.4 is evident. This relationship can be described by the following empirical expression [7.11]

$$t_s = 8.5 \cdot 10^{-7} j_{as}^{-0.42} . \qquad (7.9)$$

In Fig.7.4, the solid circles represent two points obtained after processing the waveforms from [7.3,4]. The dashed section is the region of

143

the most probable (probability 80%) values of j_{as} and t_s. This region appears to be transitory. With the obtained data one can estimate the duration of the stable electron extraction in a diode with a single CF.

7.3 Current-Voltage Characteristics of a Single-CF Diode

Let us consider a vacuum diode with a point or plane cathode, on which a single CF is operating. We shall assume that the CF plasma is at the cathode potential and has a spherical emitting surface of radius r = $v_c t \ll d$ with an unlimited emissivity. There is no known relationship to describe the current-voltage characteristic of such diodes. It is qualitatively clear that with a limited cathode emitting surface the electronic space charge will disturb the potential at any point of the interelectrode space more weakly than with an unlimited one. This should result in an increase of the limiting current density in the vacuum gap. Given in [7.13] are the results of an approximate calculation of the current density for a diode with infinite plane-parallel electrodes in the case when a small area with radius r possesses an unlimited emissivity and there is no radial repulsion of the beam by the space charge. The smaller the ratio r/d, the greater is $k_0 = j_r/j_\infty$ (j_r is the real current density; j_∞ is the current density predicted by the Child-Langmuir law for infinite plane electrodes).

Using such an approach and defining the electron emission area as $\pi v_c^2 t^2$ and the distance between the plasma and the anode as d - $v_c t$, we can write [7.14, 15]

$$i(t) = A_1 V^{3/2}(t) \frac{v_c^2 t^2}{(d - v_c t)^2} k_0 \frac{v_c t}{d - v_c t} , \qquad (7.10)$$

where A_1 = $2.33 \cdot 10^{-6}$ [AV$^{-3/2}$]. From (7.10) it follows that the electronic flux perveance P = $i(t)/V^{3/2}(t)$ should be a single-valued function of the ratio $v_c t/(d-v_c t)$. To verify the validity of this statement a great number of current waveforms were processed. These were obtained for steady-state emission with good amplitude and time resolution under the following experimental conditions: d = 1 to 4 mm, V_0 = 20 to 40 kV and d = 6 to 10 mm, V_0 = 80 to 120 kV; the cathode having been in the shape of a needle. Some experimental points chosen arbitrarily from a great number of points ($\simeq$300) are depicted in Fig.7.5 (Curve 1). The value of the velocity v_c is assumed to be $2 \cdot 10^6$

144

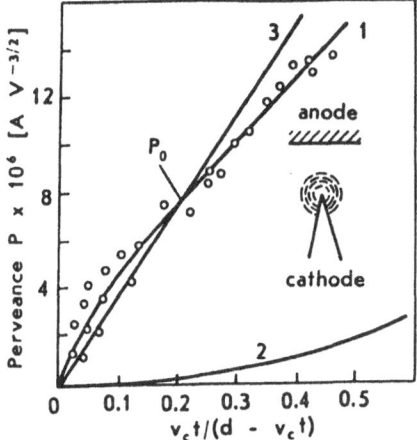

Fig.7.5. Electronic flux perveance as a function of the ratio $v_c t/(d-v_c t)$: (1) - experiment; (2) - calculation for a plane cathode; (3) - calculation according to [7.16]

cm/s. It can be seen that the majority of points fit to a single smooth curve. It applies equally to all the remaining data processed (unplotted) for values of $v_c t/(d-v_c t) \leq 0.5$. The dependence of P on $v_c t\,(d-v_c t)$ for $k_0 = 1$ is also plotted (Curve 2). One can easily notice that the "3/2 power" law, which is not valid for a finite emitting surface, underestimates the electron current values by one or two orders of magnitude [7.14, 15].

For plane electrodes and a single CF in place of an artificially produced micropoint ($V_0 = 20$ to 40 kV, d = 0.3 to 1.0 mm) the experimental points fit well onto a single curve $P = f(v_c t/d)$ (Fig.7.6).

Later on an analytical expression similar to (7.10) was obtained [7.16]. A method suggested in [7.17] was used according to which the current density at any point of a spherical emitting surface can be ob-

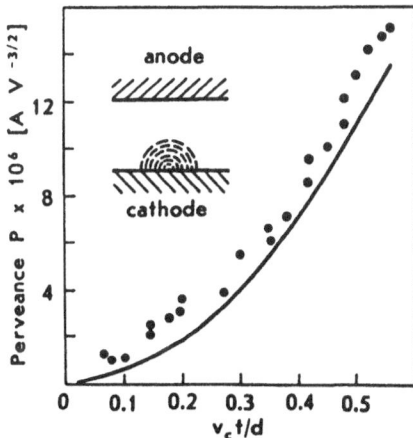

Fig.7.6. Electronic flux perveance as a function of the ratio $v_c t/d$ for the case of a flare on a plane cathode: dots - experiment; solid line - calculation according to [7.18]

145

tained using (5.6). The length of an effective gap (the effective path length of an electron coming out from the given point of the emitting surface) can be determined from the expression d_{eff} = V/E, where E is the electric field at a given point of the sphere where there is no electronic space charge. Therefore the problem is reduced to a determination of the electrostatic field at the cathode in an arbitrary electrode system. Then the total gap current can be determined by integrating the current density over the emitting surface. For a CF at the tip point [7.16]

$$i(t) = 37 \cdot 10^{-6} V^{3/2} \frac{v_c t}{d - v_c t} . \tag{7.11}$$

This relationship is reproduced in Fig. 7.5 by Curve 3; it is in good agreement with the experimental results. With this $k_0 \simeq 5(d-v_c t)/(v_c t)$.

In a similar way, for a hemispherical emitting surface on a plane cathode [7.18]

$$i(t) = 44.4 \cdot 10^{-6} V^{3/2} (v_c t/d)^2 . \tag{7.12}$$

This function is plotted in Fig. 7.6, and it also agrees with the experimental data for $k_0 \simeq 6$.

It should be noted that since the electric field in the effective diode is found from the expressions which are valid under the condition that $v_c t \ll d$, Eqs. (7.11 and 12) might be used, if only this condition is satisfied.

7.4 Dynamics of the CF Electron Emission Boundary

From (7.11) it follows that the electronic flux perveance should increase linearly with $v_c t/(d-v_c t)$. A comparison with experimental data indicates the existence of two regions (Fig. 7.5), which lie to the left and right of the point P, which is the intersection of these curves. On the left, the experimental values of perveance are greater than the calculated ones, while on the right the situation is reversed. Note that when plotting both the experimental and the theoretical function of P on $v_c t/(d-v_c t)$ the propagation velocity of the boundary of the electron emission from the plasma was assumed to be constant and equal to v_c. It seems that the difference in these plots can be attributed to the non-constant real propagation velocity of the emission boundary: during the initial time after the CF occurrence this velocity is greater than $2 \cdot 10^6$ cm/s; it decreases further and becomes less than $2 \cdot 10^6$ cm/s. For instance, the decrease in v_c soon after the EEE initiation can explain the data obtained at the explosion of field-electron emitters with no overvoltage and in operation with explosively emitting cathodes with relatively large gap spacings d. Shown in Fig.5.2 by a dashed curve, together with real current waveforms, is a plot of the dependence i(t) defined by (7.2) with $v_c = 2 \cdot 10^6$ cm/s.

New evidence concerning the change in the emission boundary velocity during the period of electron extraction from the CF plasma was obtained in the investigation described in [7.10]. Having recorded the EEE current with good time and amplitude resolution in relatively large gaps (d = 1 to 2 cm) at rather high voltages (V_0 = 10 to 30 kV) *Bazhenov* et al. [7.10] analyzed in detail the plot of i(t) at the initial period of the current passage in a diode with CF. They established that initially the EEE current rapidly increases, but after $t_s^* = 20$ to 50 ns its growth decelerates. In their opinion, by the time t_s^* the plasma concentration at the emission boundary is so high that the emission current density is in excess of the current density, which can be extracted from the emitting surface in accordance with the "3/2 power" law. A virtual cathode exists near the emission boundary. The space charge field accelerates the ions and consequently the emission boundary. The velocity increment found from (7.6) for typical experimental conditions is $\Delta v_c \leq 10^7$ cm/s, whereas experimentally [7.10] v_c = (2 to 4)$\cdot 10^8$ cm/s was obtained.

After the time t_s the emission is unable to support the current that the vacuum part of the gap can allow to pass. When this occurs the electronic space charge does not screen the emission boundary and ions begin to be "pressed" towards the cathode by the external electric field pressure. As a result, the emission boundary starts to decelerate. For typical experimental conditions we have $10^{-3} \geq p_E/p_{pl} \geq 10^{-5}$. With $\gamma = 5/3$ we obtain $v_c' = (0.9$ to $0.7)v_c$ from (7.8). From the experiments [7.10, 19] it follows that as saturation is reached, the propagation velocity of the emission boundary can decrease to 10^6 cm/s and less.

7.5 CF Plasma Potential Distribution and Plasma Emissive Properties

As shown above, in the initial stage of CF plasma expansion the current passage through the gap is highly stable, the current-voltage characteristic of the diode being well described by the modified "3/2 p⌐wer" law. Later, the current extraction from the plasma becomes unstable, which is indicated by the appearance of transient spikes in the current waveforms. In the unstable current extraction some interesting phenomena are revealed: the acceleration of positive ions to high energies toward the anode [7.3,4], the formation of thin electron jets with a current density five to ten times greater than the average one [7.3,4,6,11], the change in the character of the cathode erosion [7.11], and the appearance of new ECs in the plasma [7.20-22]. *Corop* and *Plyutto* [7.3,4] assumed that in the regime of unstable current extraction the CF plasma is charged to a high positive potential. They believed that the processes which maintain the plasma cathode conductance are similar in their nature to the processes occurring in vacuum arc cathode spots, but, in contrast to a common arc discharge, the increase in the plasma potential can be significant. In his later work *Corop* [7.23] recorded the positive CF ions moving toward the cathode with energies as high as 60-80% of the potential difference applied across the diode. In this connection it is concluded that in the oscillation regime of current passage the CF potential increases periodically up to a value which is comparable with the anode potential. *Plyutto* and co-workers [7.12] also proposed that discontinuities are possible in the CF plasma with a high potential drop. However, no direct measurements of the CF plasma potential distribution in both the stable and the unstable regimes were carried out. Therefore, in the CF plasma the potential at different distances from EC was first measured by probes at the Institute of High Current Electronics and the results were compared with the corresponding waveforms of the gap current [7.22,24,25].

7.5.1 Probe Measurements of the CF Plasma Potential

The experimental arrangement is schematically shown in Fig.7.7. Primarily, the cathode was a 2 cm long needle (molybdenum, copper). At a distance $r = 0.1$-0.5 mm from the needle tip the end of the probe (a copper wire with 50μm diameter) was placed. When a pulse with $V_0 = 20$ to 40 kV ($d = 2$-5mm) was applied to the anode, in general no self-maintained breakdown of the probe occurred. A resistor with $R_p = 75 \, \Omega$ was connected to the probe circuit, which ensured the "floating" regime and, at the same time, it served as match for the bias pulse arriving at the probe. This pulse of negative polarity with an amplitude $V_b = 40$ to 400 V was applied to the anode through a cable ($\rho_w = 75 \Omega$) simultaneously with the voltage pulse. After the breakdown

148

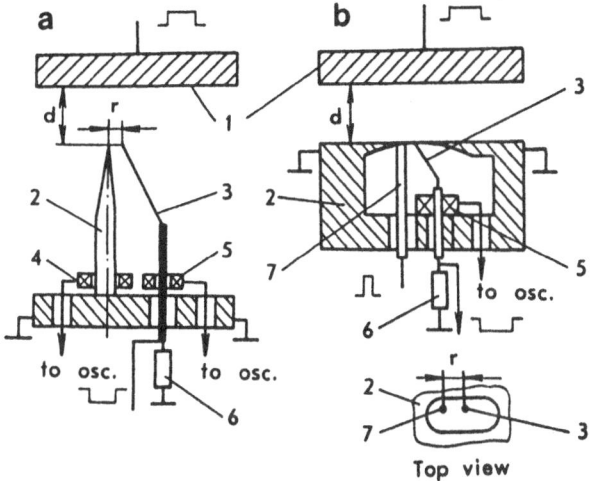

Fig.7.7a,b. Schematic representation of the experimental set-up with a needle cath-ode (a) and plane anode (b): 1 - anode; (2) - cathode; (3) - probe; (4,5) - Rogowski coils; (6) - matching resistor; (7) - igniting electrode

of the probe the limiting resistance of its circuit became equal to $\rho_w/2$ = 37.5 Ω. In terms of the voltage, the probe sensitivity was 50 V/mm.

Analysis of typical waveforms of the probe current which were recorded simultaneously with the diode current waveforms (Fig.7.8) has shown that at a probe-to-plasma bias voltage V_b = 0 the probe current represents a series of bursts occurring simultaneously with the spikes in the diode current waveforms (Fig.7.8c). At the instants of the abrupt rise in diode current, bursts of reverse polarity have been

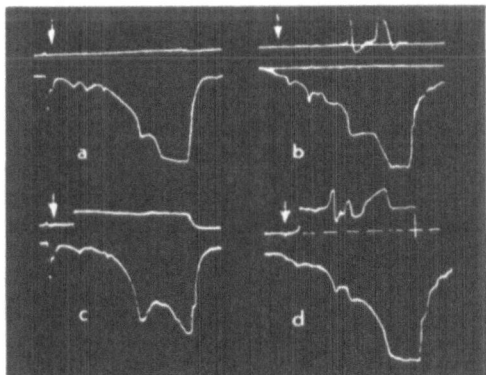

Fig.7.8a-d. Waveforms of the probe current (upper traces) and the diode current (lower traces) recorded at V_0 = 30 kV for the diode with a plane cathode (a,b) and for the diode with a needle cathode (c,d). V_b = 0 (a,b), -400 V (c), and -200 V (d). Arrows indicate the moments for dipping the probe into plasma

149

observed in the probe circuit. The first current burst does not necessarily coincide with the appearance of the first spike in the corresponding discharge current waveform. This is due to the absence of breakdown between the probe and the plasma. However, the appearance of the first burst (breakdown has occurred between probe and plasma, and EC has appeared on the former) inevitably leads to the appearance of subsequent bursts. A certain time after the application of the bias voltage V_b breakdown occurs between probe and plasma and a constant component modulated by spikes of both polarities appears on the probe current waveform (Fig.7.8d). Maxima and minima of the current waveform i_p coincide with large current rise and decay rates, respectively.

Analysis of the results showed that the probe current bursts are due to the inductance of the cathode needle. The total inductance of the cathode needle and the holder is $L = 40$ nH. With stable emission the rate of current rise di/dt is equal to about 10^9 A/s, thus corresponding to the voltage drop across the needle $V_L = L(di/dt) = 40$ V. The voltage V_L appears between the probe and the flare plasma, but its value in the stable emission is, as a rule, insufficient for EC to occur on the probe. At the time of current bursts, di/dt increases by a factor of 5 to 10, so does V_L. The probe readings coincide with V_L values calculated from the diode current waveforms. The probe current at the bias voltage V_b is $i_p = 2 V_b/\rho_w$, thus confirming that the probe operates in the "floating" regime, i.e., the resistance of the wire - cathode spot - plasma section is significantly lower than $R_p = \rho_w/2$.

To eliminate the inductive effect, the design of the cathode unit was changed (Fig.7.7b) ($L<5\cdot10^{-10}$H). EEE was initiated by application of a voltage pulse with the parameters $V = 30$ kV, $t_p = 5$ ns, $i_{max} = 5$ A to an igniting electrode (a copper wire with $0.1\,\mu$m diameter and $\simeq10^{-3}$cm away from the cathode edge). Oscillographic traces recorded under these conditions are illustrated by Fig.7.8a,b. Their analysis made it possible to establish the following: (i) the breakdown of the probe-CF plasma layer could never be accomplished without a negative bias voltage $V_b = 50$ to 200 V applied to the probe; (ii) the probe-plasma layer breakdown causes a stepwise appearance of the current i_p, which is unchanged at current bursts in the gap. These facts suggest that the CF plasma potential, at least at 0.2 to 1 mm from the EC (the probe is positioned at such distances) is not higher than 50 V and shows no significant variations at the moments of current bursts in the diode. In other words, the cathode voltage drop is not larger than 50 V, i.e., it is at the level characteristic of a vacuum arc. The conductance between the cathode and the flare plasma is high in both the stable and the unstable phases. The emissive properties of the cathode ensure the rate of the current rise to be $di/dt \simeq 10^{10}$ A/s without significantly raising the potential V_c. Thus, the process initiation of oscillations in a high-voltage diode operating in the EEE regime is not associated with a sharp increase of the potential V_c up to a value comparable with the voltage applied to the gap. Therefore we proposed [7.24] that the appearance of current bursts is associated with the phenomena developing in peripheral regions of the CF plasma, from which the extraction of electrons toward the anode takes place.

The CF plasma potential at a distance $r > 1$ mm from the cathode was measured using probes ($d = 5$ to 10 mm) [7.22,25]. The corresponding waveforms are shown in Fig.7.9. From these it is quite clear that the appearance of current bursts is associated with the increase of the potentials of CF plasma peripheral layers up to several kilovolts. Thus, the supposition that in the expanding CF plasma discontinuities with a high voltage drop occur [7.12] is confirmed. Therefore, the diode acts like a triode with a plasma grid being at high positive potential with respect to the CF part attached to the cathode. It is evident that such a system possesses a much greater current-carrying capacity than a diode.

7.5.2 The Nature of the Instability of CF Emission

Bazhenov et al. [7.10] hypothesized that a correlation exists between the EEE current instability and the irregularity of the cathode material supply to the CF plasma. To verify this, an experiment was carried out, in which the irregularity of the cathode material supply to the CF plasma was created artificially. It confirmed the hypothesis convincingly. In the opinion of *Bazhenov* et al. [7.10], under the usual conditions the current bursts occur in the following manner (Fig.7.10). After the onset of saturation the emission boundary moves at a velocity $v_c' < v_c$. If the cathode material input to the CF plasma is irregular, high-density plasmoids appear which will catch up with the decelerated emission boundary. When the next front plasmoid reaches the flare emission boundary, saturation is disturbed and the current begins to increase at a high rate. Behind the plasmoid, plasma of lower density

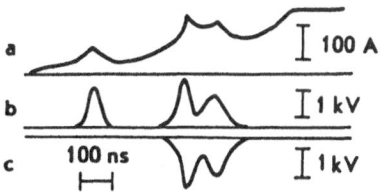

Fig.7.9a–c. Waveforms of the diode current (a) and the plasma potential (b,c) recorded with $d = 1$ cm, $V_0 = 30$ kV, $r = 4$ (b) and 8 mm (c)

151

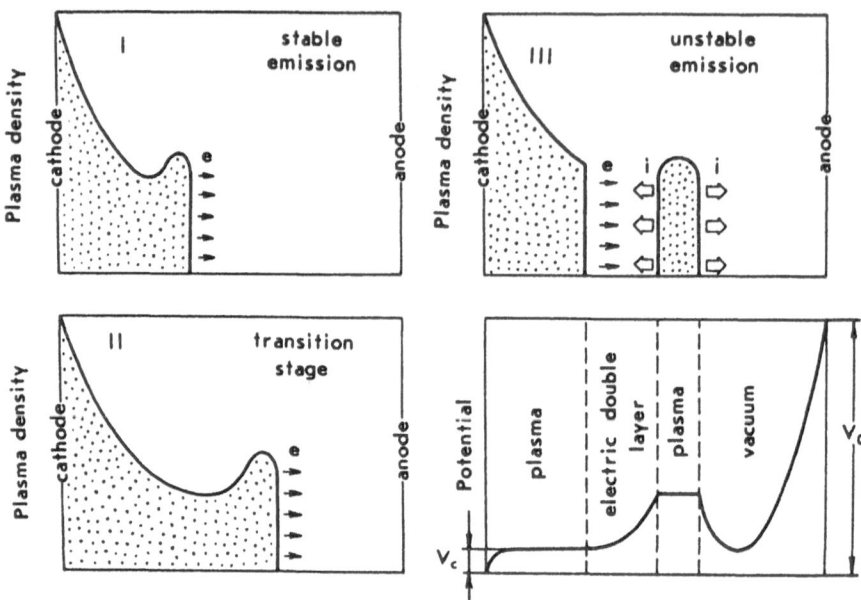

Fig.7.10. Qualitative pattern of transformation of a stable to an unstable electron emission from CF plasma

exists. If the plasmoid extension and density are such that the plasma formed behind it is not sufficient to suppose current continuity earlier than the plasmoid region of maximum density goes over to the saturation regime, then a discontinuity with a high potential drop will occur in the plasma. As a consequence, a spike will appear in the current waveform with an amplitude greater than that predicted by the "3/2 power" law for diode. If the plasmoid spreads in space, no discontinuity will form in the plasma and the burst amplitude will not exceed the value predicted by the "3/2 power" law. *Bazhenov* et al. [7.10] also showed that the appearance of the first spike in the current waveform should give rise to a series of subsequent spikes, because each current burst is followed by an additional quantity of plasma coming to a CF, so creating a nonmonotonic radial distribution of the plasma density.

According to *Bazhenov* et al. [7.10], the occurrence of the first current burst is random in nature. The data we obtained (Fig.7.4) indicate, however, a non-spontaneous character of the first current burst. Staying with the concept of varying material input to the plasma, we believe that the first intense plasma burst takes place at the moment of EEE initiation. This is because the cathode erosion starts from the explosion of sharp microprotrusions which cause the greatest rate of loss of material from the cathode (Sect.5.3). Hence, at the front of an expanding CF a denser plasmoid moves covering in a time of the order of 10^{-9} s about $2\cdot10^{-3}$ cm. Therefore, the transition to the saturation regime will coincide in time, within a nanosecond, with the appearance of the first current burst accompanied by an electron jet.

Let us assume that by the time t_s the current density extracted from the plasma boundary is equal to the saturation current density ($j_e = j_s$). Then we obtain from (7.7) [7.11]

$$t_s = \sqrt{\frac{e\dot{z}M}{4\pi v_c^3 m_i} \frac{1}{j_s} \sqrt{\frac{kT_e}{2\pi m_e}}} \quad . \tag{7.13}$$

With the rate of loss of mass from the cathode $\dot{M} = 2\cdot10^{-4}$ g/s and $kT_e = 4.5$ eV we have

$$t_s \simeq 4.3\cdot10^{-7} j_s^{-1/2} \quad . \tag{7.14}$$

Under the conditions of the experiment described in [7.11] the CF plasma filled about a half of the gap in t_s. This suggests that on the discharge axis $j_s \simeq j_a$. Thus, in spite of the essential spread in the experimental t_d values, caused by pulse-to-pulse variation of the rate of loss of mass from the cathode $\dot{M}$, the resemblance between (7.9 and 14) supports the concept of unstable current emission associated with the saturation of the CF plasma emitting ability and its irregular generation by the cathode.

Knowing the electron jet current density j_j (j_j is approximately equal to the current density j_a at the moments of bursts) and the front plasmoid potential V_f (V_f is measured by the probe when it is within the front), one can estimate the thickness of the electric double layer between the CF plasma body and the front plasmoid

$$d_f \simeq \sqrt{\frac{1.86\cdot2.33\cdot10^{-6} V_f^{3/2}}{j_j}} \quad . \tag{7.15}$$

For a known spark current and current density at the system axis one can find the location of the CF plasma emission boundary prior to and after the formation of the front plasmoid. This enables one to estimate the front plasmoid thickness d_f provided that the double layer thickness is known.

The time during which the front plasmoid dissolves due to the escape of ions (Fig.7.10), i.e., the time of existence of the electron jet, can be estimated from the relation

$$t_j \simeq \frac{q_f}{i_i} = (d_f) \left[z \frac{kT_e}{2\pi m_i} \right]^{-1/2} . \tag{7.16}$$

For instance, for the conditions $d_f \leq 10^{-2}$ cm, $z \simeq 1.5$, $m_i \simeq 10^{-22}$ g (Cu) of [7.6,11], we find from (7.16) that $t_j \stackrel{\sim}{<} 10^{-8}$ s.

Thus, a certain time after the conditions for the electron jet formation have been created, the conditions for its decay are produced as a consequence of the decomposition of the front plasmoid. Periodical escape of dense plasmoids to the emission boundary results in the occurrence of current bursts and electron jets. It is evident that the conditions for the creation of a double layer and the formation of an electron jet are first created on the system axis, where the electron current density extracted from the CF plasma is maximum.

A qualitative explanation for the cause of the appearance of ions with abnormally high energies accelerated toward the anode was given in [7.12]. The electron beam, having been formed and accelerated in the electric double layer, comes out from the positively charged front plasmoid and creates, through its space charge, an electric field providing favorable conditions for the collective acceleration of ions (Fig. 7.10).

The formation of the positively charged front plasmoids also explains [7.12,23] the reasons for the appearance of the ion flow from the cathode plasma back to the cathode, accelerated to an energy level which constitutes a noticeable part of the potential difference applied to the diode.

7.6 Spark Current Between Broad-Area Electrodes

In the above discussion it was shown that, in spite of the non-stationary nature of CF plasma generation, the electron emission from CF can be described by the modified "3/2 power" law. Based on this fact and on the results of investigations into the breakdown kinetics, we have calculated the time behavior of the vacuum spark current in a breakdown between broad-area electrodes [7.5,26,27].

7.6.1 Calculation of the Spark Current Rise

As a simplification it is assumed that the onset of the spark current rise is associated with simultaneous formation of n cathode flares, emission boundaries of which move at the same velocity which is independent of time. The opposing motion of the anode flare is not taken into account. The effect of the restricted area of the emitting surface on every CF is accounted for by using (7.10). The relation between the current and the voltage for a discharge circuit with an active impedance R has the form

$$V_0 = Ri(t) + V(t) . \qquad (7.17)$$

Substituting (7.10) into (7.17) and denoting $\tau = v_c t/d$ and $I = Ri(t)/V_0$, we obtain after transformation

$$1 = I + \frac{I^{2/3}}{(A\pi nR)^{2/3} V_0^{1/3}} \left(\frac{1-\tau}{\tau}\right)^{4/3} \left[k_0 \frac{\tau}{1-\tau}\right]^{-2/3} . \qquad (7.18)$$

On substituting $(A\pi nR)^{-2/3} V_0^{-1/3}]^{-1} = B^{4/3}$ a little manipulation yields

$$\frac{(1-I)^{3/4}}{BI^{1/2}} = \left(\frac{1-\tau}{\tau}\right) \left[k_0 \frac{\tau}{1-\tau}\right]^{-1/2} , \qquad \text{where} \qquad (7.19)$$

$$B = 370 n^{-1/2} R^{-1/2} V_0^{-1/4} . \qquad (7.20)$$

In Fig.7.11a the function $I(\tau)$ is plotted for several values of the parameter B, which correspond to real values of n, R, and V_0. The time of the current rise from 0.1 to 0.9 ($\tau_c = \tau_{0.9} - \tau_{0.1}$) as a function of B is shown in Fig.7.11b. The function $\tau_c(B)$ has a maximum at B $\simeq$ 0.8. Using this relationship one can easily obtain the commutation time as

$$t_c = d\tau_c(B)/v_c . \qquad (7.21)$$

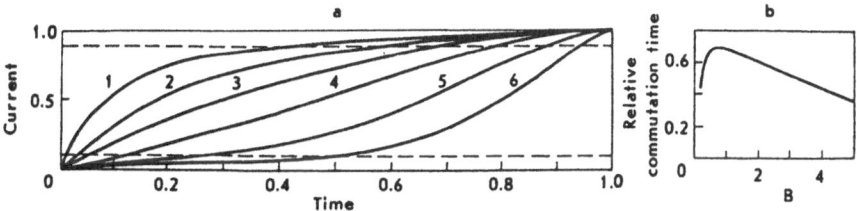

Fig.7.11a,b. Diode current time behavior (in relative units) (a) and the relative commutation time as a function of the parameter B (b)

Let us analyze the results:

(i) The function $I(\tau)$ is a plot which agrees well in form with the breakdown current waveform (Fig.4.1).

(ii) The rise time of the spark current from zero to its peak value is independent of the external circuit parameters and it is directly proportional to the gap spacing and inversely proportional to the velocity of propagation of the emitting plasma front.

(iii) The commutation time τ_c (or t_c) is measured between the levels of 0.9 to 0.1 from the peak value of current (or voltage). Therefore, a change in τ_c (or t_c) is only related to the deformation of the $I(\tau)$ - or $i(t)$ - curve.

(iv) The deformation of $I(\tau)$ and $i(t)$ curves is due to variations of the number n of plasmoids appearing at the cathode, the circuit resistance R, and the applied voltage V_0.

(v) Eqs.(7.20,21) predict a weak dependence of the commutation time t_c on the aforementioned quantities. Taking this and experimental data (Fig.7.11) into account, the function $\tau_c(B)$ allows a little increase as well as a decrease of the time t_c with an increase of the voltage V_0. The reported calculations on the current rise are compared in detail with the data of [7.28] and with our data in [7.26,27].

7.6.2 The Role of Cathode and Anode Flares

We shall now discuss one more important aspect associated with the role played by cathode and anode flares in the process of the spark current rise. In the model considered the velocity of propagation of the emission boundary is assumed to be constant. However, as the CF plasma is expanding, its density at the front turns out, at a certain time, to be lower than necessary to ensure the electron current passage to the anode (the transition to the saturation regime). As a result, the emission boundary is "decelerated" by the external field and the time t_c increases, while the ratio d/t_c decreases. (We first pointed out this aspect in [7.29]; further investigations aimed at lengthening the stage of the high-energy beam have been described in [7.30]). If the conducting medium necessary for the voltage decay is neither due to the ionization of the residual atmosphere nor due to the filling of the gap with plasma from anode, the duration of the breakdown spark stage can be considerably increased. However, the analysis of the current waveforms obtained at the pulsed breakdown of mm- [7.31] and cm- [7.32] wide gaps suggests that, provided no special measures have been taken to obtain a uniform current distribution over the broad-area electrode or to diminish the rate of the cathode plasma generation, the ratio d/t_c

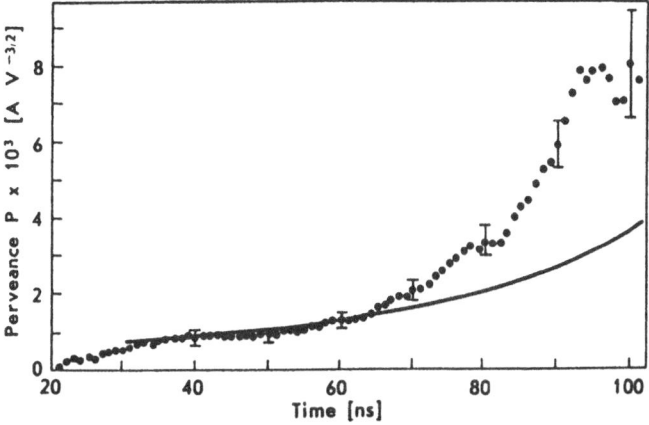

Fig.7.12. Electron flux perveance as a function of time: the dots - experiment; the solid line - calculation with $v_c = 1.8 \cdot 10^6$ cm/s [7.33]

remains in the range $(2 \text{ to } 3.5) \cdot 10^6$ cm/s. Hence, it ought to be assumed that even if a certain "deceleration" of the boundary of electron emission from CF takes place, the current rise is ensured by the opposing motion of CF and AF. In terms of the model of the current rise considered above this means that one should substitute in the formulae for determining the current (7.10-12) the sum of the cathode and the anode plasma velocities rather than the velocity of propagation of the cathode plasma emission boundary. This was demonstrated by *Parker* et al. [7.33] in their study of the pulsed breakdown phenomenon between plane graphite electrodes in an attempt to obtain a high-current pulsed electron beam. Measuring the gap current and voltage in the process of breakdown they were able to observe the variation of the electron flux perveance $P(t) \propto D^2/(d-v_c t)$ (the electrode diameter $D = 5$cm) and from the variation of $P(t)$ they could judge the change in the speed of shortening of the gap vacuum part during breakdown. The results obtained are presented in Fig.7.12. A period 30 ns after the arrival of the voltage pulse at the gap the whole surface of the graphite cathode was covered with plasma and the perveance variation could be well accounted for in terms of the CF plasma expansion toward the anode at a velocity $v_c = 1.8 \cdot 10^6$ cm/s. However, after 65 ns the perveance started to rise more rapidly, than it should follow if we assume a constant v_c, i.e., it is necessary to assume that from this moment on an anode flare appeared and moved in the opposite direction with approximately the same velocity. *Parker* et al. [7.33] checked that this took place by carrying out an electro-optical observation of the light emission occurring at the breakdown (Fig.4.8).

Table 7.1. Data for the characteristics of the anode processes, taken from [7.34]

D_a [mm]	d [mm]	v_a [cm·s⁻¹]	t_X [μs]	$v_a t_X$ [mm]	$v_a t_X/d$
30	15	$3 \cdot 10^5$	1.0	3	0.2
20	20	$6 \cdot 10^5$	0.9	5.4	0.27
10	25	$9 \cdot 10^5$	0.75	6.75	0.27
5	27.5	$1.1 \cdot 10^6$	0.6	6.6	0.24
0.3	30	$2 \cdot 10^6$	0.4	8	0.27

The fact that the vacuum gap commutation at the breakdown is determined by the cathode plasma expansion and that only the opposing motion of the AF shortens the time t_c also follows from the experiments of *Tsukerman* and co-workers [7.34] who investigated anode processes in a high-power X-ray tube of a reversed configuration. Table 7.1 lists the data of [7.34] on the electrode separation d = $(D_c - D_a)/2$ (D_c = 60 mm is the cathode diameter, D_a is the anode diameter), the anode vapor velocity v_a, and the X-ray pulse duration t_X. If one assumes in the first approximation that $t_c = t_X$ [7.25, 35], it turns out that for the time t_X the anode vapor would cover not more than a quarter of the gap spacing.

8. Formation of New Emission Centers on the Cathode

The formation of new explosive emission centers on a cathode is one of the most important processes accompanying the development of vacuum breakdown and discharge. The dominant role is played by the plasma of primary CFs. On the one hand, in the cathode regions covered with plasma the formation of new ECs is stimulated and thus the plasma-cathode interaction ensures the maintenance of explosive emission. On the other hand, the initial CFs hinder the spontaneous formation of new CFs in the cathode regions not covered with plasma due to the decrease of the applied electric field (the "screening" effect).

In the present chapter we shall first discuss the conditions of formation and subsequent operation of new ECs under the cathode plasma. The importance of studying this problem is evident from the fact that the preferential direction of new EC formation dominates the direction and the velocity of the cathode spot "motion" in both the sparking and the arcing phases. The later sections will deal with the screening effect and its influence on the structure of the electron flow formed in the breakdown spark phase.

8.1 Mechanisms of New EC Formation Under the Plasma

It is well known [8.1-3] that during the flow of electric current in a vacuum, cathode spots form and operate on the cathode surface. Their number depends on both the magnitude and the rise rate of the discharge current, and the cathode surface conditions. The cathode spot operates at the same place for 10^{-7} to 10^{-6} s. However, as shown in Chap.5, during this period a rather large number of explosive emission centers can appear and disappear under the quasi-stationary cathode plasma ball and leave craters several micrometers in diameter on the cathode surface. Moreover, it has become clear that there are two ways in which new ECs form. Firstly, the characteristic crater substructure suggests that new ECs may build up on the crater rim. Secondly, there is a real opportunity for new craters to form at distances a few or some tens of times greater than the size of the molten metal zone.

In the literature, two possible reasons for the cathode spot formation have been discussed. They are associated with of a strong electric field at the cathode. One of these is the explosion of micropoints under the action of a high-density field-assisted thermionic emission current. This conclusion is drawn as a whole from studies on the electron emission-to-vacuum arc transition [8.4], and has been discussed more than once [8.2,5-7]. Another mechanism which was mainly studied in connection with the glow-to-arc transition [8.8] relates the appearance of cathode spots with the charging of non-metallic inclusions and films on the cathode surface by the ion current from the cathode plasma and subsequent explosive-like breakdown of these impurities. It is supposed that the same reasons dominate the formation of new ECs under the plasma. One more possible mechanism associated with the formation of drops is discussed in Chap.5. Firstly, when the neck between the drop and the related point becomes sufficiently thin, it can explode under the action of the current closing onto the drop [8.9-11]. Secondly, just after the drop breaks off, a strong electric field arises between the drop and the liquid micropoint tip, which gives rise to the neck explosion [8.12]. Experimental results on the contact initiation of CS on a mercury cathode at a supply voltage of 1 to 2 V and a current of the order of 10^{-3} A [8.13] may substantiate such a mechanism of new EC formation.

In order to assess the likelihood of the various ways for EC formation one has to know the ion concentration n_i, the electric field in the near-cathode layer E_c and the layer thickness L_c as a function of the distance r from the primary EC. Assuming that the plasma is totally ionized and that there are no ion collisions in the near-cathode layer, E_c and L_c can be determined (as long as the ion temperature T_i is known) from

$$E_c = E_{D_i} \left(\frac{eV_c}{kT_i} \right)^{1/4} \quad \text{and} \tag{8.1}$$

$$L_c = L_{D_i} \left(\frac{eV_c}{kT_i} \right)^{3/4}, \tag{8.2}$$

where E_{D_i} and L_{D_i} are the characteristics of the Debye layer. Plots of the functions $n(r)$, $E_c(r)$, and $L_c(r)$ for typical values $v_c = 2 \cdot 10^6$ cm/s, $kT_i = 2$ eV, $M = 5 \cdot 10^{-4}$ g/s and for three values of potential drop V_c

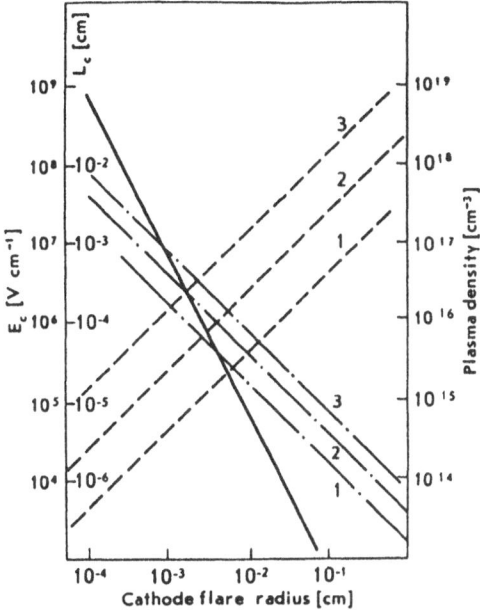

Fig.8.1. Distribution of the near-cathode layer parameters L_c (dash-lines) and E_c (dash-dot line) and the plasma density n (solid line) along the cathode flare radius at $V_c = 15$ (*1*), 100 (*2*), and 1000 V (*3*)

are given in Fig.8.1. We shall employ these data in analyzing the mechanisms described above.

8.1.1 Mechanism of the Explosion of Micropoints

As shown in Sect.5.4, the formation of new ECs under the CF plasma at small distances from the primary EC ($r \leq 10^{-4}$ cm) occurs within 10^{-9} to 10^{-8} s. If this process is caused by explosions of micropoints, the ion space charge should produce an electric field of the order of 10^8 V/cm at the micropoint tips. From Fig.8.1 it is clear that for a smooth surface at $V_c = 15$ V the electric field $E_c = 10^8$ V/cm can be attained at $r \leq 10^{-4}$ cm, i.e., just in the emission zone. For the range of $10^{-4} \leq r \leq 1$ [cm] it is impossible to find real parameters of micropoints such that the local electric field at them, βE_c, would reach the value of 10^8 V/cm. Conditions under which exploding micropoints have optimum geometrical parameters, can be realized only at much higher values of V_c (several hundred volts) and with $r \leq 10^{-3}$ cm.

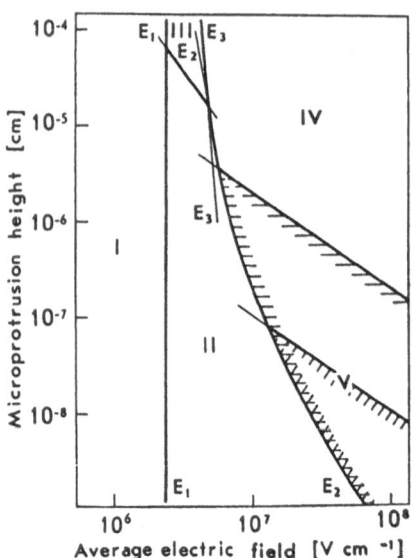

Fig.8.2. The effects of an individual explosion of an ellipsoid micropoint. The ratio of the axes is 5:1 and the long axis is perpendicular to the surface (*I*: no field effect on surface structure; *II*: surface tension deformation; *III*: field rupture; *IV*: evaporation, no individual effects; *V*: individual explosive evaporation)

The possibility of the explosion of micropoints under cathode plasma was analyzed in detail by *Ecker* [8.7]. In order that individual micropoints become new sites of intense emission and evaporation, certain conditions must be satisfied. Firstly, the micropoint tip should rapidly go to the state of intense vaporization. Secondly, the onset of this state should occur earlier than the deformation of the liquid tip under the action of surface tension or an electric field. Thirdly, the micropoint height should be noticeably less than the thickness of the near–cathode space-charge layer L_c. *Ecker* employed the method of existence diagrams which he described in [8.7]. In the case mentioned the existence diagrams were plots of the micropoint height h versus the electric field at the micropoint tip E. Such a diagram is given in Fig.8.2. Curve E_1 corresponds to the condition that the steady-state temperature of the tip is equal to the boiling point. The regions lying to the right of curves E_2 and E_3 correspond to the second condition. Curve E_4 is associated with the condition $L_c = 3h$ for $V_c = 10$ and 100 V. Thus the values of the height h and the electric field E, where individual micropoints are potential explosive emission centers, are established. From Fig.8.2 it is clear that if $V_c = 10$ V the contribution of explosions of individual micropoints to the formation of new ECs should be scarcely taken into consideration, since with their height of about 10^{-7} cm the total number of atoms contained in them amounts only to several thousand. An essentially different situation arises, if the cathode drop has a chance to rise up to $V_c \geq 100$ V. In this case the micropoints make a real contribution to the explosive emission. *Ecker's* conclusions [8.7] correspond, in many respects, to our ideas [8.9, 10, 14-17]. It is important to note that since the plasma density sharply decreases as the distance from the emission center increases, the maximum distance at which the micropoints present under the plasma are still able to explode is not in excess of about 10^{-3} cm.

8.1.2 Mechanism of the Explosion of the Liquid Neck

During the detachment of a liquid drop it may occur that the ratio of the drop surface area S_d to the neck cross-sectional area S_n reaches a value of the order of 10^2 to 10^3. In Fig.5.20 the distribution maximum corresponds to the drops with a surface area $S_d \simeq 10^{-9}$ cm^2. The neck cross-sectional area can be estimated from the dimensions of the tips of the points which appear after the termination of the current pulse. Thus, according to the data of [8.18], $S_n \simeq 10^{-11}$ to 10^{-12} cm^2.

In order that the liquid neck could explode within 10^{-9} s, the current density passing through it should be $\sim 10^9$ A/cm^2 (Sect.5.1). Hence, the density of the current closing on the drop should be 10^6 to 10^7 A/cm^2.

The question of how such a high current density could be produced has been discussed in [8.10, 11, 17]. Since such a drop is in the environment of dense plasma, it can be treated as an electric plasma probe. In doing so the following current components should be involved (Fig.8.3): (a) field-assisted thermionic emission current from the drop surface; (b) ion current from the plasma to the drop; (c) electron current from the plasma to the drop.

The contribution of the field-assisted thermionic emission from the drop surface can be estimated from ranges of the probable drop temperature T_d and the electric field at its surface E_d. Curves illustrating the relation between T_d and E_d for two values of the field-assisted thermionic emission current density $j_e = 10^6$ and 10^7 A/cm^2, calculated from the Richardson-Shottky equation, are shown in Fig.8.4. It can be suggested that the drop temperature T_d is scarcely above the boiling point, hence the field-assisted thermionic emission current density should not be greater than 10^6 A/cm^2.

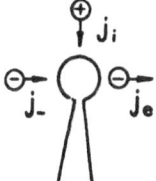

Fig.8.3. Components of the current closing on a drop

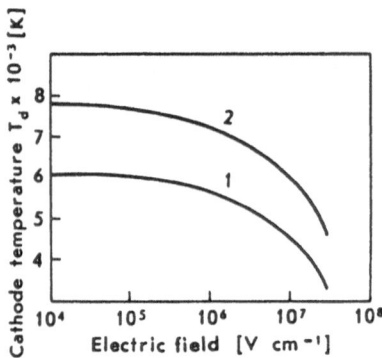

Fig.8.4. Cathode (drop) temperature T_d as a function of the electric field E_d at the metal surface for the field-assisted thermionic emission current density 10^6 A/cm^2 (1) and 10^7 A/cm^2 (2)

The ion current density (to the probe) can be defined as follows

$$j_i = \bar{z}en_i \sqrt{\frac{kT_e}{2\pi m_i}} . \tag{8.3}$$

For $n_i = 10^{19}$ cm^{-3} (at the crater edge) $j_i \leq 10^6$ A/cm^2.
The electron current density (from the plasma to the probe) is

$$j_e = \bar{z}en_e \sqrt{\frac{kT_e}{2m_e}} \left[1 - f(T_e) \right] \exp\left(-\frac{eV_c}{kT_e}\right), \tag{8.4}$$

where $f(T_e)$ is a function accounting for the plasma cooling due to the electron flow to the cathode. Using (8.4) gives for $V_c = 15$ V an estimate $j_e = 10^6$ to 10^7 A/cm^2. *Mesyats* and co-workers [8.10, 11, 17] concluded that the plasma electrons associated with the "tail" of the Maxwellian distribution provide the dominant contribution to the current closing on the drop.

The neck will be heated if the voltage drop across it is low compared with $|V_c - V_f|$, where V_f is the floating potential of the probe drop. The condition for the onset of the neck heating is expressed by [8.11]

$$T_n \sqrt{\lambda\kappa_o} < |V_c - V_f| , \tag{8.5}$$

where T_n is the neck temperature. If we assume $|V_c - V_f| \simeq V_c$ and set $T_n \simeq 3 \cdot 10^3$ K (the boiling point of copper), then the condition (8.5) turns out to be fulfilled within an order of magnitude. It is important, however, to note that the neck can explode, if the plasma density only ensures that $j_e = 10^6$ to 10^7 A/cm^2, i.e., at $r < 10^{-3}$ cm.

8.1.3 Mechanism of the Breakdown of Non-metallic Inclusions

On a real cathode surface non-metallic inclusions and films are usually present (Sect.2.1). The charged plasma particles, having passed through the near-cathode layer, are deposited on these inclusions and films, thus creating a surface charge of density $\sigma_q = jt$. If the surface and space leakages are relatively small, the electric field in the dielectric will increase with time up to the breakdown electric field value E_{br}. The breakdown delay time, i.e., the time prior to the appearance of a new emission center, will be

$$t_{EC} = \frac{\epsilon \epsilon_o E_{br}}{j(r)} , \qquad (8.6)$$

where ϵ and ϵ_o are the relative and absolute dielectric constants, respectively. It is assumed that the time the breakdown develops within the dielectric itself is negligible compared with the time t_{EC}. Moreover, the condition should be satisfied that

$$E_{br} \, \Delta d < |V_c - V_f| , \qquad (8.7)$$

where Δd is the thickness of the inclusion or the film. Since the breakdown electric field E_{br} for thin films amounts to 10^6 to 10^7 V/cm [8.19], the condition (8.7) is satisfied with $\Delta d \leq 10^{-6}$ to 10^{-5} cm. Inclusions and films of such thickness are typical for experimental conditions. It should also be noted that a new explosive emission center can arise as a result of the development of sliding discharge along the dielectric inclusion.

Estimates obtained with (8.6) show that the breakdown of dielectrical inclusions under the plasma may result in the new EC formation not only at small distances from the primary EC, but also at those distances which are an order of magnitude larger than the crater radius. Thus, at $r \leq 10^{-3}$ cm $t_{EC} \leq 10^{-9}$ s, while at $r = (1$ to $5) \cdot 10^{-2}$ cm $t_{EC} \leq 10^{-6}$ s.

From the above analysis it is obvious that well planned experimental investigations into the formation and operation of such ECs at various distances from the primary centers are important.

8.2 New EC Formation and Operation Under Cathode Plasma

One of the most successful ways of approaching this problem is measuring the current in different cathode regions [8.20-22]. We carried out two sets of experiments using this method [8.15,23]. The second set differed from the first one in that a transverse magnetic field was applied across the discharge gap. These experiments and the results obtained are discussed below.

8.2.1 Experiments Without Application of a Magnetic Field

The main part of the experiment was carried out in a "commercial" vacuum of ~10^{-3} Pa with a low-inductance cathode assembly (Fig. 7.7b) [8.15]. An igniting electrode and a probe, both made of wire $5 \cdot 10^{-3}$ cm thick, were mounted in a radial slit at the level of the cathode surface. A rectangular voltage pulse of up to 30 kV and 1.3 μs long was applied to the anode simultaneously with a low-power igniting pulse ($t_p = 5$ns, $i_i = 5$A). The electrode separation d was varied in the range of 0.5 to 2 cm. The average rate of current rise in the diode amounted to 10^8 to 10^9 A/s. The inductance of the cathode assembly was about 10^{-9} H, so the potential difference between the plasma and the probe did exceed a few volts even at $di/dt = 5 \cdot 10^9$ A/s (current bursts). The probe potential relative to the cathode and plasma could be lowered artificially.

As the igniting pulse arrived at the anode, a single CF arose and the diode current started to rise. After the time r/v_c the plasma reached the probe. After a time t_{EC} from the moment when the probe had been immersed into the plasma, ECs appeared on its surface. The idea that the new ECs are EEE centers was supported by the appearance of several local luminous figures, the radiation spectrum of which showed the lines of the probe material. The glow appeared within ~10^{-9} s in synchronism with the probe current i_p [8.22]. As a result of EC operation, microcraters with solidified micropoints on their rims formed on its surface. This, however, did not give rise to an intensification of the process of EC formation at subsequent discharges. Three characteristic ranges for the distance r from a primary EC, which differed in the conditions under which the EC occurred [8.15,24], were observed.

(i) In the range $40 \leq r \leq 150$ [μm] a nonzero probability for the appearance of EC existed, which increased with decreasing r and in-

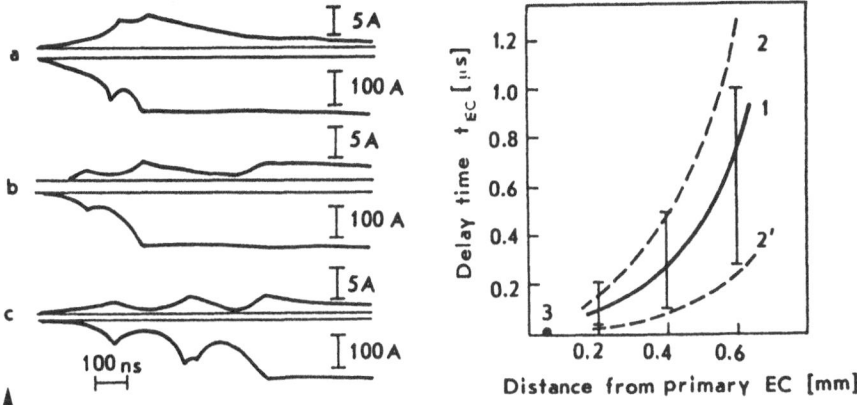

Fig.8.5a-c. Waveforms of the diode current (lower traces) and the probe current (upper traces) recorded with d = 0.5 (a, b) and 1 cm (c), r = 50 μm.

Fig.8.6. Delay time of the appearance of new ECs as a function of the distance from the primary EC obtained experimentally (1) and by calculating with E_{br} = $5 \cdot 10^6$ (2) and $1 \cdot 10^6$ V/cm (2')

creasing di/dt, and reached 0.5 to 0.8 at r ≤ 50 μm. The most likely delay time for these conditions was $t_{EC} \leq 10^{-8}$ s, but a few cases were observed with $t_{EC} = 10^{-7}$ to 10^{-6} s. The probe current correlated with the diode current (Fig.8.5), the amplitude of i_p increasing with di/dt. After the current rise stopped, the vacuum gap turned out to be completely shortened by plasma, and the arc stage of discharge began. The current i_p gradually decreased and sometimes cut off before the end of the diode current pulse. In some cases, a new EC occurred during the arc stage.

(ii) In the range 0.15 ≤ r ≤ 2 [mm] the current i_p appeared only with an artificially created negative bias at the probe relative to the plasma not less than $-V_b$ = 30 to 40 V. Increasing V_b magnified the probability of EC appearance. The function t_{EC} = f(r) obtained with V_b = -70 V is plotted in Fig.8.6. Every point on the distribution curve represents the most probable t_{EC} value.

(iii) In the range r > 2 mm the current i_p did not appear before time r/v_c, and then it appeared only at that moment when a spike appeared in the diode current waveform (Fig.8.7). The amplitude of i_p just after its appearance was much greater than for the range (i) and amounted to 5 to 20 A. The appearance of the spike was associated with an increase of the potential of the plasma boundary layer around the probe up to several kilovolts (Fig.8.8). The probe placed 4 mm away from the EC recorded potentials not only at the first current burst, but also at subsequent ones.

167

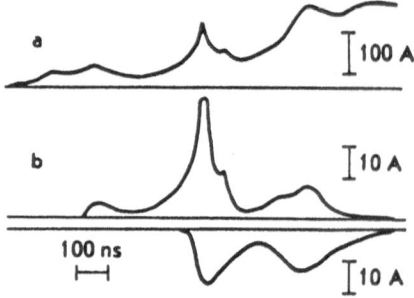

Fig.8.7a,b. Waveforms of the diode current (a) and the probe current (b) recorded with r = 4 (the upper trace) and 8 mm (the lower trace)

The results reported show convincingly that in the spark stage of discharge the peripheral part of the cathode plasma acquires a high potential relative to the cathode at the moments of current bursts due to the formation of a double layer. This situation is qualitatively illustrated by Fig.8.8.

In addition to the experiments already described, another one was carried out to determine the conditions for the appearance of EC on the probe in a vacuum arc discharge, when $di/dt = 0$. A voltage pulse of 1.3 μs duration was applied to a 20 μm igniting gap. The transition to the arc regime occurred within a time of the order of 10^{-9} s. The arc current was limited by a level of 10 to 50 A. The bias was provided by connecting a low-inductance resistor into the cathode circuit. The mechanisms of EC occurrence turned out to be similar to those that took place in the ranges (i) and (ii), and the probability of EC formation increased with arc current.

A study of EC appearance at an arc discharge in "oil-free" vacuum with the use of heated probes, made of pure materials, yielded results somewhat different from those described above. At $40 \leq r \leq 70$ [μm] EC appeared on a probe only when V_b was not less than -100 V with a probability of about 0.1 in a time t_{EC} of 10^{-7} s in the average. As the tungsten probe temperature was increased to 1500 K, the probability remained almost unchanged, while the average time t_{EC} incre-

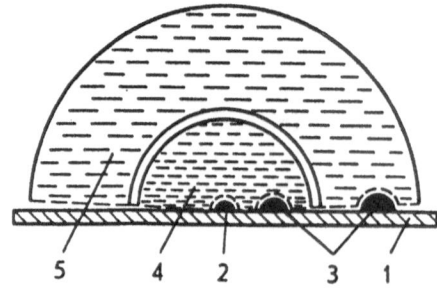

Fig.8.8. Schematic illustration of the plasma regions present on the cathode (1: cathode, 2: primary EC, 3: new ECs, 4: low-potential plasma region, 5: abnormally-high-potential plasma region)

168

ased to $8 \cdot 10^{-7}$ s. The appearance of erosion traces on the probe did not facilitate the process of EC formation at subsequent discharges. With $-V_b = 1$ to 2 kV the probability of the appearance of current in the probe circuit approached 1, while the delay time decreased. EC did not always occur on the probe at a minimum distance from the site of ignition. With $r > 100$ μm and $V_b = -100$ V no EC appeared on the probe.

Under the same conditions experiments were carried out with a probe made of aluminum wire. On its surface an Al_2O_3 layer of a certain thickness with a known breakdown voltage was formed. The time characteristics of the process under investigation for the ranges (i) and (ii), as well as the spread in t_{EC} were not significantly different from those observed for the case of uncontrolled non-metallic inclusions and films.

8.2.2 Effect of Transverse Magnetic Field on New EC Formation

In the presence of a magnetic field tangential to the cathode surface new ECs occur in an arc discharge preferentially in the direction opposite to that of the magnetic force, which predetermines the abnormal character of the cathode spot motion. However, not a single investigation on cathode spots in a magnetic field took into consideration the initial (spark) stage of the vacuum arc formation, i.e., the case when the current flow is determined by the expansion of the plasma in transverse electric and magnetic fields.

We studied the dynamics of new EC formation on a cathode under these conditions [8.23]. At the same time the opportunity arose to compare the behavior of cathode spots in arc and spark discharges. Experiments were carried out using a coaxial diode shown schematically in Fig.8.9. The cathode was a copper disc 12 mm in diameter and

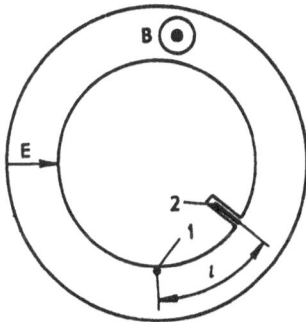

Fig.8.9. Schematic representation of the coaxial diode (1: site of ignition, 2: probe)

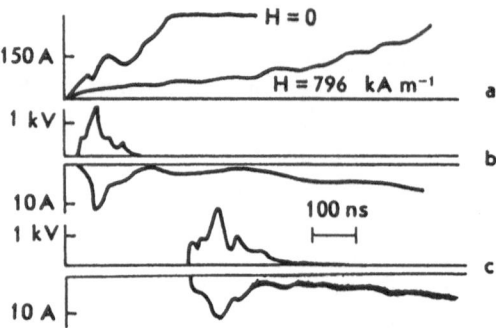

Fig. 8.10a–c. Waveforms of the diode current (a), the plasma potential (upper traces), and the current of the probes placed at a distance of 2 (b) and 10 mm (c) from the site of ignition (lower traces)

0.5 mm thick. The electrode separation was 5 mm. Probes made of 0.5 mm copper wires were placed at a distance 1 from the site of ignition in the cathode slits, level with the edge of the disc. The diode was placed in the pulsed magnetic field of a solenoid. The magnetic field amplitude H was varied in the range of $7.9 \cdot 10^4$ to $7.9 \cdot 10^5$ A/m. As the magnetic field reached its maximum, a rectangular voltage pulse of duration t_p = 0.3 or 1.3 μs and amplitude of up to 30 kV was applied to the anode. At the same time, a pulse with t_p = 5 ns and i = 5 A was applied to the igniting electrode. The site of CF formation was indicated by the igniting electrode within ±0.1 mm. Without an igniting pulse there was no breakdown of the gap. When photographing the glow at the cathode, a negative voltage pulse was applied to it. A primary CF was initiated by the igniting electrode connected to the ground through a 5 kΩ limiting resistor. The rate of current rise in the diode, di/dt, was determined by the velocity of plasma expanding in the magnetic field. It varied from pulse to pulse in the range of 0.7 to $2 \cdot 10^8$ A/s (Fig. 8.10). The cathode assembly inductance was $2 \cdot 10^{-8}$ H. For measuring the plasma potential a "plasma" probe was used as previously. Experiments were carried out in "commercial" vacuum of about 10^{-3} Pa.

Figure 8.11a,b illustrate the cathode light emission for two pulse durations with an electrode separation of 5 mm and a magnetic field of $7.96 \cdot 10^5$ A/m. It can be seen that the EC multiplication occurs in the direction of the magnetic force. In order to establish in which stage of the discharge (spark or arc) the EC multiplication took place, the electrode separation was decreased to 1 mm. Then, with a pulse duration of 1.3 μs, a vacuum arc operated for most of the time, the commutation time being as short as 100 to 150 ns. It turned out that in this case new ECs appeared at a distance not more than 1 to 2 mm

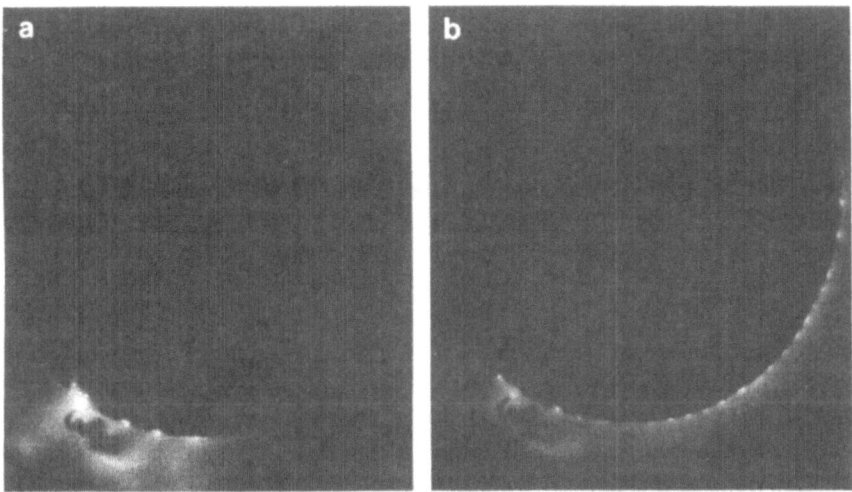

Fig.8.11a,b. Photographs illustrating how ECs can spread along the cathode edge in the presence of magnetic field. $t_p = 0.3$ (a) and 1.3 μs (b); 1 - site of ignition

from the primary one. Hence, the EC formation takes place virtually only during current rise. As the magnetic field is increased from (1.6 to 8)$\cdot 10^5$ A/m, the commutation time increases twofold and therefore the maximum distance at which new ECs appear also increases approximately twofold.

With the use of a probe placed at a distance 1 from the site of discharge initiation we investigated the delay time t_{EC} of the emission from the probe and the plasma potential at given cathode sites. A plot of $t_{EC}(l)$ is given in Fig.8.12. The tangent $1/t_{EC}$ defines the propagation velocity of the EC formation boundary, which is about $1.6 \cdot 10^6$ cm/s for this case.

Shown in Fig.8.10b,c are the typical probe current waveforms (lower traces). The emission occurs due to an increase of the plasma

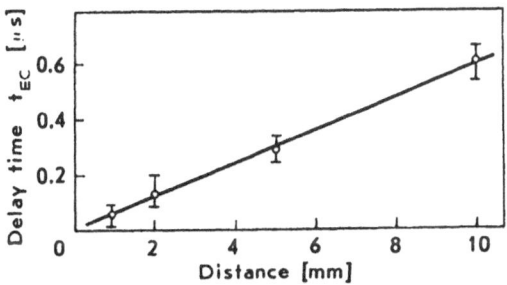

Fig.8.12. Delay time of the emission appearance from the probe as a function of the distance form the site of discharge ignition

potential just above the given cathode site, indicated by the potential waveforms (upper traces). Sometimes, the probes placed 2 and 5 mm from the site of initiation emitted practically simultaneously. From the waveforms shown in Fig.8.10 it is clear that the time of existence of the heightened potential of up to several kilovolts, and consequently of more intense emission from the given cathode sites, ranges from 50 to 150 ns. The dimension of the charged plasma section was not above 3 mm. From the data obtained one can also conclude that new ECs occur in a "relaywise" way, as the diode current rises and the plasma expands.

After the action of five pulses the cathode surface was examined by Scanning Electron Microscopy (SEM) [8.25]. On the cathode section where the light emission had been observed a great number of craters not more than 1 μm in diameter were found along with a much smaller number of craters with a characteristic substructure 2 to 5 μm in diameter.

8.2.3 Results and Discussion

Let us first consider one of the mechanisms of the new EC formation under the plasma at the cathode regions with a low potential drop (r<0.2cm). If new ECs were to form due to the explosion of micro-points under the action of field emission current of a high density, the relationship $t_{EC}(r)$ would have the form $t_{EC} \propto r^2 \exp(2r^{1/2})$, since the field emission current density $j \propto \beta^2 E_c^2 \exp[-1/(\beta E_c)]$, $j^2 t_d$ = const., and $\beta \propto L_c^{1/2}$ [8.26]. If the same process should occur due to the explosion of liquid necks, this dependence would be as follows: $t_{EC} \propto r^4$, because $t_{EC} \propto j^{-2}$ and $j \propto n \propto r^{-2}$. The experimentally obtained relationship $t_{EC}(r)$ is not as pronounced (Fig.8.6). Suppose that new ECs show up at breakdown of non-conducting inclusions and films. Then the function $t_{EC}(r)$ will have the form $t_{EC} \propto r^2$. Given in Fig.8.6 are the curves of the function $t_{EC}(r)$, calculated according to (8.6) for two possible values of E_{br}, for a dielectric film. The satisfactory agreement between theory and experiment is evidence that, within the investigated range of distances r, new ECs form preferentially as a result of breakdown of non-conducting thin films and small inclusions. The large spread in t_{EC} values can not only be due to a spread in E_{br} values, but also due to pulse-to-pulse fluctuations of the plasma density, to wandering of the site of EC formation, or to a wide range of effects due to non-metallic films and inclusions, as well as to the possibility that breakdown is the result of a different mechanism [8.19] (including the surface flashover [8.8]). The fact that the breakdown voltage of thin dielectric films increases with their thickness (from $\simeq 10$ V for 10^{-6}cm up to $\simeq 200$V for 10^{-4}cm) [8.19] allows one to explain the increasing probability of EC formation with an increase of plasma potential. Actually, at $r \leq 100$ μm the emission centers seem to appear due to the breakdown of relatively small non-metallic inclusions and films. To charge them up to the breakdown voltage, the usual cathode potential drop is sufficient. As r increases, small inclusions become unable to be charged up to such an extent because of a relatively stronge influence from the leakage current. For breakdown of the larger inclusions a higher voltage is necessary (in our case 50 to 70 V was sufficient). Also given in Fig.8.6 is an experimental point which is related to the region (i) ($r = 50 \mu$m). The calculated time t_{EC} for this case is about 10^{-9} s. Increasing the purity of the original material and improving the vacuum condition may only impede the EC formation according to the mechansim under consideration, which was also observed in the experiment. Increasing the probe temperature results in an increase of the conductivity of the non-metallic inclusions and, consequently, in a lengthening of the time t_{EC}. These results do not disagree with the data of [8.27], where it is shown that the transition probability of a high-current glow discharge to an arc is significantly reduced when a purer material is utilized for the cathode. The fact that the appearance of erosion traces on the probe does not give rise to further EC formation also indicates that this may be the mechanism of breakdown. The fact that emission is initiated more efficiently in the presence of organic contaminations was confirmed by simple observation that when the probe was touched with a finger, this caused the appearance of a great number of ECs, although there was no emission from the probe, even when its potential relative to the plasma was lowered [8.15, 24].

We shall discuss the formation of new ECs at those cathode sections with a relatively high potential drop, which lie under the plasma. At distances $r > 0.2$ cm the plasma density becomes less than 10^{13} cm^{-3} and for $V_c = 10^3$ V we have $L_c = 10^{-2}$ cm, $E_c = 10^5$ V/cm. In order that micropoints can explode under these conditions, it is necessary that the enhancement factor β exceeds 10^2. As has been shown more than once [8.28-32], high β values ($>10^2$) usually correspond to those cathode areas where non-metallic inclusions and films are present. The calculated time for charging these inclusions and films at such distances is above 10^{-6} s. Nevertheless, experiments show that ECs appear about 10 to 20 ns after the onset of the plasma potential increase. For charging and breakdown of inclusions in such short periods a plasma density of about $5 \cdot 10^{15}$ cm^{-3} is necessary. It can be assumed that this plasma density is produced as a result of absorbed gas desorption and ionization by ion bombardment. It can be easily estimated that 10 ns after the plasma has arrived at the cathode, a gas layer of thickness 10^{-3} cm ($<L_c$) and density of the order of 10^{19} cm^{-3} forms. At the bursts the ionization probability due to ion bombardment increases. Newly produced electrons and the UV radiation give an additional contribution to the ionization. To charge inclusions efficiently, an ionization degree of the desorbed gas of not more than 10^{-3} may be sufficient.

It should be noted that dielectric inclusions can initiate the breakdown of a vacuum gap under the action of an applied electric field, as has been observed in experiment [8.30, 32, 33]. In this case, the breakdown delay time can be about 10^{-9} s with a macroscopic electric field at the cathode of $(1 \text{ to } 5) \cdot 10^6$ V/cm [8.34]. Thus, in our case, with $r \leq 30$ μm, when the ion-space-charge field reaches a value of about 10^6 V/cm, ECs can occur under the plasma as a result of the breakdown of non-metallic inclusions and films solely under the influence of the electric field, only if the breakdown delay time is less than that predicted by (8.6).

Also to be noted is the correlation between the probe current and the total spark current (Fig.8.10). Such a correlation can be explained qualitatively as follows. At high rates of current rise both the near-cathode potential drop and the plasma potential drop increase (Chap.11). As the discharge goes over to the arc phase, the plasma resistance decreases and, consequently, conditions are created for EC to be destructed, the probe current to decrease and even to be cut off. Such a large current value for the probe placed within the range (iii) is due to a high voltage drop across the double layer formed between the probe plasma and the charged plasma coming from the cathode (Chap.11). When the thickness of the double layer increases due to scattering of the charged plasma, the probe current falls abruptly.

The process of EC formation in the discharge spark stage in the presence of a transverse magnetic field has some pronounced peculiarities when compared to the behavior of an arc spot. It is observed that in the discharge spark stage the high propagation velocity of the boundary of the region within which ECs form, of about $2 \cdot 10^6$ cm/s (in an arc it is 10^4 cm/s) is in the magnetic-force direction. These peculiarities are specific to the conditions under which CFs existed in our experiments. After the EC formation, the plasma filters through the magnetic field and drifts in the magnetic-force direction, giving rise to the diode current. The plasma drift velocity, judging by photographs and oscillographic waveforms, amounts to about $2 \cdot 10^6$ cm/s. The new ECs only occur during the time when the CF plasma is expanding and remains isolated from the anode. An essential condition that ensures an extensive EC formation in a magentic field is fast charging of the peripheral regions of the plasma drifting in the Ampere direction up to a high potential. This suggests that EC formation proceeds in the same way as in the case of plasma expansion in a high-voltage diode without a magnetic field. It has been shown that in such a diode new ECs appear simultaneously with bursts of the plasma boundary layer potential and the propagation velocity of the boundary of the region of EC formation is about $2 \cdot 10^6$ cm/s. Charging of the plasma's peripheral layer to a high potential can occur as a result of the formation of plasma discontinuities due to the random nature of cathode erosion. The random emission of cathode material into the plasma is due to the "relaywise" occurrence of new ECs. Indeed, from Fig.8.10 it can be seen that the current waveform has the shape of a pulse, i.e., it rapidly decays after the EC appearance. Hence, a situation is created periodically where the plasma density becomes higher in front of the expanding plasma than behind it, and this results in a discontinuity. Since the plasma does not expand in a magnetic field with spherical symmetry, but rather expands into a half-space from the EC, the discontinuity with a high voltage drop occurs in the same direction. Thus, the EC appearance causes, in a certain time (20 to 100 ns), the occurrence of a new EC or a group of ECs along the direction of plasma drift.

After shortening of the interelectrode gap by plasma, the diode voltage decreases to the level of the arc voltage drop and the "relaywise" EC multiplication becomes inefficient. From this time onwards the formation of the new ECs seems to proceed in the same way as in an arc discharge, i.e., in the reverse direction, but as the rate of this process is of the order of 10^4 cm/s, it is difficult to record it reliably with the pulse duration used (1.3μm). Careful analysis of photographs of the glow has shown that in the presence of a magnetic field, ECs were often observed in the "reverse direction" at a distance not less than 1 mm from their site of the initiation. This corresponds to a velocity of the EC "motion" of less than $5 \cdot 10^4$ cm/s.

8.3 "Screening" Effect and Electron Beam Structure in a Diode

8.3.1 "Screening" Effect

Let us now analyze the situation occurring when a high electric field is created instantly at the surface of a real cathode. In such a situation the sharpest microprotrusions will be the first ones to explode. The electric field at the cathode areas adjoining the CF formed will decrease for two reasons. Firstly, the plasma hemispheres will acquire a potential close to the cathode potential. Secondly, a strong effect will be exerted by the space charge of the electrons emitted by the plasma. Since the period for a micropoint explosion to occur is strongly dependent on the electric field at the cathode (Chap.5), even a small decrease in the electric field strength sharply reduces the probability of the appearance of new ECs in the neighbourhood of the plasma of an operating CF.

We now consider how the screening effect manifests itself in a vacuum diode with plane-parallel electrodes [8.35]. Assume that a plasma hemisphere of radius $v_c t$ emitting electrons is present on the cathode. The difference between the forces created by the cathode and the anode electric fields is equal to the electron beam momentum per unit time, i.e.,

$$\frac{1}{8\pi} (\int E_a^2 dS - \int E_c^2 dS) = \frac{mc}{e} i \sqrt{\gamma^2 - 1} , \qquad (8.8)$$

where E_a and E_c are the electric fields at the anode and at the cathode, respectively; c is the velocity of light; $\gamma = 1 + eV/(mc^2)$ is the relativistic factor. It is assumed that the incidence of the beam electrons on the anode is normal. Integration is carried out over the electrode surface.

In order to estimate the maximum radius of the cathode surface area screened by the electron beam we shall suppose that the space charge of the beam electrons reduces the electric field to zero at the cathode area of radius r_{scr} and that $E_a = V/d$. Then the left side of (8.8) will be equal to $V^2 r_{scr}^2/(8d^2)$ and therefore

$$r_{scr} = \sqrt{\frac{8mc^2 d^2 i(\gamma^2 - 1)^{1/2}}{eV^2}} . \qquad (8.9)$$

From (8.9) it follows that for nonrelativistic electrons

$$r_{scr} \simeq 5 \cdot 10^2 V^{-3/4} i^{1/2} d , \qquad (8.10)$$

and for ultrarelativistic electrons

$$r_{scr} \simeq 15.5 V^{-1/2} i^{1/2} d , \qquad (8.11)$$

where V is measured in Volts, i in Amperes, and d in centimeters.

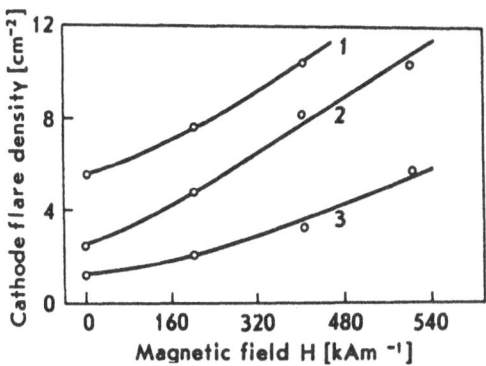

Fig.8.13. Cathode flare distribution density on a plane graphite cathode as a function of magnetic field at E_{av} = 4.7 (1), 4.0 (2), and 3.4·10⁵ V·cm⁻¹ (3) with d = 0.8 mm, t_p = 10 ns

To evaluate how the screening effect manifests itself in a multi-point cathode, the influence of space charge on the electric field of neighboring points is calculated [8.35]. It is assumed that at the tip of each point (at height h) there is a plasma ball of radius $v_c t$, which emits electrons. A self-consistent solution with the space charge taken into account has shown that with the plasma ball radius of 0.01h or 0.28h at h/2, the electric field decreases by 21% and 65%, respectively.

The screening effect was experimentally observed in various diodes. In a diode with a multipoint cathode [8.36] the density of points was varied (one point per 0.16 or 1.25 cm², respectively). The 250 kV pulse had a duration of 4 μs. Reducing the electrode separation from 10 to 4 cm resulted in an increase of the fraction of the exploded points from 1% to 4% in the first case and from 25% to 60% in the second, the current drawn from one point having been 30 and 90 A, respectively.

Application of a magnetic field along the beam reduces the screening effect due to diminishing of the beam spread, and makes it possible to increase the distribution density of simultaneously operating CFs. The CF distribution density as a function of electric and magnetic fields obtained in a plane-parallel vacuum diode with a graphite cathode 1.2 cm in diameter and d = 0.8 mm apart is given in Fig.8.13 [8.37]. Estimating the screened area radius r_{scr} using (8.10) provides satisfactory agreement with the characteristic extent of the area on which one CF falls.

The same phenomenon was observed in a magnetically insulated coaxial diode [8.35]. Given in Fig.8.14 are photographs of the cathode obtained using pulses of duration t_p = 5 ns. The graphite cathode was a hollow cylinder 2 cm in diameter with a width of the operating edge

177

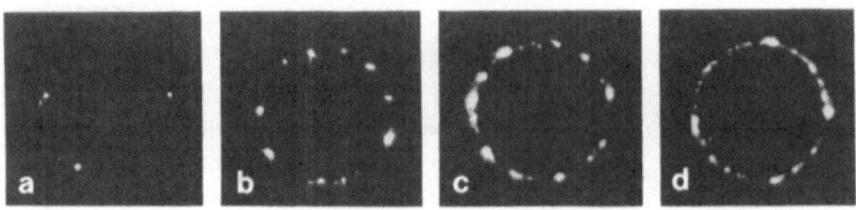

Fig.8.14a–d. Illustration of the magnetic field effect on the cathode flare distribution density in a coaxial diode with a cylindrical cathode. H = 0 (a), 79.6 (b), 239 (c), and 796 kA·m⁻¹ (d)

of 0.5 mm. The EC number increased from 3 to 5 at H = 0 up to 20 to 30 at H = $7.96 \cdot 10^5$ A/m. It is characteristic that this growth occurs mainly in the range of small values of magnetic field, at which the Larmor radius of the electrons is still much greater than the CF radius. When these radii become comparable to each other, the growth becomes slower.

8.3.2 Influence of Neighbouring CFs on the Electron Beam Structure in the Diode

The screening effect results in uniform electron beams in diodes with explosively emitting cathodes. This was first revealed experimetally and explained qualitatively by *Bazhenov* et al. [8.38]. They established that in the spark stage of vacuum breakdown between broad-area electrodes erosion traces often appear on the anode in the form of circular erosion marks which are a short distance apart, and have a clearly outlined stretched trace of stronger erosion between them called a "touch" (Fig.3.7). They further assumed that such erosion traces appear because of an increase of the current density at certain anode sections, which is due to the distortion of electron trajectories as a result of the interaction of electron flows emitted by two closely situated CFs. This supposition was substantiated by a direct experiment using two points as a cathode. The "touch" appeared as a strip erosion perpendicular to the plane in which the points lie (Fig.8.15). An experimental study of the "touch" development revealed that, at first, the edges appear in the form of sharp lines and then the residual part shows up. Thus it has been illustrated that a "touch" is a result of the interaction of the neighboring electron flows. It is worth noting that the beam inhomogeneities in the form of the prolonged "strata" of heightened current density were observed more than once in the study of the electron-beam structure in explosive emission diodes [8.39-42].

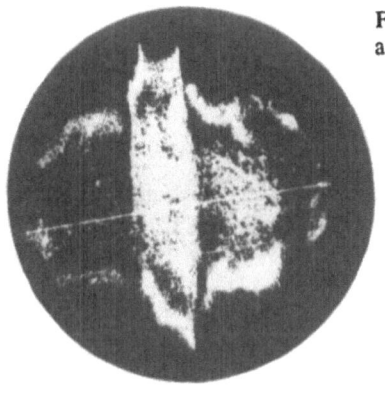

Fig.8.15. Erosion trace formed on a copper anode opposite a two-needle cathode

To explain the appearance of "touches" and "strata" the following problem was solved numerically [8.43]. It is assumed that on a plane cathode there are two plasma semicylinders, from which electrons are emitted (Fig.8.16a). The effect of the beam's magnetic field is neglected. The relative dimensions of the system necessary to solve the problem by the method of current "tubes" (Fig.8.16) were chosen close to those at which the "touch" was observed in [8.38]. From each semicylinder 50 current "tubes" were taken. The region under calculation was covered by a network with 30 equal steps along the y-axis. Along the x-axis, the step was 1 or 3 for $0 \le x \le 75$ and $75 \le x \le 150$, respectively. As boundary condition at $x = 150$ it was assumed that the normal derivative of the potential vanishes. The boundary problem for the Poisson equation was solved using difference methods with a rectangular network.

According to the calculation, if electrons escape from the semicylinder at a certain angle ϕ, they reach the anode at a maximum deviation toward the neighboring semicylinder (Fig.8.16a). For our case $\phi = 68°$. At the boundary, electron trajectories are closer to each other. Seen in the figure is the region of a "touch" ("stratum") with a greatly

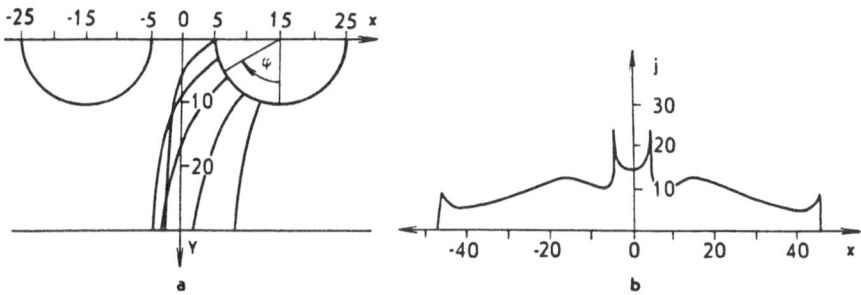

Fig.8.16. A diagram illustrating the appearance of "touches"

179

increased current density at its boundaries (Fig.8.16b). The current density in the "touch" region is higher than that opposite the semicylinder tops. The convergence of the electron trajectories is similar to the convergence at the "touch" boundary and takes place at the beam edge.

To establish how the "touch" width changes in time a calculation was carried out with the semicylinder radius equal to 12 (in relative units). It was found that for the given parameters the "touch" width decreases at a rate approximately equal to the velocity of the flare expansion.

The calculated current density at the "touch" boundaries and that of the beam itself (the height of lines) may involve a large error, since the line width is less than the network step. It is quite evident that the current density increases in this region. Thus, a "touch" is a section of overlapping electron beams emitted from closely spaced cathode flares. Sharp erosion lines at the "touch" boundaries can be attributed to the convergence, on the beam edge, of the trajectories of electrons emitted from one of the flares.

Topfer and *Bradly* [8.39] also explained the beam structure by the spatial periodicity of the plasma at the edge of the cathode blade. However, they attributed the occurrence of this periodicity to the development of an instability in the current-carrying cathode plasma. In our opinion, the geometrical periodicity of the CF plasma should be attributed to the screening effect, i.e., to the fact that the earlier CFs weaken the electric field in the neighborhood of the blade edge.

9. Anode Processes in the Spark Stage of Vacuum Breakdown

The investigations described in Chap.3 made it possible to analyze anode processes at vacuum breakdown qualitatively. As the qualitative pattern was being established, it was necessary to observe the time behavior of anode processes to obtain new data and compare them with those available in the literature. The most accurate way of obtaining quantitative information is the conditioning of the energy and the power transferred to the anode by an electron flux. The simplest conditioning technique consists of controlling the amplitude and the duration of a voltage pulse applied across the gap, and varying the gap spacing. To analyze anode processes a simple gap geometry was chosen (apoint cathode and a plane anode), that provided for a wide range of electron beam parameters and also allowed well reproducible results to be obtained.

The aim was to study:

1) The energy characteristics of the electron flow affecting the anode and the heating of the anode;

2) the pattern of damage to the anode and the structure of the material in the irradiated zone;

3) the conditions under which one or another type of erosion occurs; and

4) the conditions for the formation of anode flares.

It was also intended to analyze the parameters of the X-radiation pulse at the anode.

9.1 Anode Heat Conditions

9.1.1 Power Density Deposited at the Anode

To study the heat conditions at the anode one needs to obtain some data on the power density q_a of the electron flux bombarding the anode. The value of q_a can be easily found from the anode-current density j_a and the gap voltage V, since $q_a = j_a(t)V(t)$. Analysis of typ-

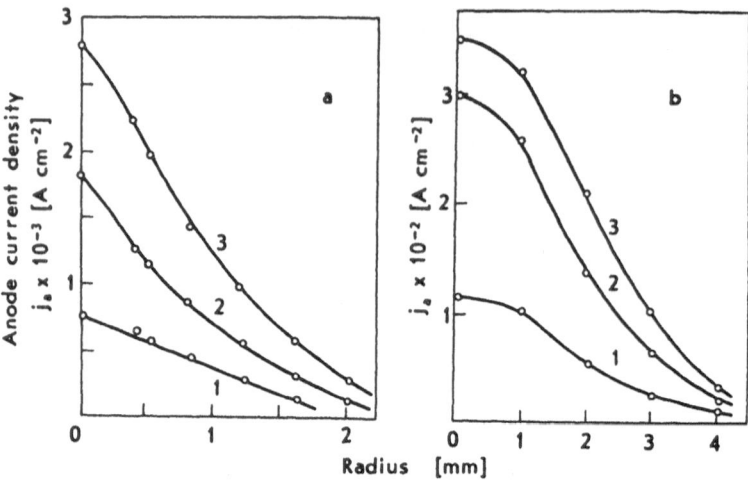

Fig.9.1a,b. Anode current density distribution at V_0 = 30 kV for d = 1 mm, t = 12 (*1*), 24 (*2*), and 36 ns (*3*) (a) and d = 2 mm, t = 24 (*1*), 48 (*2*), and 72 ns (*3*) (b)

ical anode current density distributions (Fig.9.1) [9.1,2] showed that at a given moment about 80% of the electron current falls within a circle of radius r = 2d and that at r = d the current density j_a is about one-half of its magnitude on the discharge axis. As the gap spacing is increased, the current density j_a abruptly decreases. Some plots of the function $q_a(t)$ on the system axis where the current density is highest are given in Fig.9.2. These curves have a characteristic bell-like shape

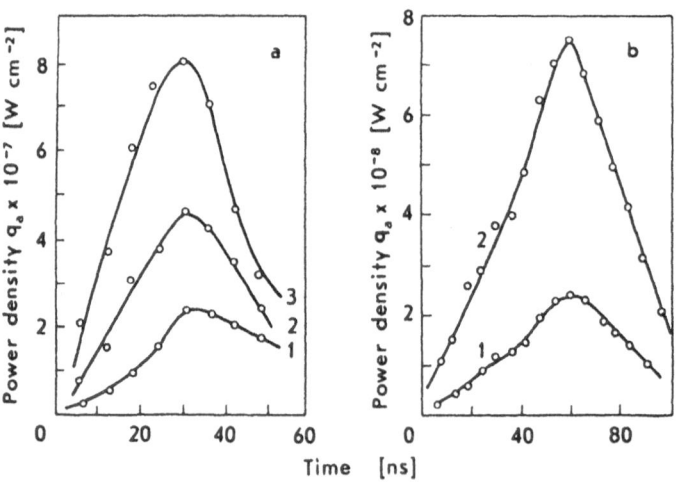

Fig.9.2a,b. Time behavior of the electron flux power density on the anode at the discharge axis for d = 1 mm, V_0 = 20 (*1*), 30 (*2*) and 40 kV (*3*) (a) and d = 2 mm, V_0 = 20 (*1*) and 30 kV (*2*) (b)

with a maximum at $t_{max} \simeq t_c/2$. The average rate of power density rise on the growth part of the function $q_a(t)$ can be defined as $q_a \simeq q_{a_{max}}/t_{max}$. Measurements have shown that in the voltage ranges examined ($20 \leq V_0 \leq 40$ [kV]) with spacings $0.5 \leq d \leq 4.0$ [mm] the maximum power density changes from 10^6 to 10^9 W/cm^2 and q_a varies from 10^{13} to 10^{17} W/cm$^2\cdot$s. Thus, during the breakdown spark stage, the electron flux provides a high energy influx density to the anode with a large rise rate. The energetic parameters of the flux can easily be controlled over a wide range by varying the electrode separation and the voltage amplitude.

The energetic parameters are close to the radiation parameters of Q-switched lasers (radiation fluxes with power densities $q = 10^7$ to 10^9 W/cm^2 are considered "moderate" [9.3]). The interaction of laser radiation with metals, the structure and properties of metals and alloys in the zone of beam action and the production of laser plasma have been extensively covered [9.3-6]. However, until recently there have been no attempts to study the phenomena occurring in the interaction between nanosecond electron beams of moderate power density and metals systematically. Although of significant interest, these problems exceed the scope of this monograph.

9.1.2 The Anode Temperature

To find out more about the conditions at which one or another type of anode erosion takes place, it is necessary to consider anode thermal processes. We may write a one-dimensional heat equation (the x-axis is normal to the anode surface and directed into it) [9.7]

$$\rho c \, \frac{dT}{dt} = \lambda \, \frac{\partial^2 T}{\partial x^2} + q_a(t)f(x) \ . \tag{9.1}$$

We assume, for simplicity, that electrons lose their energy with a constant rate over a distance $\bar{x}$. The electron energy loss distribution at a depth $f(x)$ will then take the form

$$f(x) = \begin{cases} 1/\bar{x} & \text{for } x < \bar{x} \\ \\ 0 & \text{for } x > \bar{x} \ . \end{cases} \tag{9.2}$$

Since most of the anode erosion experiments were carried out for $t_p \leq t_c/2 = t_{max}$, one can use for the power density

$$q_a(t) \simeq \dot{q}_a t . \tag{9.3}$$

Without any noticeable loss due to evaporation the boundary conditions for (9.1) have the form

$$T_{t=0} = T_0 ; \quad T\Big|_{x\to 0} = T_0 ; \quad \frac{dt}{dx}\Big|_{x=0} = 0 . \tag{9.4}$$

On solving (9.1) by taking into account (9.2-4), we obtain an expression for the anode surface temperature

$$T(x=0;t) = T_0 + \tag{9.5}$$

$$+ \frac{(\dot{q}_a t^2)}{\chi\rho c}\left[\frac{1}{2} \, \text{erf}(k) + \frac{k}{3\sqrt{\pi}} (5 + 2k^2) \, e^{-k^2} - 2k^2\left(1 - \frac{k^2}{3}\right) \text{erfc}(k)\right]$$

where $k = (\chi/2) [\rho c/(\lambda t_p)]^{1/2} = \chi[2(at_p)^{-1/2}]$.

In analyzing the solution (9.5) it is necessary to consider three cases which differ in the relation between the electron penetration depth and the thermal field penetration depth $(at_p)^{1/2}$, namely

$$\text{(i)} \quad \chi \ll (at_p)^{1/2} , \tag{9.6}$$

$$\text{(ii)} \quad \chi \simeq (at_p)^{1/2} , \tag{9.7}$$

$$\text{(iii)} \quad \chi \gg (at_p)^{1/2} . \tag{9.8}$$

In the first case a volume heat source can be exchanged for a surface one and the solution (9.5) takes the form

$$T(x=0;t) = T_0 + \dot{q}_a t^{3/2}(\pi\rho c\lambda)^{-1/2} . \tag{9.9}$$

When the condition (9.7) is satisfied, one can use the results of the analysis performed in [9.8]. There it was shown that in this case the surface temperature may be calculated by considering the reduced power release in an infinitesimally thin surface layer. The reduction factor, which is smaller than unity, is a function of the ratio of the thermal field penetration depth to the electron penetration depth.

If condition (9.8) is satisfied, then the temperature may be calculated neglecting the heat transport into the anode, and the solution of (9.5) is

$$T(x=0;t) = T_0 + \dot{q}_a t^2/(2\varkappa\rho c) \ . \tag{9.10}$$

It should be noted here that the assumption about the regular depth distribution of the electron energy loss may give a somewhat overestimated value of the anode surface temperature, since according to [9.9] electrons lose most of their energy near the surface.

After the pulse ends, the anode surface temperature begins to decrease according to [9.7]

$$T(x=0;t>t_p) \simeq T(x=0; \ t_p) \ \text{erf}\sqrt{t_p/(t-t_p)} \ . \tag{9.11}$$

It follows that the surface cools after irradiation much more slowly than it heats up. For instance, at the time $t = 5t_p$ the anode surface temperature is halved. Regions deep in the anode $(x>0)$ initially heat up due to the heat removal from the surface (i.e., after the pulse ends the dimensions of the heated-up layer increase). There is subsequent cooling.

It was shown in Chap.4 that with a power density $q_a \simeq 10^9$ W/cm^2 conditions are created for intense evaporation of the cathode material. In analyzing this process, it is more convenient to solve the heat equation (9.1) in a moving coordinate system with its origin on the evaporation surface. The boundary conditions may then be written as

$$T\Big|_{t=0} = T_0 \ , \quad T\Big|_{x\to\infty} = T_0 \ ,$$

$$\frac{dt}{dx}\Big|_{x=0} = \frac{\rho\epsilon_s}{\lambda} v_s \exp\left(-\frac{\epsilon_s m_a}{kT(0;t)}\right) \ . \tag{9.12}$$

Solving (9.1) by taking into account (9.2,3 and 12) leads to either the transcendental equation [9.10]

$$T + \epsilon_s V_s \sqrt{\frac{\rho t}{\lambda c}} \exp\left(-\frac{\epsilon_s m_a}{kT}\right) = T_0 + \frac{\dot{q}_a t^{3/2}}{(\pi\rho c\lambda)^{1/2}} \tag{9.13}$$

185

with the condition (9.6) or the transcendental equation

$$T + \epsilon_s V_s \sqrt{\frac{\rho t}{\lambda c}} \, \exp\left[-\frac{\epsilon_s m_a}{kT}\right] = T_0 + \frac{\dot{q}_a t^2}{2x\rho c} \tag{9.14}$$

with the condition (9.8). From (9.13, 14) it follows that with sufficiently intense evaporation, the evaporation front temperature reaches a certain critical value T_p^*. After reaching the critical temperature, the front temperature remains practically unchanged because the evaporation heat loss becomes approximately equal to the energy delivered by the electrons. A more detailed calculation of the anode heating with evaporation taken into account was given in [9.11].

9.2 Surface Structure of the Anode in the Discharge Zone

9.2.1 Summary of Previous Work

The first attempts to study the anode damage at nanosecond pulsed breakdown of short vacuum gaps were made in the late sixties [9.12, 13]. It was shown that the initial anode erosion in the breakdown spark stage occurred in the form of numerous craters 5 to 10 μm in diameter. Similar craters, caused by microsecond pulsed breakdowns [9.14, 15] and dc breakdowns were observed in studies of the erosion of anodes made of different materials. In the case of dc breakdown, the majority of craters were grouped along the boundaries of grains and twins [9.16, 17].

A further investigation of the anode surface damage in the initial stage of vacuum breakdown was carried out using point cathodes and plane anodes [9.18, 19]. In these experiments it was found that there are three types of anode erosion, which occurred one after another as the energy transferred to the anode by electrons was increased. Initially, some surface roughness appeared (erosion "marks"), the height of the irregularities increasing with t_p. In the next stage craters formed in the center of the erosion mark. As the current was increased, a melt zone appeared in the center of the crater zone. The dimenisons of all these zones increased with the spark current pulse duration. The same sequence of erosion zones on metal surfaces was observed after irradiation of targets by laser pulses [9.20].

Perhaps the least understood aspect was the occurrence of the erosion phenomena which preceded the surface melting. In [9.19,20] it was suggested that these phenomena are related to various types of defects in the metallic crystal structure itself. Therefore, it was of interest to compare the erosion pattern on anode samples having surface layers with greatly different properties. It is well known that defects of any kind (point defects, dislocations, packing defects, etc.) can be introduced into metal by conventional mechanical treatment which gives rise to a significant plastic deformation. In order to diminish the defect concentration, high-temperature annealing is applied which causes a decrease in the dislocation density due to recrystallization.

9.2.2 Metallographic Studies

The results of previous studies indicated the need of carrying out special experiments to elucidate the character of anode damage in the spark stage of vacuum breakdown. To make the conditions of irradiation identical, a disc consisting of two semi-discs with different properties of the surface layer was used as anode. A point cathode was mounted so that its axis (the beam axis) always crossed the contact line between the two anode halves. The semi-discs were made of oxygen-free copper. One of them was pre-annealed at 1200 K in high vacuum for two hours. Both anode halves were electrochemically polished. The experimental conditions were as follows: $d = 0.5$ mm, $t_p = 20$ ns, $V_0 =$ 20 to 40 kV. After each pulse the cathode was moved along the semi-disc boundary to a distance more than the size of the erosion zone.

The results obtained are displayed in Fig.9.3. The melt zone usually covered both the anode halves equally, a result which shows the

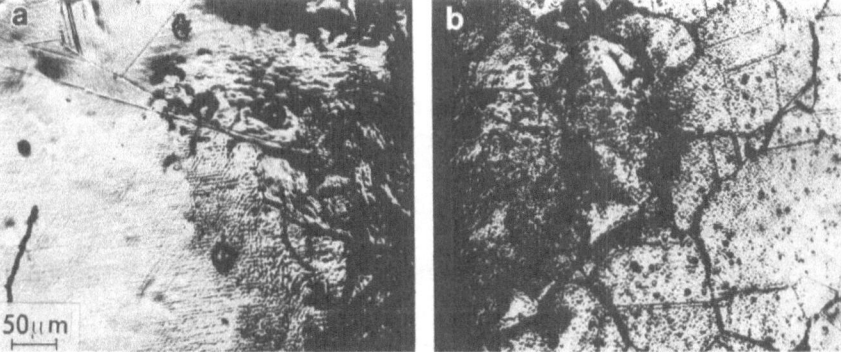

Fig.9.3a,b. Microphotographs illustrating the surface damage pattern of a copper anode consisting of an annealed (a) and an unannealed (b) part

187

identity of irradiation conditions for the two different samples and also the independence of macroscopic thermal properties of the surface layer on its defect density. On the unannealed sample, a crater zone was present behind the melting zone. The crater size and density decreased with the distance from the melting zone. This was evidently associated with a decrease in the beam power density. There was no crater zone on the annealed sample. In this case, zone a rough strain relief was seen behind the melting zone. This relief became more and more regular with increasing distance from the erosion mark and assumed a form typical for plastic shear deformation (Fig.9.3a). (It sould be noted that plastic deformation has been observed after irradiation of metals with high-power pulses of duration $\sim 10^{-8}$ s [9.21]). The rough strain relief adjoining the melting zone seems to represent a structure which has occurred as a result of melting of steps (separate stacks of crystallographic planes which emerge on the surfaces along which slipping occurs). Melting of these steps (about 1μm in height) takes place at a beam density somewhat lower than that for a smooth surface, since heat is removed less easily from steps than from a plane. The degree of deformation, as seen from Fig.9.3a, decreases with the distance from the center of the mark.

The external diameter of the plastic deformation zone which is resolvable using conventional optical metallography, is about 1.5 times larger than the crater zone diameter. Therefore, the formation of craters requires a higher beam power density than the minimum shear deformation.

The depth of the strained layer was determined by the appearance of bands of slipping on the back of annealed electropolished foils of various thickness. Under the conditions when only such bands were observed on the irradiated side and the penetration depth x was less than 10^{-4} cm, the thiickness of the strained layer was evaluated as 10 to 20 μm. As the power density was increased, the layer became deeper and by the time a significant melt zone could be seen on the irradiated side, it was as thick as 30 to 40 μm.

The effect of plastic deformation was more pronounced at electron energies almost an order of magnitude greater ($V_0 = 100$ to 200 kV). In this case a volume heat source plays a more essential role (at $V_0 = 200$kV for copper $x \simeq 45\mu$m, $(at_p)^{1/2} \simeq 2\mu$m) than a surface one (as at $V_0 = 10$ to 30 kV).

After irradiating a cold-strained copper sample by an electron beam with an energy of 200 keV, craters also formed on the sample surface but they were larger and their distribution density was almost an order of magnitude greater than at 20 keV. This seems to be because a significant contribution comes from defects which are deeper within the metal body. The volume heat source operating at $V_0 = 200$ kV favored the activation of these defects and the consequent formation of craters.

It should be noted that the structure of both the plastic deformation zone and the crater zone formed after a single period of irradiation under the same conditions. Such behavior of the plastic deformation zone can be attributed to a considerable hardening of the strain layer by the primary irradiation.

Experiments have shown that the region adjoining the melting zone has the greatest microhardness. It is about 15% higher than the initial value for annealed, and 20% for unannealed samples. The microhardness is observed to be somewhat less in the melting zone than in both the neighbouring crater zone and the plastic deformation zone.

On increasing the beam power density a melt zone occurs after the formation of craters. A further increase in the power density leads to the formation of a hole from which ridges of solidified metal diverge radially. This is evidence of the transition to a regime of intense evaporation of the anode material. Under the action of the vapor pressure the liquid metal is ejected from the hole in different directions. When operating in such a regime the cathode holder already appears to be covered with a solid copper layer after several pulses. It has therefore been shown experimentally that the destruction of the metal under the action of short-time intense electron bombardment proceeds in several characteristic stages (plastic deformation, crater formation, melting, evaporation). This sequence is associated with an increase of the beam power density. Therefore, it may be said that the onset of each stage is due to the achievement of a certain critical temperature on the anode surface, i.e., that the destruction has a "threshold" nature. This threshold nature was observed with laser irradiation of metals at the power density of 10^6 to 10^8 W/cm^2 [9.3,4].

Special experiments were carried out to verify the threshold nature of the appearance of different types of anode surface damage and to find the critical values of the power density [9.22,23]. It was established that for pulse durations of 5 to 80 ns and gap spacings of 0.4 to 2 mm a metallographically resolvable plastic deformation of copper appears at a power density $q_a = (0.8 \text{ to } 2.5) \cdot 10^7$ W/cm^2 and increases with diminishing t_p. According to calculations, the anode temperature reaches 700 to 800 K at the end of the pulse. Similar measurements for the crater zone have shown that in this case the critical power density is about 1.5 times higher than the q_a values necessary for the plastic deformation zone to occur. It was estimated to be $(1.2 \text{ to } 4) \cdot 10^7$ W/cm^2, with the attainable anode temperature being 1300 K. The critical power density at which the melt zone could be observed to appear was about $6 \cdot 10^7$ W/cm^2 (t_p=20ns, d=0.8mm). According to calculations (not taking into account the phase transition energy) the anode temperature by the end of the pulse amounts to 1800 K.

9.2.3 Electron-Microscopic Studies

The dislocation structure of the destroyed layer under various types of erosion of copper anodes in the breakdown spark stage was investigated by using transmission electron microscopy [9.24]. Up until this investigation there were practically no data on the dislocation structure of metals deformed by irradiation with a short-duration intense electron beam.

The electron-microscopy study has shown that the dislocation density is observed to increase significantly in the zone of the electron beam action. A general character of the dislocation structure of samples irradiated in a regime characterized by distinct plastic deformation but with no indications of melting is that the pattern corresponds roughly to that observed at a slow tensile strain of 3% to 10%. At the minimum electron flux intensity the dislocation distribution in the exploded zone is characterized by a relatively low density. In this case individual dislocations and their clumps are observed. The structure formed is usually called "cellular", as it consists of cells with a dislocation density comparatively low on the inside and high on the boundaries. A typical cell size is 0.5 μm. The majority of them are open and the dislocation density is of the order of 10^9 cm^{-2}. Increasing the beam power density results in a stronger deformation, thus affecting the dislocation structure. The cell size decreases and the dislocation density increases accordingly. The cell boundaries are very smeared. At a maximum beam power density when there are still no indications of melting on the slipping pattern, a dislocation structure characteristic of a strongly strained metal forms. In this regime a significant increase of the dislocation density up to about 10^{10} cm^{-2}, on average, is observed. The cell size considerably decreases reaching a certain limiting value of about 0.1 μm. Within the cell boundaries, individual dislocations are almost unresolvable. This indicates that the dislocation density there is close to 10^{11} cm^{-2}.

In samples irradiated by an electron flux with a power density sufficient for the formation of a solidified-melt zone ($5 \cdot 10^7$ W/cm^2), in spite of the high rate of cooling after pulse, some indications of tempering are clearly visible in the dislocation structure (a significant decrease of the dislocation density, an increase of the cell size). The dislocation density in the crater zone is, as a rule, much lower than in the neighboring plastic deformation zone; it is close to the dislocation density in the melt. This suggests that the crater formation is associated with the heating up to the melting point of local metal microsections; hence it is of purely thermal nature.

9.2.4 Mechanisms of the Anode Surface Damage

Plastic deformation of the anode surface layers under the action of high-power pulsed electron irradiation is caused by the following. As calculations show, the rate of heating of the anode surface in our experiments reaches $\sim 10^{10}$ K/s. The electron energy is absorbed in a layer of a few micrometers thick. The temperature gradient in the surface layer exceeds 10^6 K/cm. With such a rapid increase in temperature occurs thermal expansion of the heated layer, which in turn leads to the appearance of significant thermal stresses in the layer. These stresses relax due to plastic flow - formation, movement and interaction of dislocations.

The mechanism of the plastic deformation of the metal under the action of a high-power nanosecond electron beam is discussed in [9.23]. Using the dislocation theory and the theory of elasticity one can calculate that in our experiments the deformation time amounts to 10^{-7} to 10^{-6} s, i.e., the shearing occurs mostly after irradiation. The residual shear deformation, under the conditions described above, can vary from 10 to 100. The same order deformation was observed in our experiments.

The craters formed in the breakdown spark stage may be thought of as solidified liquid-phase elements. This is justified not only by a purely external analogy, but also by the pattern of the dislocation structure in both the melting and the crater zones. The appearance of craters should be related to the presence in the metal of impurities (carbides, oxides, sulphides, among others). Since the impurities have, as a rule, less heat capacity, heat conductivity and density than metals, localized areas of high temperature will occur in the surface layer even during regular irradiation. This is the reason for the appearance of a liquid phase. Based on these ideas, one can easily explain the origin of craters on steel (containing carbide inclusions [9.25]) and copper (containing a great deal of oxides). Abrasive materials also form alien inclusions. This is one of the reasons why craters may occur near the grain boundaries [9.16, 17], as this is where the impurities diffuse in the process of metal recrystallization. If due to the existence of an inclusion, a disc-shaped liquid element forms there at the end of a pulse, liquid metal will flow due to thermal expansion and solidify, forming a ridge on the edge of the crater. Estimates show that the ridge formation can be accounted for by the thermal expansion of the metal which is heated up near the inclusion.

9.3 Formation of Anode Flares

The exposure of the anode to an electron flux in the breakdown spark stage results not only in erosion of the anode, but also in the formation of anode flares. To understand the breakdown spark stage processes, one needs to know the background of anode flare formation, its composition, temperature, and velocity of propagation into the gap.

9.3.1 Conditions for AF Formation, Its Composition and Temperature

A study of the conditions for AF formation from the matter adsorbed on the anode surface was carried out at the Institute of High Current Electronics [9.26, 27]. The occurrence of ion current from the anode served as an indication for the appearance of AFs. Experiments were carried out under the same conditions at which we studied anode erosion. Breakdown was initiated between a point cathode inclined at an angle of 10° to the system axis and a disc or ring anode made of aluminum, copper, molybdenum, tantalum or tungsten with an electrode separation d = 0.3 to 5 mm. Time, mass and energy characteristics of anode ions were studied with the use of an electrostatic analyzer and a Thomson parabolic spectrometer. It was established that the ion beam mass spectrum depends on the voltage pulse amplitude and duration and the electrode separation. Changing the electrode material did not give rise to any change in the mass spectrum of ion beams. The following ions were seen under various conditions: H^+, H_2^+, C^+ (m_i/z = 1 to 4), O^+, CO^+, CO_2^+, as well as ions with m_i/z = 60 to 190. It is interesting that at frist only H^+, H_2^+, C^+, and O^+ ions appeared, then followed by ions of a greater mass. It was not possible to detect the anode material ions because a high proportion of desorbed gas was present in the background. Some data on the relationship between the delay time of ion current appearance and the gap spacing were given in [9.27]. This, together with (8.9, 10), leads to the conclusion that the anode is heated by an electron beam which causes thermal desorption.

Estimates obtained in [9.28] for d = 14 cm and $V_0 = 10^6$ V have shown that about 10^{-6} s after the onset of the current rise the anode surface temperature reaches 520 to 570 K. This temperature is insufficient for the anode material to evaporate. The AF appearance at this moment should be attributed to the desorption and the subsequent ionization of the gas present on the anode surface. Evidence for the formation of AFs from the desorbed gas was given in [9.26, 27, 29] which deal with the intense ion fluxes (H^+, C^+) found in the vacuum breakdown spark stage. The theoretical results for calculation of the energy absorbed by a graphite anode on the formation of anode flares were given in [9.30]. It has been shown that this energy is an order of magnitude less than that necessary for evaporation of graphite. This suggests that under certain conditions a desorption mechanism for AF formation is possible.

Additional information on the role of electron-stimulated desorption from the anode may be found in [9.31, 32]. The efficiency of the desorption was determined in [9.32]. An electron beam was injected through a thin foil into a space containing an ionization gauge. For electron beams with $V_0 = (1$ to $2) \cdot 10^5$ V, j = 0.3 to 9.0 A/cm^2, t_p = 0.3 to 5 μs a significant increase in pressure was recorded. The evolution of gas from the beam-irradiated surface was due to electron-stimulated gas desorption rather than heating of the surface. The desorption yield turned out to be 7 to 12 molecules per electron.

The delay of the onset of gas liberation relative to the beginning of the beam action, the velocity of propagation of the desorbed gas front into the vacuum gap and the gas concentration in the flow were obtained [9.31]. The gas release characteristics were measured using a plane triode ionization gauge installed on the base of an explosive-emission electron source producing an electron beam with the cross-sectional area of about 100 cm^2. At an electron energy of 10 keV an increase of the ion current to the gauge collector corresponding to the expansion of the gas flow from the anode with a front velocity of $(3.5$ to $5) \cdot 10^4$ cm/s was observed. With $V_0 = 10$ kV the desorption yield turned out to be 1 to 3 molecules per electron. The gas evolution started within 10^{-6} s with the beginning of the electron bombardment of the anode.

A study of the formation and composition of AFs at $q_a = 10^7$ to 10^8 W/cm^2 is described in [9.33]. A pulse generator with the parameters $V_0 = 20$ to 40 kV, $\rho_w = 5$ Ω (i_{max}=4 to 8kA), and $t_p = 10^{-7}$ s was used. The electrode gap with a spacing d = 0.5 to 3 mm was formed by a plane copper anode and a multipoint copper cathode. The anode current density was varied by varying the number of cathode points (n=1, 10, and 62). The velocity of motion of the luminous boundary of

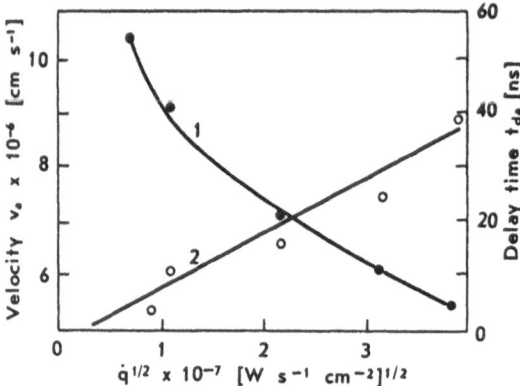

Fig.9.4. Delay time of the appearance of AF (*1*) and the velocity of AF expansion (*2*) as a function of the rate of rise of the power density at the anode

an AF and its radiation spectrum were investigated using the method described in Chap.6. The AF spectrum was characterized by the existence of a continuum on the background of which the most intense of the CuI lines could be seen. The CuI line of wavelength of 515.3 nm appeared simultaneously with the occurrence of the AF. The velocity of AF motion was measured by this line only. Both the velocity of propagation of the AF boundary, v_a, and the delay time of the AF appearance, t_{da}, were dependent on the rise rate of the power density at the anode (Fig.9.4). *Baksht* et al. [9.33] suggested that the AF velocity is directly proportional to the quantity $\dot{q}_a^{1/2}$ since the anode surface temperature $T_a \propto \dot{q}_a$, see (9.9), and $v_a \propto (kT_a/m_a)^{1/2}$. However, because we are dealing with a system where evaporation occurs, (9.13) should be used with $T_a \simeq T_{ev}^{*} \propto \ln \dot{q}_a$. However, even calculations using (9.9) show that the anode temperature, with the $\dot{q}_a$ values given in the figure, reaches about 800 K within the time t_{da}. It seems that because of the irregularity of the emission from the points, some anode

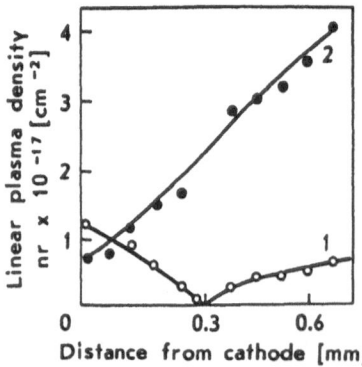

Fig.9.5. Linear plasma density as a function of the distance from the cathode for d= 0.7 mm, t = 10^{-8} (*1*) and $3 \cdot 10^{-8}$ s (*2*)

areas were heated up to a temperature at which noticeable evaporation took place.

Under approximately the same experimental conditions (the cathode being a single needle) the formation of a high-current vacuum spark was investigated using the high-speed interferometry method. *Bugaev* et al. [9.9] were able to obtain clear interferograms of AF in the final stage of transfer towards the cathode at a velocity $v_a \leq 10^6$ cm/s. The results for d = 0.7 mm (t_c=35ns) are presented in Fig.9.5. It is clear that in the beginning of the flow ($t<t_c$) filling of the gap with conducting medium occurred from the cathode, while in the break-down terminating stage ($t>t_c$) the bulk of the material came into the gap from the anode. The electron density near the anode reached $3 \cdot 10^{19}$ cm^{-3}. According to the estimates of *Bugaev* et al. with $q_{a_{max}} \simeq 10^9$ W/cm^2 and eV=10keV, the anode temperature reached about 7800 K. Calculated values of the volume of the evaporated anode material agreed with their experimental results to within about $5 \cdot 10^{-7}$ g.

9.3.2 The Expansion Velocity of AF

The expansion velocity of AF was treated in [9.11]. According to [9.3] the metal destruction occurs in a purely thermal way provided that the specific energy delivered to the metal is not in excess of (2 to 3)ϵ_s. In order to check whether this condition is satisfied, it is necessary to es-timate the highest possible specific energy delivered to the anode sur-face for a characteristic time t, assuming t_c/2. To obtain an upper limit the heat conduction loss is assumed to be negligible. The specific energy introduced to the anode will then be

$$\epsilon_a = \int_0^{t_c/2} [j_a V/(\rho x)]dt .$$

To obtain this upper limit it is necessary to impose restructions on the anode current density j_a. In accordance with (7.12), $j_a = AV^{3/2}/d^2$ (A = 44.4·10^{-6}AV$^{-3/2}$. To find ρx Widdington's formula $\rho x \simeq B_1 V^2$ can be used. Then

$$\epsilon_a \simeq AV^{1/2}/(2B_1 v_c d) . \tag{9.15}$$

From (9.15) it follows that the specific energy delivered to the anode decreases with increasing gap spacing and is proportional to d$^{-0.75}$ to

$d^{-0.5}$, since $V_{br} \simeq d^{0.5}$ to d^1. For most conventional electrode materials, with $d < 1$ mm, (9.15) predicts $\epsilon_a \simeq 2 \cdot 10^4$ J/g.

Estimates have shown that the interactions of the electron flux, secondary electrons, and X-radiation with the AF matter are negligible. Hence, the AF matter internal energy is in fact determined by the preceding development of the AF associated with the action of the primary electron flux on the anode. Assuming that AFs consist in the whole of anode material vapor and plasma one can determine the velocity of the AF expansion in an adiabatic approximation using $\gamma = 5/3$ as for a monatomic gas. From (6.3) with $\epsilon_a = 2 \cdot 10^4$ J/g we obtain $v_a \simeq 10^6$ cm/s. We have observed about the same velocity in experiments with pulsed vacuum breakdown of short vacuum gaps ($d < 1$ mm) [9.1, 12, 34, 35].

For a better determination of the AF expansion velocity it is necessary to examine the heating of the anode by an electron beam taking into account the evaporation and heat-conduction energy losses. In [9.11] it was shown that with the anode power density $q_a > 10^8$ W/cm^2 the evaporation losses are insignificant for a certain time and the anode surface temperature can be calculated from (9.5). However, from a certain moment, the increase of the surface temperature slows down abruptly because of the exponential dependence of the evaporation energy loss on temperature (9.13, 14). From this moment the surface temperature remains practically unchanged and for $10^8 \leq q \leq 10^{10}$ [W/cm^2] it can be found from the relation

$$kT_{ev}^{\,*}/\epsilon_s \simeq 0.14 \quad \text{to} \quad 0.23 \, . \tag{9.16}$$

In [9.3] it was shown that under such conditions the parameters of the vapor moving from the surface acquire equilibrium values at a distance of several mean free paths. The vapor equilibrium temperature for $kT_{ev}^{\,*}/\epsilon_s \simeq 0.2$ is $T_{eq} \simeq 0.82 \, T_{ev}^{\,*}$.

To estimate the velocity, it is necessary to find that part of the vapor internal energy which changes into the kinetic energy of the matter expanding into vacuum. As shown in [9.3], the terminal velocity of the expansion of the evaporation products is determined by their initial total energy together with the evaporation potential energy, which changes over to the flux kinetic energy in the process of condensation. This process can also play a certain role in the AF vapor expansion. In this case the fraction of the sublimation energy which is changed to kinetic energy is determined by the degree of vapor condensation $\alpha_c = \rho_1/(\rho_1 + \rho_v)$, where ρ_1 and ρ_v are the liquid and vapor density, respectively. Below a certain value of the gas density, the degree of condensation ceases to follow the vapor expansion law and the α_c value becomes constant [9.36]. Thus, it can be assumed that the internal energy of the anode-vapor gas phase in [9.11] is defined as

$$\epsilon_a = (\alpha_c \epsilon_s + 3/2 \; kT_{eq} - 3 \; \alpha_c kT_{eq})/(1 - \alpha_c) \; . \qquad (9.17)$$

The AF velocity, with the help of (9.15-17), can be expressed as

$$v_a = \epsilon_s \sqrt{\frac{4\gamma}{\gamma - 1} \frac{0.5\alpha_c + 0.24}{1 - \alpha_c}} \; . \qquad (9.18)$$

For most metals evaporation is maintained until the discharge changes to arcing. With this, the velocity of anode vapor expansion is very weakly dependent on the energy absorbed by the anode surface. The only exceptions are metals with a low sublimation energy, for which a hydrodynamic regime of evaporation is possible at an energy delivered to the anode as low as $(2 \text{ to } 4) \cdot 10^4$ J/g.

The relationship between the velocity of AF expansion and the anode power density was measured for a copper anode using the photoelectrical technique described in [9.11]. The investigation was carried out in "commercial" vacuum with an electrode separation of 1.5 mm. The anode was a circular plate 5 mm in diameter, and the face of a hollow cylinder made of copper foil 20 μm thick served as the cathode. A 50 kV, 75 ns voltage pulse generated by a cable generator with the resistance of 150 Ω was applied across the gap. The electron beam density at the anode was varied by varying the diameter of the cylindrical cathode. It is known that when using a cylindrical cathode, a sharp burst of current density occurs at its axis. A maximum current density was reached when the cathode diameter was about one third of the electrode separation [9.37]. A relative change in the current density was indicated by a change in the erosion pattern and by the appearance of light emitted from copper vapor at the anode. The earliest appearance of copper vapor corresponded to the minimum size of the erosion spot.

The procedure was as follows. Using a photoelectric spectrometer the instant when the CuI ($\lambda = 521.8$ nm and $\lambda = 515.3$ nm) lines appeared was recorded. This indicated the beginning of evaporation at the anode. The entrance slit of the spectrometer was then moved along the gap image from anode to cathode. The velocity of propagation of the anode vapor was determined by observing the time at which light appeared at a given distance from the anode. The distance covered by AF versus time is given in Fig.9.6. It is quite clear that, in spite of the fact that the time of anode vapor appearance varied by a factor of 3.5,

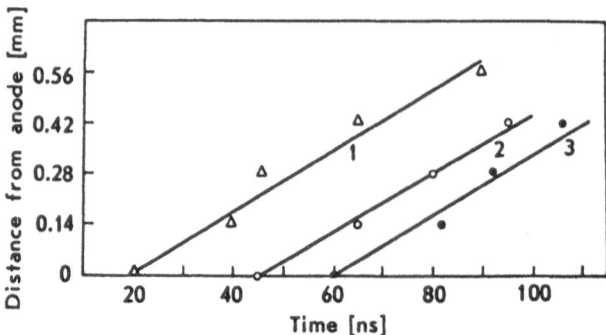

Fig.9.6. Distance from the anode covered by the anode flare plasma as a function of time for cathode diameters of 1.5 (*1*), 1.0 (*2*), and 0.5 mm (*3*)

its propagation velocity turned out to be practically unchanged and was estimated as $(7 \text{ to } 8) \cdot 10^5$ cm/s, which confirms the above conclusions.

Estimates of v_a obtained using (9.18) and experimental data in [9.38, 39] are compared in [9.11]. The condensation degree α_c at the time of its stabilization is usually in the range 0.2 to 0.5. For metals such as aluminum, copper, tin, iron, and tungsten v_a is $(4 \text{ to } 9) \cdot 10^5$ cm/s. The experimentally obtained AF velocities for these metals are in the same range.

9.4 X-Radiation Generated at the Anode

The action of an electron flux on the anode in the spark stage of breakdown results not only in anode erosion and AF formation, but also in the appearance of a high-power X-ray flash. Although the first diode-type X-ray pulse tubes based on the principle of vacuum pulsed breakdown were developed as early as in the late 1930s [9.40-42], the mechanism behind the high electron currents remained unclear for a long time. The high-current electron flow in the tube at breakdown can be understood only on the basis of the studies of the breakdown kinetics described in Chap.4. This, in turn, made it possible to interpret correctly the data on the time of appearance, the waveform and the duration of the X-ray pulse [9.43]. Later on, following a detailed investigation into diode current-voltage characteristics, we were able to interpret the main mechanisms of the X-ray pulse generation in X-ray tubes [9.44].

The X-ray pulse parameters are dependent on both the parameters of the discharge circuit and the diode gap geometry. In every calculation the problem is reduced to measuring the functions i(t) and V(t) correctly and determining from them the X-radiation intensity $J_X(t)$, the conventional X-ray pulse duration t_X (for example, on the level of 0.1 or 0.5 from $J_{X_{max}}$) and the dose per pulse D_X, etc.

9.4.1 X-Radiation on Discharging a Line

For the case of discharging a line with a resistive impedance R on a diode tube one may use the i(t) and V(t) functions obtained analytically (Sect.7.3). In a simple estimation, however, it can be assumed that the tube current increases linearly, i.e.,

$$i(t) \simeq \frac{V_0}{R} \frac{v_c t}{d} = \frac{V_0}{R} \tau .$$

The gap voltage can be expressed as

$$V(t) \simeq V_0 \left(1 - \frac{v_c t}{d} \right) = V_0 (1 - \tau) .$$

Then the X-radiation intensity in the breakdown spark stage may be written as

$$J_X(t) \propto i(t) V^2(t) \simeq \frac{V_0^3}{R} \frac{v_c t}{d} \left[1 - \frac{v_c t}{d} \right]^2 = \frac{V_0^3}{R} \tau (1 - \tau)^2 . \qquad (9.19)$$

The maximum intensity $J_{X_{max}}$ is reached at the time $t_{X_M} \simeq d/(3v_c)$ after the beginning of current rise, the corresponding current and voltage being $i_M \simeq V_0/(3R)$ and $V_{max} \simeq (2/3)V_0$, respectively. The radiation dose per breakdown

$$D_X \simeq \int_0^{t_c} J_X(t) dt \simeq \frac{V_0^3 d}{12 R v_c} . \qquad (9.20)$$

From (9.19,20) it follows that both the intensity and the dose are proportional to the cube of the voltage pulse amplitude V_0 and the dose is also proportional to the gap spacing d. In Fig.9.7 the results of the X-radiation dose measured at the breakdown of millimeter gaps are given. They show satisfactory agreement with an approximate calcula-

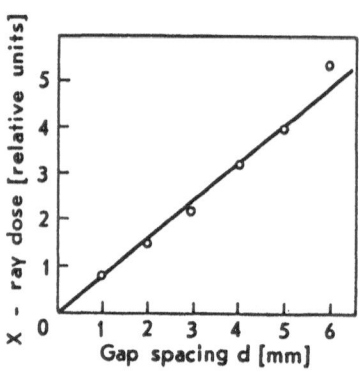

Fig.9.7. X-radiation dose as a function of gap spacing determined for breakdown of mm gaps

tion using (9.20) [9.45]. The hypothesis that the X-ray pulse duration is proportional to the gap length and is practically equal to the commutation time t_c (Fig.4.12) was also confirmed experimentally in [9.45].

The conclusion that the duration of the X-radiation pulse $t_x \propto d$ is also in agreement with the oscilloscopic records in [9.46], where X-radiation generated in discharging a line (t_p=250ns) on a diode-type tube with a needle cathode was studied. As the gap spacing d was increased, the X-radiation pulse duration t_x lengthened, while the pulse duration t_p was in excess of the commutation time t_c. With a further increase of d the X-ray flash duration t_x remained constant and equal to about t_p, while the amplitude decreased in accordance with (9.19).

9.4.2 X-Radiation on Discharging a Capacitor

The most widespread method for generating high-power, short X-ray pulses is that where a capacitor is discharged onto an X-ray tube through a spark gap filled with compressed gas [9.42,44]. The spark gap enables the capacitor to be connected to the tube within ~10^{-9} s. If the circuit inductance and the spark gap resistance are neglected, the relation between the tube current and voltage may be defined as

$$i(t) + C \frac{dV(t)}{dt} = 0 \ , \tag{9.21}$$

where C is the capacitance. The problem now is to describe approximately the current-voltage characteristic for this tube geometry. Difficulties may arise as none of the expressions given in Sect.7.3 rigorously describes the tube current-voltage characteristic throughout the period of the shortening of the gap by plasma. Therefore, we analyze here a simple case of a cathode consisting of n points operating simultaneously, the current-voltage characteristic of each point being described by (7.11). Moreover, for a relatively low capacitance the condition $t_X \ll d/v_c$ is usually valid. The system of (9.21) and (7.11) with the initial conditions $V=V_0$, $i=0$ at $t=0$ can be reduced to the equation [9.44]

$$\frac{dY}{dt} + \frac{\tau}{1-\tau} BY^{3/2} = 0 \quad , \tag{9.22}$$

where $Y=V/V_0$, $\tau=v_c t/d$, $I=i/i_0=id/(CV_0 v_c)$, $B=A_1 ndV_0^{1/2}/(Cv_c)$. If $\tau \ll 1$, then (9.22) can be simplified and a little manipulation yields the tube current and voltage

$$I = \frac{64\tau B}{(4 + B\tau^2)^3} \quad ; \quad Y = \frac{16}{(4 + B\tau^2)^2} \quad . \tag{9.23}$$

The time at which the current maximum occurs and the maximum current are determined from the formulae

$$t_{i_M} \simeq \frac{2(Cd)^{1/2}}{(5A_1 nv_c \sqrt{V_0})^{1/2}} \quad , \tag{9.24}$$

$$i_M \simeq 0.52V_0^{5/4} \sqrt{CA_1 nv_c/d} \quad .$$

The time to the attainment of the maximum X-radiation intensity and the intensity value may be expressed as

$$t_{X_M} \simeq \sqrt{\frac{4Cd}{BA_1 nv_c \sqrt{V_0}}} \simeq 0.6t_{i_M} \quad , \tag{9.25}$$

$$J_{X_{max}} \propto V_0^{13/4} \sqrt{A_1 nv_c C/d} \quad . \tag{9.26}$$

Essentially the same calculation of the capacitor discharge on the X-ray diode tube was carried out later by *Jamet* and *Thomer* [9.42]. They approximated the tube resistance by the relation

$$R_d(t) = \frac{A}{t} \quad . \tag{9.27}$$

It can easily be shown that this relationship follows from (7.11). Indeed,

$$R_d(t) = \frac{V(t)}{i(t)} = \frac{d - v_c t}{A_1 v_c t \sqrt{V(t)}} .$$

(9.28)

Provided that $v_c t \ll d$ and $\sqrt{V(t)} \simeq$ const., we obtain

$$R_d(t) = \frac{d}{A_1 v_c \sqrt{V}} \frac{1}{t} = \frac{A}{t} .$$

(9.29)

In the range $0.4 \leq d \leq 1.0$ [cm] with $V_0 < 10^5$ V the constant A is close to the value $25 \cdot 10^{-6}$ $\Omega \cdot$s as assumed in [9.42]. From the calculation of [9.42] it also follows that $t_{i_M} \propto \sqrt{(AC)}$ and $t_{X_M} \simeq 0.6 t_{i_M}$.

Jamet and *Thomer* measured the half-width of the X-ray pulse duration t_X as a function of electrode separation in coaxial three-electrode X-ray tubes with the anode placed at the end of the tube axis [9.42]. The tube voltage was varied in the range $20 \leq V_0 \leq 35$ [kV], while the electrode separation was d = 4 to 9 mm and the capacitance C = 7000 pF. The X-ray pulse duration turned out to increase almost linearly with d and to decrease slightly with increasing V_0. The slope of the curve $d(t_X)$ lies mainly in the range $(1.5$ to $3) \cdot 10^6$ cm/s. The variation of the duration t_X with V_0 can be explained in terms of the analysis of current waveforms carried out in Sect. 7.6.

Description of technical developments of pulsed X-ray tubes exceeds the scope of this monograph. The reader may refer, for example, to [9.40-42].

10. Fast Processes at DC Breakdown of Vacuum Gaps

As a result of the studies described in the previous chapters the processes responsible for pulsed breakdown initiation and development were properly defined, and this made it possible to classify the successive stages of this phenomenon. However, while we were establishing the mechanism of the development of the vacuum breakdown, it could not be asserted that, in principle, the same breakdown stages take place at a slowly increasing or dc voltage. When a high voltage is applied across the gap for a long time, a great number of different processes occur at the electrodes and in the gap [10.1, 2]. It is of interest to establish which processes are directly responsible for the breakdown initiation and dominate the further development of the breakdown. Since, in practice, most breakdown processes are either uncontrolled or controlled with insufficient accuracy and spatial-temporal resolution, it is often difficult to distinguish the most important of them. There are a number of indications that the transition to a high conductivity of the gap occurs extremely rapidly at dc breakdown. Therefore, it is necessary to use instruments with sufficiently high temporal and spatial resolution under these conditions. Independently and virtually simultaneously, direct observations of the origin of the conducting medium in the gap at dc breakdown were carried out at the Institute of High Current Electronics [10.3-9], and by *Davies* and *Biondi* [10.10-12]. Their results are analyzed below.

We have already noted that in studies of dc breakdown with conventional temporal resolution it was impossible to observe the phase in which the gap fills up with the conducting medium. It was therefore impossible to determine how the cathode and the anode were involved in the generation of the conducting medium and the moment when the cathode spot appeared. The techniques which we developed to investigate these problems are based on pulsed breakdown studies and those of the fast stage of dc breakdown in vacuum gaps with $d \leq 1$ mm. The investigation included: (i) a study of the prebreakdown conductivity and the breakdown voltage; (ii) an oscillographic study of the current growth and the X-ray generation in the fast stage of the breakdown; (iii) an investigation of the space-time pattern of the light emission accompanying a breakdown event; (iv) an examination of the electrode erosion in the fast stage of the breakdown.

10.1 Electrical Study of DC Breakdown

10.1.1 Electric Circuit

To solve the problem stated it was necessary to work out a special scheme which would allow simultaneous measurement of dc breakdown characteristics, together with investigations of the fast breakdown stage at high temporal resolution. Additionally, the installation should permit on-line control of the measurements of the breakdown voltage at certain times and allow one to carry out experiments on pulsed nanosecond breakdown. For this purpose a discharge circuit with long coaxial lines similar to those used in the study of the pulsed breakdown was used. The vacuum gap under investigation acts as a switch which separates the storage line from the transmission line and allows the voltage V_0 to rise gradually until the occurrence of the breakdown. The time and the rate of current rise in the transmission line at breakdown are then entirely determined by the processes taking place in the vacuum gap. This principle was used in the construction of the discharge device (Fig.3.7).

Problems arose due to two contradictory requirements for the charging resistance R_0. On the one hand, to provide a single breakdown, the R_0 value should be large, because otherwise the pulse-forming line which discharges after the first breakdown is charged rapidly again and the next breakdown occurs, etc. until the voltage V_0 is eliminated by some external influence. On the other hand, R_0 should not be large enough to provide a high prebreakdown voltage drop across it. The problem was solved by using an element with a strongly nonlinear current-voltage characteristic - a high-voltage, high-vacuum, hot-cathode rectifier tube. By varying the filament voltage, a regime was chosen, in which the tube saturation current was somewhat in excess of the maximum prebreakdown current. In this case the tube resistance is low (about $1 M\Omega$). After breakdown and discharge of the pulse-forming line the resistance of the high-vacuum hot-cathode rectifier tube increased by almost two orders of magnitude. This permitted a significant lengthening of the charging time of the line (up to 10^{-2} to 10^{-1} s). During this time the output voltage of the high-voltage power source was sharply decreased and the source was turned off with the use of a special scheme started by the breakdown current. This power source ensured that the amplitude of bursts was not more than 200 V on a level of 30 kV with the load current 0.5 mA. The maximum source voltage was 60 kV.

To measure the breakdown current a 50 Ω load resistor connected to the end of the transmission line was grounded (by a reverse current line) through a small air gap which broke a few nanoseconds after the voltage wave had arrived at it. A low-inductance capacitor was connected in parallel with the air gap. The capacitor and the resistor protected the measuring device from being overloaded.

10.1.2 Prebreakdown Current and Breakdown Voltage

As noted in Chap.2, there are a great deal of data which indicate that there is an increase of the vacuum gap electric strength at high discharge currents. In order to see whether there was any significant difference between our experimental data and the results of studies with low discharge currents, it was necessary to examine the process of the electrode conditioning by discharges and also to clarify the nature of the breakdown currents as well as to measure the parameters of cathode micropoints, the breakdown voltage and electric field.

Figure 10.1 shows the progression of the electrode conditioning by subsequent breakdowns. For electrochemically polished electrodes, the maximum electric strength is reached with fewer breakdowns. In this case a smaller spread (<10%) in the breakdown voltages is observed after the discharge conditioning is completed. The breakdown voltage values attained suggest that, in spite of high discharge currents, the gap electric strength is not lower than when using high limiting resistances.

During the conditioning and subsequent testing the prebreakdown current was measured and the gap was examined using a telescop device. In the case of electrochemically polished electrodes, the prebreakdown current was much more stable. As the gap conditioning

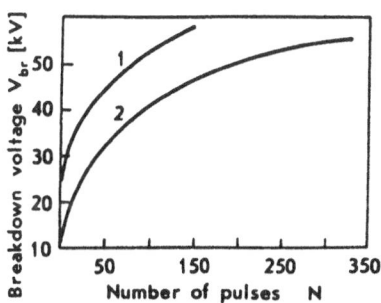

Fig.10.1. A plot of the function $V_{br}(N)$ for electropolished (1) and paste-polished (2) copper electrodes with d = 0.6 mm

Table 10.1. V_{br} [kV] and d/t_c [cm/s] values for various eletrode materials

Breakdown characteristics	Material														
	W	Ta	Mo	Ni	Ti	Nb	Cu	Al	Zn	Pb	In	Cd	Bi	Sn	C
V_{br}	73	68	62	56	55	48	46	45	26	24	23	22	21	19	16
$d/t_c \times 10^6$	2.7	3.5	2.6	1.9	1.9	3.0	2.5	2.6	1.6	1.3	1.3	1.3	1.1	1.5	2.0

proceeded, the breakdown current increased. The number of ECs on the cathode also increased as discharges occurred over a larger area on the electrode surfaces. Breakdowns often occurred away from the electrode axis, i.e., where the surface had not been cleaned so well by the discharges. The surface peripheries were cleaned by appropriate changes in the polarities of the voltage applied. Immediately after the change of polarity the breakdown voltage decreased about half, but after several dozens of breakdowns it was restored to the intial level. For well-conditioned electrodes, the current before the breakdown was 5 to 7 mA. A great number of brightly luminous local spots could be observed at the anode. Such spots cause a sharp increase in the scatterd light in the discharge chamber which hampers optical measurements. Under these conditions even cooling of the anode with water was not always sufficient to maintain the gap spacing constant.

The Fowler-Nordheim plots of the prebreakdown current versus the gap voltage were straight lines. This suggested that the current had a field-emission nature. As the conditioning progressed the micropoint-field enhancement factor decreased while the emitting area increased. The experimentally determined critical electric field at cathode micropoints $E_{cr} = \beta V_{cr}/d = \beta E_{av}$ for copper was $(6.2\pm1)\cdot10^7$ V/cm which agrees with the data of [10.13].

It should be noted that most of the experiments (especially those investigating light phenomena) were carried out under "intermediate" vacuum conditions. Prebreakdown dark currents were about two orders lower than normal and the anode microspot brightness was also greatly reduced. Therefore it was possible to make the required optical measurements. The conditioning by discharges resulted in some increase of the breakdown voltage and a decrease of the spread in its values. However, V_{br} values in this case were about half as compared to those obtained under "pure" vacuum conditions. The results obtained [10.8] are listed in Table 10.1 where the breakdown voltages are given in terms of the gap spacing d = 1 mm. Both the dependence of V_{br} on the electrode material and the absolute values of V_{br} are in accordance with the data available in literature.

10.1.3 The Current Rise Time at Breakdown

As shown above, much of the information on the mechanism of pulsed breakdown was obtained in studies of the characteristic features of current rise. Therefore, the intention was to investigate the current rise at dc breakdown and to compare the results with data for pulsed breakdown. For this purpose, control tests on pulsed breakdown were carried out, in addition to dc voltage measurements. The voltage pulse amplitude was chosen to be 10% higher than the average V_{br} dc value, so that the breakdown delay time was less than 10^{-7} s. Analysis of the data obtained has shown that the character of the current rise is practically the same for both dc and pulsed voltages, the absolute values of the current rise time being close to each other. Shown in Fig.10.2 is the commutation time t_c versus the gap spacing d for a breakdown between copper electrodes. The time t_c is the same for both cases and independent of the "purity" of the vacuum. As in the case of pulsed breakdown, the spread of t_c values from discharge to discharge is due to the fact that breakdowns not only occur at the electrode axis where the field strength is at its maximum, but also at a certain distance from the axis. For dc breakdowns, the spread is somewhat greater than at pulsed breakdowns under the same conditions.

To investigate the processes responsible for the initiation of dc breakdown, it was of interest to study the character of the current rise in the initial stage of breakdown. Using a pulse oscilloscope with the input connected directly to the electrode through a matching cable, current waveforms were obtained with a resolution of about 0.07 A/mm [10.4,9]. Some typical current waveforms are shown in Fig.10.3, where the time dependence of the EEE current, as predicted by (7.12), is plotted. Good agreement between the experimental and the calculated waveforms i(t) is evident. This result directly indicates that the

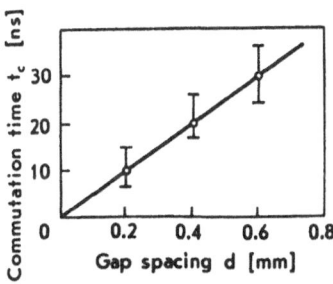

Fig.10.2. Commutation time as a function of gap spacing for a dc breakdown between copper electrodes

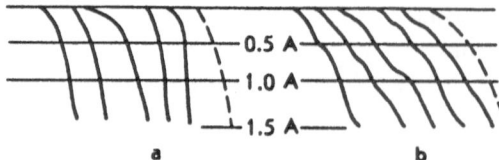

Fig.10.3a,b. Typical oscilloscope records (solid) and calculated time-dependent curves (dashed) of the current between molybdenum (a) and indium (b) electrodes obtained for the initial stage of breakdown

onset of the current rise at dc breakdown is also associated with the EEE initiation at the cathode.

Results of measurements of the dependence of the commutation time t_c (average values) on the electrode separation d for electrodes made of various metals and graphite were given in [10.8]. The dependence is linear and zero at the origin for every electrode material. The d/t_c values for different electrode materials are given in Table 10.1. It is important to emphasize that this dependence is the same for both pulsed and dc breakdowns.

The spark current rise measurements therefore showed that there is no noticeable difference between the discharge spark stage time at pulsed and dc voltage. There is also no difference in the character of the current rise. In the spark stage of the pulsed breakdown, current variations in the form of spikes in the waveforms are characteristic for some metals (for example, copper, aluminum, titanium, niobium), while for some other (indium, lead, bismuth) the absence of such variations is more typical; the same pattern is observed at dc breakdowns, too.

Of practical interest is the effect of the electrode material constants on the discharge current rise for electrodes made of different materials. Measurements of t_c were made for several electrode pairs. The graphs obtained for an indium - copper pair are plotted in Fig.10.4. From this figure it is clear that the material with the smaller d/t_c value has a dominant effect on t_c; this effect being more pronounced when the cathode is made of this material. This is also valid for other material pairs (for example, indium-aluminum, lead-copper).

Figure 10.5 shows the commutation time as a function of the number of breakdowns N for electrodes made of different materials. The data were obtained with d = 0.75 mm. At the first breakdowns (N<10) the cathode material strongly influences t_c. As the number of breakdowns is increased, the anode material has a greater effect on t_c. This effect is more pronounced when the anode electrode is made of a material with a lower d/t_c ratio. Thus, the breakdown current rise time is mainly determined by the electrode material which shows the greater

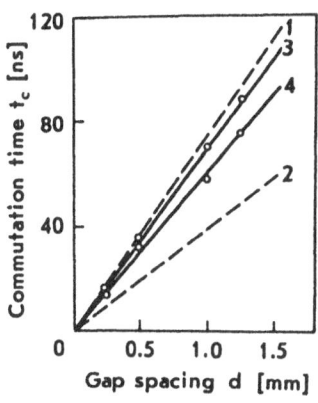

Fig.10.4. Commutation time as a function of gap spacing for the electrode pairs: In-In (*1*); Cu-Cu (*2*); Cu-In (*3*), and In-Cu (*4*)

commutation time. Even if the electrode made of this material is an anode, its influence soon becomes evident because of the anode-to-cathode material transport occurring at breakdown (Sect.10.1.4). Looking at the data given in Table 10.1, it can be seen that the d/t_c ratio increases as a rule with the breakdown voltage. This means that if the cathode material has some heterogeneous areas which differ in the breakdown electric field, breakdown starts in the area where the breakdown voltage is the lowest, and the current rise time will be determined by the irregularities in the material in that area.

These findings can be qualitatively understood by analyzing the energy release during the explosion of a micropoint and the formation of CF plasma (Sect.6.4). Materials with low melting points are characterized by comparatively small values of the d/t_c ratio, but they have, at the same time, relatively low specific sublimation energies. Therefore, for a cathode made of a material with a low melting point, as energy is first introduced into the micropoint the material disintegrates although the accumulated specific energy may be relatively small. Therefore, on the one hand, the value of the breakdown electric field is relatively small and, on the other hand, the expansion velocity of the CF plasma is rather low.

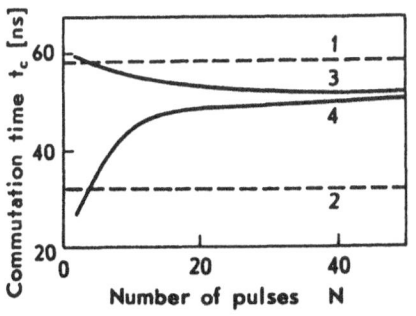

Fig.10.5. Commutation time as a function of the number of subsequent breakdowns for the electrode pairs: In-In (*1*), Al-Al (*2*), In-In (*3*), and Al-In (*4*)

209

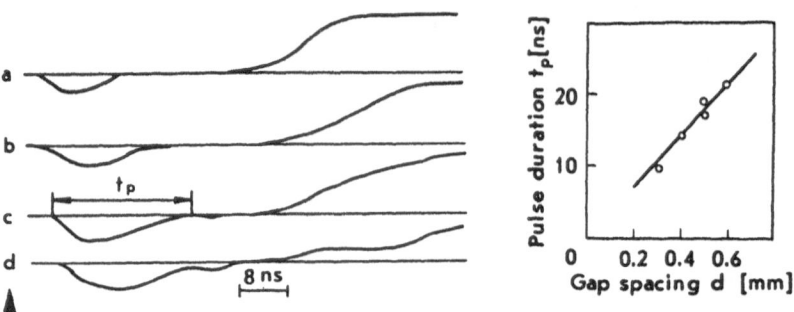

Fig.10.6a–d. Waveforms of the X-ray pulses (lower trace) and the breakdown current (upper trace) recorded with d = 0.3 mm (a), 0.4 mm (b), 0.5 mm (c) and 0.6 mm (d)

Fig.10.7. X-ray pulse duration as a function of gap spacing at dc breakdown

10.1.4 X-Radiation and Electrode Erosion at Breakdown

It is of interest to investigate the mechanism of X-ray generation at dc breakdown. Waveforms of X-ray pulses and breakdown currents for gaps with d = 0.3–0.6 mm between molybdenum electrodes are shown in Fig.10.6. Figure 10.7 illustrates the X-ray pulse duration t_X versus the gap spacing, the plot being entirely similar to that of $t_c(d)$. It is evident that at dc breakdown, an X-ray pulse occurs simultaneously with the onset of the current rise. This fact indicates that there exists a heavy electron flux which affects the anode in the fast stage of breakdown. The existence of such a flux is also confirmed by studies of the electrode erosion using the method of incomplete breakdown. Using pulse-forming lines with the "electric length" less than t_c it was established that during the current passage intense erosion of the anode and transfer of the anode material to the cathode in the form of molten drops 1 to 5 μm in diameter take place. The beginning of intense anode erosion is delayed with respect to the onset of the current rise by about 10 to 20 ns (d = 0.5–1.0 mm).

10.2 Optical Studies

10.2.1 Determination of the Time of Appearance of Light

The results described above suggest that on dc breakdown the main processes dominating the development of the fast phase of breakdown are essentially the same as at pulsed breakdown. Nevertheless, these results do not answer the question as to whether or not there was any conducting medium in the gap immediately before the onset of the current rise. Since vapor (if available) should be in the excited or ionized state, one should be able to detect it using a photoelectrical technique. Moreover, such a technique could help in obtaining some knowledge about the kinetics of the glow as early as in the stage of current rise.

Hemispherical copper or molybdenum electrodes of radius 10 mm were used in the study [10.3, 7]. Experiments were also carried out on a system with a needle cathode and a plane anode (molybdenum), where it is known that breakdown is induced by the cathode. The electrode separation was 0.7 mm (this provided a sufficiently large time of current rise at a moderate breakdown voltage). Examination of signal waveforms showed that the system was capable of detecting the appearance of single photoelectrons. Using this very sensitive technique, we found that no light appeared in the gap 100 ns before the onset of current rise under any conditions. The time at which the light appeared in any section of the gap fluctuated from discharge to discharge and seemed to be related to the electrode curvature. One hundred oscilloscope records were obtained for each position of the slit to determine the most probable time the light appeared. Statistical distributions of the delay time of the light appearance with respect to the onset of current rise for both dc and pulsed breakdowns and are given in Fig. 10.8. It can be seen that the distribution maximum coincides with the onset of the current rise only in the case when the slit is disposed at the cathode. Comparing this with the pulsed breakdown data provides some more evidence that the onset of current rise at dc breakdown can also be due to the occurrence of cathode flares.

The anode glow appears on average 10 ns later than the start of current rise. This means that the AF appearance, as it is at pulsed breakdown, is associated with electron bombardment of the anode, i.e., it is a secondary process. Light emission in the gap center is revealed on average 18 ns after the onset of current rise. If one suppose that this glow appears as a result of CF expansion in vacuum, then the velocity of propagation of the plasma front would be about $2 \cdot 10^6$ cm/s, which is close to v_c for copper and molybdenum.

10.2.2 Electro-optical Breakdown Studies

Electro-optical studies have the advantage that they make it possible to visualize a process [10.6]. At dc voltages we could only use the method of time-base sweeping which is less sensitive than the single-frame photography.

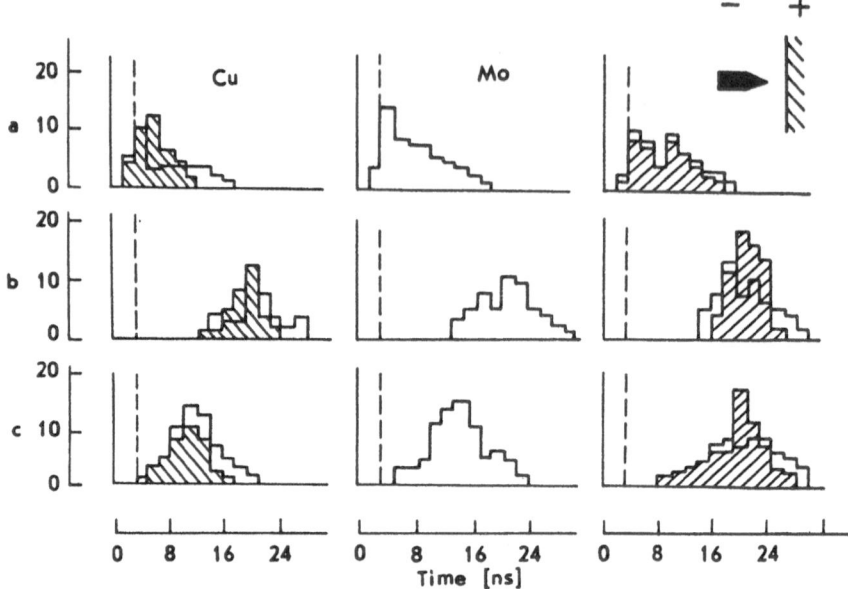

Fig.10.8a-c. Statistical distributions of the delay time of the appearance of light respective to the onset of current rise for the slit at the cathode (a), at the gap center (b) and at the anode (c). The dashed region represents pulsed breakdown. The dashed line denotes the moment of the onset of current rise

A typical EOIC recording of the sweep of the glow accompanying a breakdown in the direction perpendicular to the axis of copper electrodes is given in Fig.10.9, where the electrode arrangement and the scanning direction are shown schematically. The glow is first observed at the cathode and only after one-third of the commutation time it appears at the anode. At dc breakdown the glow always appears as a single luminous point at the cathode, the brightness of which increases with time. No significant differences were observed in the characteristics of the glow development for other materials, with the exception of brightness. It is worth noting the broken character of the cathode glow in the discharge initial period. This seems to reflect the cyclical nature of the cathode processes.

10.3 Comparison with Results of Other Investigations

In Chap.2 we have discussed briefly the hypothesis of the dc breakdown suggested by *Davies* and *Biondi* [10.14-16]. The main point of

212

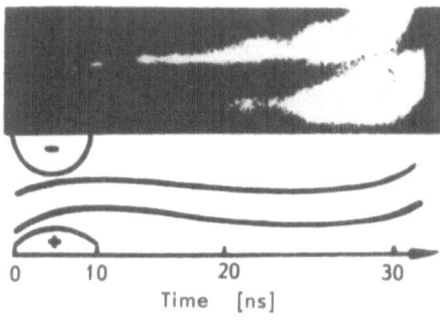

Fig.10.9. A typical electro-optical sweep of the glow accompanying dc breakdown. Copper electrodes, d=0.7 mm

this hypothesis is that a breakdown develops as a result of evaporation and ionization of the vapor formed during the heating by a field emission current of a molten anode particle in flight from anode to cathode. To verify this hypothesis *Davies* and *Biondi* [10.10-12] attempted to reveal the existence of vapor in the gap just before breakdown. Primarily, they used the method of determination of the density of electrode material vapor (copper) from measurements of the absorption of resonance radiation. The light from the resonance radiation source passed through the vacuum gap, got into a monochromator and was recorded by a photomultiplier tube (PMT). The PMT signal was applied to one of the scanners of a double-beam oscilloscope while the other scanner served for recording the breakdown current pulse. If copper vapor with a density of $\geq 10^{16}$ cm^3 were to exist in the gap prior to breakdown, then the signal amplitude should decrease as a consequence of the absorption of radiation by the vapor. The experiment [10.10] showed that about 0.8 μs before the onset of current rise a noticeable absorption of the resonance radiation starts. This was considered by *Davies* and *Biondi* [10.10] to be proof of the existence in the gap of copper vapor just before breakdown.

In subsequent years the same investigators carried out a number of new experiments aimed at explaining from which electrode - cathode or anode - the vapor enters the gap just before the breakdown [10.11, 12]. The experimental set-up was the following. The radiation emitted from the gap was collected by a lens, split into two beams and foccused at the input slits of two monochromators adjusted to the resonance lines of copper and chromium. At the output of each monochromator were situated photomultiplier tubes. The signals, having been amplified, were displayed on a double-beam oscilloscopes simultaneously with the breakdown current signal. The optical and electrical paths for both the channels were identical. Under investigation were discharges between a copper anode and chromium cathode as well as between a chromium anode and copper cathode. The following results

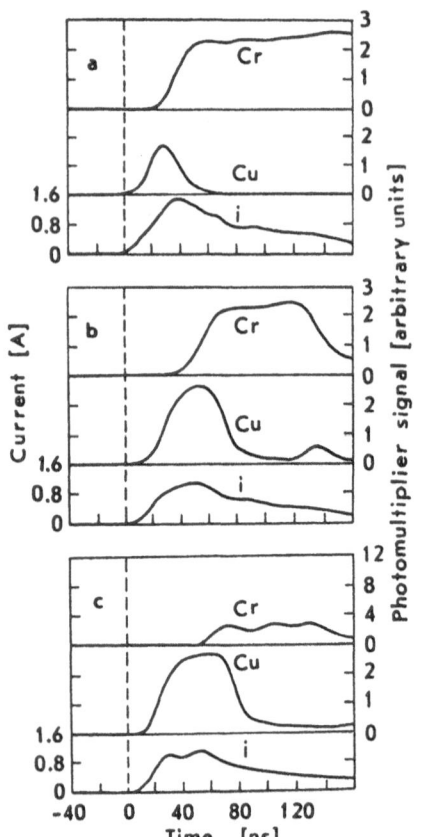

Fig.10.10a-c. Waveforms of the photo-multiplier signal and the breakdown current at dc voltage recorded for a chromium cathode - copper anode system with the slit disposed at the cathode (a), in the middle of the gap (b), and at the anode (c) [10.12]

were obtained (Fig.10.10): i) the resonance radiation of the electrode vapor does not occur earlier than the start of the breakdown current rise; ii) in any section of the gap, the resonance radiation of the anode material vapor occurs first and the radiation of the cathode material vapor can be observed only after 10 to 20 ns; iii) the beginning of the anode vapor radiation may be in step with the onset of the current rise, but on average it was delayed by about 10 ns; iv) the radiation of the anode material vapor usually lasts about 60 ns and falls abruptly before the passage of breakdown current comes to an end; v) the duration of the radiation of the cathode material vapor usually corresponds to the time of the breakdown current passage, the radiation peak falling at the breakdown current peak.

Davies and *Biondi* stated that the vapor radiation always occurs initially from a region on the cathode and this is in good agreement with our results, as discussed above [10.12]. However, in contrast to our results, they showed that the radiation of the anode vapor appears from the region near the cathode. Thus, they believe that the gap is already filled with anode vapor before the onset of current rise and this does not contradict their model of breakdown. They wrote: "The emission measurements demonstrate that the initial current amplification occurs in anode vapor, since excitation and ionization of the atomic vapor go hand in hand. At present it is not clear whether this vapor is the result of evaporation of the initiating particle which produces the microplasma or of further anode evaporation due to enhanced bombardment of the anode by electrons produced in the initial microplasma". As to the time of appearance of the cathode vapor radiation in different regions of the gap, in the opinion of *Davies* and *Biondi*, it can be interpreted, in accordance with our model, as the motion of cathode plasma with a velocity of $2 \cdot 10^6$ cm/s. "Further measurements are required, however, to provide a more complete description both of the mechanisms of production of the vapor sources and of the transition phase of the discharge".

From the work described in [10.12] the following conclusions can be drawn:

1) The main result which agrees entirely with our data is that no light emission can be recorded in the vacuum gap until the onset of current rise. In our opinion, this contradicts the measurements in [10.10]. *Davies* and *Biondi*, however, considered that there is a complete agreement between their early and subsequent experimental data, i.e., the anode vapor is recorded earlier than the cathode vapor, but while in the early study [10.10] anode vapor could be recorded before the onset of current rise, in the study described in [10.12] it was not observed earlier than the breakdown current began to rise. This is of principal importance.

2) While investigating the appearance of the integrated glow in the gap, we believed that besides resonance lines of the electrode material neutral vapor ion lines may also appear in the glow spectrum and that ions other than those of the electrode material may be present, such as those due to the matter absorbed at the electrodes. Moreover, the appearance of a continuum (bremsstrahlung and recombination radiation) should not be excluded. Since the question of whether or not the glow of conducting medium in the gap occurred just before the beginning of current rise was of principle interest in the initial stage of the study, the measurement of the integral radiation was justified. Firstly, a maximum quantity of photons related to the sensitivity range of the photomultiplier photocathode was "collected". Secondly, the light losses due to the connection of a monochromator in the optical path were reduced. This enabled us to obtain better spatial resolution by

using an optical slit with a width of one seventh of the gap spacing (in [10.10-12] the slit width was 0.3d to 0.5d, d being the gap width).[1]

3) The fact that anode glow was detected first may be associated with the transfer of anode material to the cathode. Firstly, such a transfer could take place as early as in the prebreakdown stage. Secondly, as follows from the calculation in [10.16], the particle initiating breakdown does not evaporate completely until impact with the cathode. Thirdly, the anode material transfer seems to take place in the breakdown spark stage by discharging the stray capacitance of a high-voltage electrode into the earthed vacuum chamber metal case and other earthed units of the installation. It should be noted that *Davies* and *Biondi* took measures to reduce this capacitance and to limit the energy deposited at the electrodes by breakdown.

Estimates obtained from the current waveforms and other data of [10.12] have shown that the stray capacitance amounted to about 4.5 pF. Half of the energy stored in it released into the gap (about $1 \cdot 10^{-3}$ J), and about $0.5 \cdot 10^{-3}$ J released in discharging the interelectrode capacitance ($\simeq 1$ pF). If one assumes that practically all this energy is absorbed by a small section of the anode surface of radius $R_a \simeq d = 1$ mm and depth $x = 10^{-4}$ cm in a time of about t_c, this energy would be sufficient to heat this section up to about 1000 K. Since, according to *Davies* and *Biondi's* calculation, the temperature at the center of this section reaches the melting point before the breakdown, it can be said that a melt zone forms on the anode and a material transfer takes place from anode to cathode.

One should also note that the measurements in [10.10] did not cover the first breakdown and the case of a single breakdown, but rather they seem to have been performed on conditioned electrodes (there is no indication of this in [10.12]). In those circumstances there is no doubt that anode material was present on the cathode. Therefore, the fact that the anode and cathode material vapors were detected after the onset of the current rise does not prove the validity of the hypotheses in [10.8-10]. Moreover, if one compares the above observation with our results, a conclusion can be drawn that in a CF, which occurs at the same time as the onset of current rise, both the cathode and the anode materials are present.

4) In the research by *Davies* and *Biondi* one more question is left unanswered: In what form (neutrals or ions) the anode and cathode materials exist in the gap just before breakdown. *Davies* and *Biondi* [10.10-12] attempted to detect the radiation of neutrals in the gap.

---

[1] As described in [10.10], the bandwidth of the electric line was limited by the oscilloscope bandwidth and was equal to 25 MHz, i.e., narrower by a factor of 6 than that in [10.5]. The gain factor of the electric line in [10.10] was about one-third of that in [10.5]

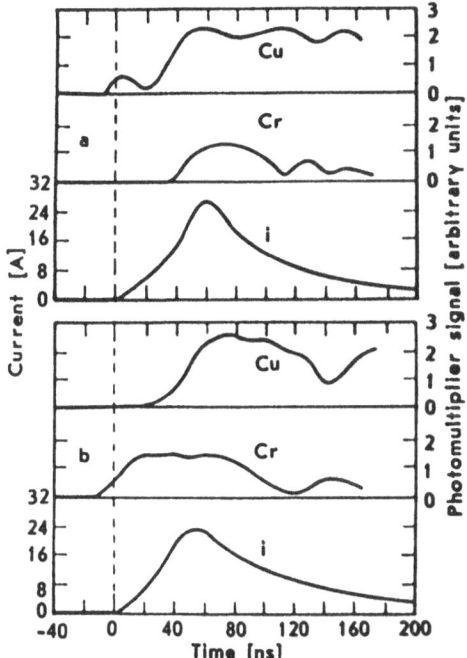

Fig.10.11a,b. Waveforms of the photomultiplier signal and the breakdown current recorded for a copper cathode - chromium anode system. The delay time is 14 μs (a) and 100 μs (b) [10.17]

However, in accordance with their own estimates [10.16], the characteristic time of the vapor-to-plasma transition for a vapor cloud of radius about 10^{-3} cm available around a particle is less than 10^{-8} s.

5) The light emitted by the neutral vapor in the middle of the gap and at the anode usually appears about 10 ns after or even at the same time as the onset of the breakdown current rise. The latter disagrees with the calculations in [10.15, 16] and cannot be understood in terms of other hypotheses. It is possible that such a result is due to insufficient temporal resolution of the detector or some other unknown reasons.

Yen et al. [10.17] presented a logical continuation of the experiments described in [10.11, 12]. They improved the high-voltage power supply system. In this way it was possible to apply not only the dc voltage across the gap, but also rectangular voltage pulses of a 200 μs duration, thus creating a small overvoltage. The remaining experimental conditions, the diagnostic apparatus, and the measurement methods were pratically the same as in [10.12]. The main results of [10.17] can be summarized as follows (Fig.10.11): (i) the resonance radiation of electrode vapors appears during or sometimes some tens of nanoseconds prior to the onset of the current rise; (ii) the maximum of the intensity of this radiation usually falls on the spark current maximum; (iii) at relatively high overvoltage coefficients (≤ 1.3) when the breakdown delay time t_d is not in excess of 20 μs, the resonance radiation of cathode vapor is emitted first, while for $t_d > 20$ μs the resonance radiation of anode vapor is the first to be detected.

Yen et al. [10.17] stated that the primary appearance of the resonance radiation of the anode vapor at comparatively large delay times corresponds to the anode-initiated breakdown predicted by the theory developed earlier [10.15, 16]. On other occasions it has been assumed that the explosive electron emission mechanism suggested by the present authors is dominant.

On account of the important paper [10.17] we would like to emphasize that the interpretation of the results obtained in [10.17] is still far from being thorough. Firstly, judging from the data of [10.17], the character of spark current and current amplitude waveforms is practically the same irrespective of which electrode vapor appears initially in the gap. This shows that the breakdown development inevitably goes through a stage of explosive emission, where the gap conductivity is insignificant at the time of the emission initiation. Secondly, even with microsecond breakdown delay times the breakdown seems to be initiated in a more complicated manner than if it was due to the explosion of cathode micropoints. Indeed, for micropoints of height 10^{-4} cm the relaxation time of thermal processes within a micropoint is of the order of 10^{-8} s. Thirdly, with increased currents in the breakdown spark stage (~25A), one should take into account the noticeable erosion and the transport of erosion products from electrode to electrode (Cu - Cr).

When analyzing the data more attention has to be paid to the breakdown initiation mechanism associated with the microparticle impact. This idea is prompted by both the results of recent studies of the motion of microparticles between the electrodes [10.18] and the analysis of the influence of microparticles on breakdown [10.2, 19]. The fact that microparticles of both cathode and anode materials could be present on the electrodes in the experiments described in [10.17] is practically beyond doubt. Moreover, since in some cases the resonance lines of electrode vapors appeared some tens of nanoseconds prior to the onset of the spark current rise, it is necessary to record the behavior of the prebreakdown current in all stages preceding the occurrence of the explosive electron emission. This would allow for a better determination of the influence of microparticles on the breakdown initiation.

Some interesting results for dc breakdown of gaps up to 6 mm wide for voltages up to 120 kV were obtained by *Jüttner* and *Siemroth*

[10.20]. Discharges were initiated between molybdenum electrodes. With a very small discharge capacitance equal only to the electrode capacitance plus the stray capacitance it was found that light emission initially appeared at the cathode in the form of a single bright point. Supposing that such points were EEE centers, *Jüttner* and *Siemroth* evaluated the velocity of cathode plasma expansion at a capacitance of 100 pF from the current waveforms and obtained $v_c \simeq 2 \cdot 10^6$ cm/s. These experiments agreed well with our results on dc breakdown for up to 1 mm wide gaps, i.e., in [10.20] it was shown that the high-current discharge stage in comparatively wide gaps is due to the occurrence of explosive electron emission at the cathode. However, a question was left unanswered concerning the formation of ECs at the cathode. It was shown later by *Litvinov* et al. [10.21] that the EC formation is due to the explosion of the neck between the point and a molten drop formed on its tip as a result of heating.

10.4 EEE Initiation at DC Breakdown

The studies on dc breakdown kinetics described above show that the irreversible failure of vacuum insulation under these conditions is associated with the initiation of explosive electron emission at the cathode. From the data obtained it also follows that just before the EEE initiation there is no conducting medium in the gap which would be sufficient for a discharge to develop across it. Such a medium is generated at the electrodes at the same time as the current starts rising in the breakdown spark stage. The initiation of dc breakdown is therefore of great interest, especially with respect to the processes leading to a high energy concentration in the cathode microvolume, particularly due to a strong electric field at the cathode, and to FEE which results in EEE. Let us now consider the mechanisms leading to the initiation of explosive emission.

10.4.1 EEE Initiation under Pure Conditions

With good vacuum conditions and pure electrodes, provided that microparticles are absent and there is no evaporation from the anode, the EEE initiation is related to the development of thermal instability within the cathode micropoints. *Dyke* and co-workers were the first to deal with the steady-state problem of heating of a point cathode by a field emission current of high density [10.22]. They obtained the following expression for the emitter tip temperature:

$$T_\infty = T_0 + 2j^2 r_e^2 \kappa/(\lambda\theta^2) \; . \tag{10.1}$$

For a tungsten field-electron emitter of typical size, the current density necessary for heating up the emitter tip to T_m is, according to (10.1), 10^7 A/cm^2. *Vibrans* later improved this analysis [10.23] by introducing temperature-dependent resistivity and current density, and extending the problem to include the case of small micropoints on plane electrodes. From the data of [10.23], it follows that at temperatures much lower than T_m the emitter becomes thermally unstable because of the resisitivity increase with temperature. A number of subsequent studies [10.24-27] took into account the Nottingham effect on the micropoint heat balance and the expression for T_∞ then became [10.25]

$$T_\infty = T_0 + \frac{2j^2 r_e^2 \kappa(T_\infty)}{\lambda\theta^2} + \frac{2jr_e}{\theta} \frac{\pi k T_\infty}{\lambda} \cot(\pi p), \tag{10.2}$$

where $p \simeq 9.3 \cdot 10^{-3} \, \phi^{3/2} T_\infty /E$.

The results reported in [10.24-27] show that the Nottingham effect is important in the cathode heat balance. The steady-state energy balance in a cathode micropoint was also treated in [10.28]. The calculation for a variety of metals predicts a limiting field emission current density of 10^7 to 10^8 A/cm^2. This suggests that the critical breakdown electric fields E_{br} should not differ considerably for electrodes made of different metals, as the majority of metals have close values for the work function. The calculated values $E_{br} \simeq (5$ to $10) \cdot 10^7$ V/cm [10.28] are in good agreement with experimental data [10.14, 29, 30].[2] Thus, in this case the cathode mechanism of breakdown initiation suggests that E_{br} is independent of the vacuum gap spacing d, i.e., that V_{br} and d are directly proportional.

Let us consider a vacuum discharge between very clean electrodes, which is caused by heating of the anode by field electron current microbeams. It is commonly assumed that this mechanism becomes more likely to occur as the gap spacing is increased. The voltage necessary to attain a certain field emission current from cathode is $V \propto d^{\alpha_1}$. For a uniform field $\alpha_1 = 1$. However, since the macroscopic field enhancement effect increases with increasing d, one may conclude that the in-

---

[2] In some investigations (e.g., [10.31, 32]) it was shown that E_{br} depends on r_e and on the micropoint geometry

fluence of the points situated in the region of non-uniform field, i.e. at the cathode edge, is stronger. Usually $\alpha_1 = 0.6$ to 0.8 [10.33]. The radius of the area heated by the electron microbeam is $r_a \propto d^n$, where $n = 0.5$ for a uniform field and $n > 0.5$ for a non-uniform field, i.e., in a non-uniform field the beam divergence is greater. In a dc case, the anode temperature, according to [10.24], can be expressed as follows

$$T_a - T_0 \propto iV/r_a \propto d^{(\alpha_1 - n)} \; . \tag{10.3}$$

For sites with a non-uniform field the value of $(\alpha_1 - n) \ll 1$ or even less than zero. In other words, the likelihood of observing the manifestation of the anode mechanism is not increased. Since the accelerated electrons release their energy mainly at a certain depth $x \propto V^k$, where $k \simeq 1.5$ to 2, this also supports our conclusion. This is because as the voltage is increased, the electrons penetrate deeper and the surface heats up less. It is probable that the electron microbeam current rises irregularly [10.20]. Then

$$T_a - T_0 \propto iV/(r_a^2 x) \propto d^{(\alpha_1 - \alpha_1 k - 2n)} \; . \tag{10.4}$$

The value of $(\alpha_1 - \alpha_1 k - 2n) < 0$, so the anode temperature will decrease with increasing electrode separation d. Therefore, it is unlikely that under pure conditions the breakdown of a large gap is initiated at the anode [10.20].

10.4.2 EEE Initiation and the Total Voltage Effect

The so-called "total voltage effect" (a decrease of the breakdown electric field with increasing gap spacing) appears to contradict the cathode mechanism of breakdown initiation. This effect may be due to a number of reasons [10.34]. At present there are a great deal of experimental data which permit a qualitative explanation of it in terms of the mechanism of cathode-initiated breakdown, and these are listed below.

1) The total voltage effect may be influenced by the electrode geometry. It should be noted that the field enhancement factor β is not exclusively characterized by the cathode microgeometry. Actually, $\beta = \beta_1 \beta_2$, where β_1 and β_2 are characteristics of the cathode micro- and macrogeometry, respectively. The factor β_2 is dependent on the radius of curvature r of the surface on which the microprotrusion is present

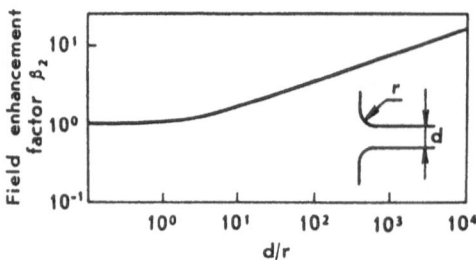

Fig.10.12. Enhancement factor β_2 as a function of the ratio d/r for a pair of semi-infinite plane electrodes with rounded edges [10.28]

and on the electrode separation d. Figure 10.12 illustrates the relationship between β_2 and the ratio d/r [10.29] for electrodes which represent a pair of semi-infinite planes with a rounded edge of radius r. The factor β_2 is determined as a ratio of the maximum electric field to the electric field inside the gap. From Fig.10.12 it is clear that $\beta_2 \simeq 1$ for d/r $\leq$ 1, i.e., only in this case it can be assumed that $E_{av} \simeq V/d$. With d/r > 1 the macrofield $E_{av} = V \beta_2/d$.

2) The total voltage effect and the decrease of prebreakdown currents with increasing electrode separation may be due to the presence of adsorbed gas and vapor on the electrode surface. If the cathode area is rather small (for example, when using a classical field emitter), it is possible to free the surface from adsorbed gases. However, even in this case and under high-vacuum conditions the gas molecules coming from the anode produce a sharp increase of the probability of the field emission-to-vacuum arc transition [10.35]. The deflection of prebreakdown field emission currents from the anode by application of a transverse magnetic field results in an increase in breakdown voltage [10.36]. The presence of surface contaminations facilitates the process of surface migration of cathode material and adsorbate atoms to the region with a strong electric field [10.37]. As a result, during a time interval of a few tens of seconds the field emitter rearranges and sharpens up. This process results, in the end, in breakdown. A significant amount of the energy necessary to intensify the migration process is supplied from the cathode bombardment by ions produced by electron-beam ionization of residual and adsorbed gases [10.37].

In [10.34, 38, 39] it was suggested that the total voltage effect can be associated with the fact that as the voltage (the gap spacing) is increased the energy and the number of ions produced from the adsorbate increase. Because of this, the tips of the microemitters become sharper and more free from adsorbed impurities (adsorbate atoms are usually electronegative). The frequency and the amplitude of the prebreakdown current bursts increase, and in the end this gives rise to the breakdown. In [10.39] experimental results indicate that the total voltage effect can be reduced if the experiment is carried out under high standards of purity. However, it is not possible to free the electrode surface from contaminations absolutely. Moreover, to secure uniform electric fields any increase in the electrode separation automatically requires an increase of the electrode surface. Therefore, as the gap spacing is increased, it becomes more and more difficult to ensure that the electrodes are pure.

It was shown in [10.38,40] that as a result of the competition of the dynamic processes of adsorbtion and desorption in ultrahigh vacuum with a strong electric field, the resonance tunelling effect can be created at some local cathode areas within a time of the order of 10^{-8} s [10.41]. The occurrence of resonance emission results in the fact that the emitting area predicted by the Fowler-Nordheim equation may turn out to be highly overestimated and, consequently, the critical value of the prebreakdown current density will be underestimated [10.40].

3) The total voltage effect may be associated with the presence of dielectric films and inclusions on the cathode surface. Dielectric films in the form of islets are produced on contaminated (for instance, with organic substances) electrode surfaces under the action of discharge processes, the prebreakdown electron current being rather significant with an average field as low as $\sim 10^4$ V/cm [10.41]. Such high conductivity is associated with electron emission from the dielectric inclusions as a result of their charging up, which leads to the enhancement of electric field both at the cathode and inside the dielectric [10.42,43]. The charging up can occur due to the processes taking place in the dielectric itself (impact ionization in the bulk and in the pores of the dielectric islet, and also escape of electrons from the dielectric into the vacuum [10.43]) and due to the bombardment of dielectric films by ions formed from the residual gas and the gas desorbed from the anode [10.42]. Surface charging of the film up to an electric field strength of 10^6 V/cm leads to its local breakdown, thus initiating explosive electron emission. As the ion energy increases, the ion-electron emission coefficient increases and charging of the islets is likely to occur. Hence the gap spacing is not necessarily directly proportional to the breakdown voltage (i.e., α_1 in (2.16) will be less than unity). This is also the case for non-metallic inclusions usually present in the electrode material as impurities.

4) With a prolonged application of voltage across the gap the anode becomes a supplier of microparticles [10.1,2]. For example, when the average electric field reached a value of the order of 10^5 V/cm, it was observed that the microparticles broke off from the electropolished

anode surface and left micron-size hollows on it. The majority of the hollows were localized at the sites of alien inclusions situated mainly along or near the grain boundaries. Non-metallic particles deposited on the cathode may become sites of breakdown initiation.

Experiments with metallic particles which were introduced artificially into the gap proved convincingly the existence of the polarity effect: Breakdown only took place in those cases when particles started from the anode and impacted onto the cathode [10.44,45]. *Little* and *Smith* [10.45] concluded that under the action of ponderomotive forces an inelastic impact of a particle onto the cathode gives rise to the formation of efficiently operating micropoints from molten metal. Although there is no mention of further development of breakdown in [10.45], the results of *Little* and *Smith*, particularly when a transverse magnetic field was used to deflect the electron beam, proved convincingly that the mechanisms of the breakdown initiation and development are dominated by cathode processes.

For inelastic collisions the velocity of the particle should be in excess of a certain value $v_p \simeq (\sigma_{ys}/\rho)^{1/2}$ [10.2] determined by the yield strength of the particle and the target material σ_{ys}. If a particle of radius r_p breaks off from the anode, its impact velocity will be [10.46,47]

$$v_p = \sqrt{9.87 \epsilon_0 V^2 \beta_a /(r_p \rho d)} , \qquad (10.5)$$

where β_a is the enhancement factor of the electric field at the anode site from which the particle breaks off. Hence, the minimum breakdown voltage necessary for micropoint formation to take place as a result of an inelastic impact can be defined as follows:

$$V_{br_{min}} = \sqrt{\sigma_{ys} r_p d/(4.94 \ \epsilon_0 \beta_a)} \propto d^{1/2} . \qquad (10.6)$$

Analysis shows [10.47,48] that for $\beta_a = 1$ and $d \geq 0.1$-1 cm (copper electrodes) the experimental data available can be understood on the assumption that micropoints are formed on the impact of particles with $r_p \leq 1$ μm onto the cathode. These conclusions were confirmed later on experimentally by laser diagnosis of microparticles using time referencing of the processes under examination [10.18]. It turned out that breakdown between copper electrodes was initiated by microparticles with radii not in excess of 0.4 μm impacting on the cathode at a velocity which was somewhat higher (about 1.5 times) than that predicted by (10.5) for $\beta_a = 1$. If one takes into account that in reality β_a can be much greater than unity, then the proposed mechanism may also be applied to particles of a larger size.

When their size is tens of micrometers or more, the velocity gained by the particles in the gap is insufficient for inelastic impact [10.1, 2, 47, 48]. However, as a particle approaches the cathode at a distance less than r_p, the electric field between the cathode and the particle can rise by several orders of magnitude in comparison to the average value. This is because of the initial and induced charges [10.1, 2, 47, 48]. If there is a point opposite the microparticle ($r_e \simeq 10^{-6}$ cm, $\beta \simeq 5$), then the electron flux emitted by the point can heat the particle surface up to the temperature of intense evaporation [10.47, 48]. It has been suggested [10.47, 48], that in such a situation a microplasmoid forms due to ionization of vapor, which causes discharge. The microparticle creates cathode micropoints or produces an igniting spark during the inelastic impact. This approach explains the total voltage effect and the dominant influence of the anode material on the dc breakdown voltage [10.49]. There is a correlation between σ_{ys} and V_{br}. V_{br} decreases with an increase of the particle size [10.50]. In order that breakdown occurs as a result of particle participation, passage of a noticeable prebreakdown current over a longer period of time is not necessary, since the time from the instant of the particle impact onto the cathode until the micropoint explosion, i.e., the time of the formation of the igniting spark, can be extremely short.

10.4.3 Criteria for Vacuum Breakdown and EEE Initiation

Summarizing the above, it can be concluded that the electric field enhancement at the cathode in the presence of a strong electric field depends on many processes. All these processes create favorable conditions for the initiation of electron emission. Thus, a vacuum breakdown is only possible when an EEE center forms on the cathode, i.e., only in this case will the discharge gap resistance tend to a low value. Other breakdown mechanisms cannot explain the reason for the appearance of strong discharge currents (tens of amperes) of duration of the order of 10^{-9} to 10^{-8} s. They only suggest possible reasons for the appearance in the gap of the electrode material vapor, organic contaminations or desorbed gases [10.51]. In our opinion, these processes should lead in the end to the occurrence of explosive emission responsible for the existence of the vacuum breakdown spark stage.

It is important to know which criteria define the occurrence of vacuum breakdown and discharge. If vacuum breakdown is initiated by EEE, then the criteria for dc and pulsed breakdown will be, respectively,

$$j = C_1 \quad \text{and} \quad j^2 t_d = C_2 \, , \tag{10.7}$$

where C_1 and C_2 are constants, j is the current density initiating EEE, and t_d is the explosion delay time. If the dependence of j on the electric field is known, one can find a value of the electric field E_{EEE} at which explosive emission occurs. As we showed above, this field is equal to E_{br} at which the vacuum breakdown is initiated. This is not always the case, since there are some processes which lead to the condition $E_{br} < E_{EEE}$. Therefore, we believe that to characterize vacuum breakdown and discharge phenomena it is necessary to establish (i) a criterion for the vacuum breakdown initiation, and (ii) a criterion for the EEE and vacuum arc occurrence.

The total voltage effect is just one of the manifestations of the existence of these two criteria. Breakdown and EEE are often initiated under the same conditions, for example at a nanosecond pulsed breakdown or at a dc breakdown under the conditions of high vacuum and well-processed electrodes. Equations (10.7), which in our opinion are the criteria for EEE appearance, will then be criteria for vacuum breakdown and will permit one to estimate the ultimate electric strength of vacuum gaps. With this there will be no total voltage effect.

11. Nonstationary Processes in the Vacuum Arc Cathode Spot

It may be inferred from the previous chapters that vacuum electrical breakdown and spark discharge are governed by intricate, often dependent nonstationary processes. The nonstationary nature is also typical for vacuum arc discharge and phenomena taking place in the cathode spot [11.1]. The experience we gained by studying explosive emission led us to investigate the nonstationary phenomena in vacuum arc discharges in more detail. Some of the experiments are described above, while others will be discussed and analyzed in the present chapter. We shall also discusss the results of numerical simulation of the processes in a single explosive center. A comprehensive analysis of the available data leads to a conclusion about the fundamental role played by explosive electron emission in the operation of vacuum arc cathode spots.

11.1 The Motion of Vacuum Arc Cathode Spots

11.1.1 The Effect of Surface Conditions

Lyubimov and *Rakhovsky* established experimentally [11.2] that a first-type cathode spot consists of several (four of five) fragments. The distance between fragments is $l_0 \simeq 10$ to 30 μm, the current per fragment ranges from 0.2 to 0.7 A and the fragment lifetime is of the order of 10^{-6} s. The apparent motion of a spot occurs due to the disappearance of some fragments and the appearance of others. When the distance between individual groups of fragments increases to a value of the order of 10^{-2} cm division of the spot occurs. The spot lifetime is (3 to 6)$\cdot 10^{-6}$ s. In the opinion of *Lyubimov* and *Rakhovsky* the explosion of the next micropoint results in the formation of a discharge channel with a greater conductivity and in a rapid decrease of voltage. In this situation the electric field and the current density at other micropoints are such that the explosion of micropoints cannot occur. As the plasma produced by explosion, expands and cools, its conductivity decreases and the near-cathode voltage drop increases. At a certain moment, some suitable microirregularity explodes, and the cycle is repeated. This model is remarkable in that the current passage is mostly due to the field electron emission from micropoints at current densities much lower than the explosive current density, thus producing an average spot current density of about $5 \cdot 10^4$ A/cm^2. In other words, the main electron current from cathode to plasma is provided by the mciropoints existing under the plasma in conditions which are far from explosive and the explosion of one of these micropoints leads to a cessation of current passage through this area of the cathode surface. *Lyubimov* and *Rakhovsky* used this model in an attempt to explain the properties of rapidly moving cathode spots.

We believe, however, that the motion of a first-type spot can be explained by the appearance of ECs (i.e., fragments) as a result of breakdown of non-metallic inclusions and films. It is shown in Chap.8 that only this mechanism can explain why new ECs can arise in an arc discharge at a distance of tens of micrometers from a primary emission center. Moreover, the minimum probe current and its duration are of the same orders as those in the case of the first-type cathode spot fragments.

From our results it follows that the delay time of the appearance of new ECs at a distance of tens of micrometers is less than 10^{-8} s and this suggests that the velocity of expansion of the region occupied by ECs is practically determined by the plasma expansion velocity (10^6 cm/s). The velocity of the first-type spots is, however, much less (10^3 to 10^4 cm/s). The apparent contradiction is quite understandable. The fact of the matter is that the expansion of the boundary of the EC-occupied region with the velocity of 10^6 cm/s proceeds only until the number of ECs on the cathode reaches a value determined by the total current and by the minimum current per EC (the size of this region, judging by the data reported in [11.3], is limited by a radius of 10^{-2} cm, i.e., the time of spot formation is of the order of 10^{-8} s). After that, for a time of the order of 10^{-8} s some ECs disappear and the current is redistributed over several more efficiently operating centers. The plasma density in the neighbourhood of these centers increases and as a result of this the probability that new ECs appear also increases. The new ECs appear mostly near the efficiently operating EC on areas of the cathode which are not clean because the breakdown mechansim of inclusions and films can occur under somewhat less strict conditions in the near-cathode layer. Thus, beginning from the moment in which the dynamic equilibrium between the number of appearing and extinguishing ECs is established, the velocity of the spot motion is determined by their lifetime t_0 rather than by the delay time of the appearance of new ECs (which involves the time of plasma propagation and the time of breakdown of a film or inclusion). The chaotic nature of the first-type spot motion along the cathode surface has been demonstrated experimentally [11.2, 4-6]. The radial velocity of expansion of the region controlled by the spot can be found from [11.7]

$$v_{EC} = \frac{\sqrt{\pi}\, l_0}{4\sqrt{t_0 t}} = K_{EC}/\sqrt{t}\,, \tag{11.1}$$

where l_0 is the average distance between neighbouring ECs. The better the conditions for the appearance of new ECs, the greater the number of ECs which can exist simultaneously. This means that for a contaminated or oxidized cathode surface the current per EC and consequently the EC lifetime t_0 may turn out to be significantly less than those parameters for a clean surface. According to (11.1), bearing in mind that under such conditions the distance between ECs can reach 10 to 30 μm, this should lead to an increase of the velocity v_{EC}. Indeed, as shown in [11.1], the cathode surface oxidation results in a three- to five-fold increase of the coefficient K_{EC} in (11.1).

As contaminations are removed from the cathode surface, a more significant role is played by field-assisted thermionic emission and explosion of the liquid neck at the detachment of a drop, i.e., new ECs will form in the neighorhood of thoses already present on the surface ($l_0 \simeq 10^{-4}$cm). They therefore are concentrated within a small region, and the velocity of the spot motion decreases.

The effect of surface contamination on the velocity of cathode spot motion was discussed by *Achtert* et al. [11.8]. Having analyzed the composition of the cathode surface layer using Auger and secondary-ion mass spectrometry, they showed that first-type spots appear only on an uncleaned cathode surface. Using an AlN dielectric film 0.2 μm thick, specially deposited on the surface of a molybdenum cathode, they made sure that under such conditions the cathode spot moved rapidly along the surface, consuming the film and practically leaving the molybdenum base undamaged (Fig. 11.1). After conditioning of the cathode by arc discharges in ultrahigh vacuum, cathode spots did not become mobile, in spite of the existence on the cathode of a great number of micropoints of various sizes formed by discharges (Fig. 11.2).

Bushik et al. [11.9] studied the influence of the cathode surface condition on the dynamics of cathode spots and have shown, in particular, that (i) the first-type cathode spots occur on oxidized cathode surfaces, including previously cleaned, eroded surfaces after oxidation

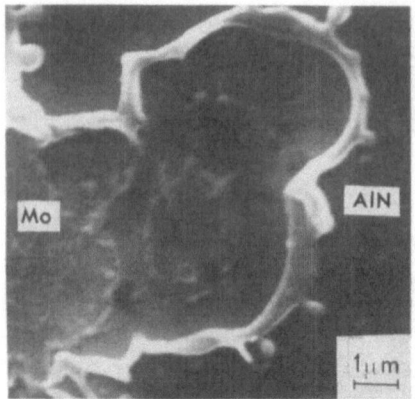

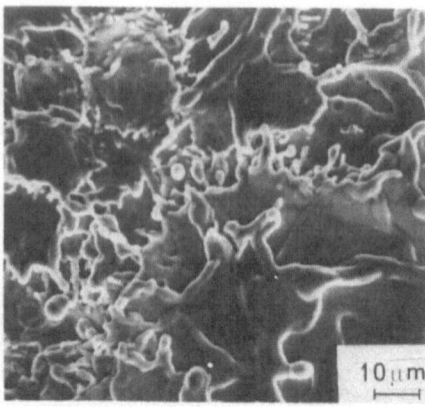

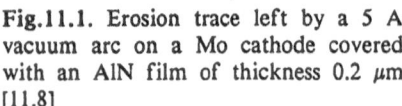

Fig.11.1. Erosion trace left by a 5 A vacuum arc on a Mo cathode covered with an AlN film of thickness 0.2 μm [11.8]

Fig.11.2. Erosion trace left by a 20 A vacuum arc on a clean Mo cathode [11.5]

by air; (ii) the cathode material constants have a weak effect on the development of first-type cathode spots; (iii) on cathodes cleaned with arc discharges only second-type spots are observed; (iv) the development of second-type cathode spots is strongly dependent on the cathode material; (v) the motion of second-type cathode spots along a clean cathode surface is accompanied by explosion with ejection of microparticles and plasma jets.

Observations of the erosion traces left by a single spark or arc discharge of duration 10^{-7} to 10^{-6} s on clean and smooth virgin surfaces of cathodes made of high-melting-point materials, which were obtained by melting the wire in ultrahigh vacuum were described in [11.5, 10]. In Fig.11.3a more than ten separate craters can be clearly seen. It can be assumed that the emitting zone stayed in one place for less than 10 ns. The pictures reproduced here indicate that the nature of the cathode spot processes is dynamic. If there are no contaminations on the cathode surface, the emission centers never move further away than the melt region and, as a rule, they form on the crater edges. As can be seen from Figs.11.3b,c, the cathode spot is not attached to the grain boundaries of molybdenum microcrystals and this suggests that the spot migration is mainly affected by the regeneration of the microprotrusions on the crater edges. Hence, the cathode spot produces craters and micropoints, which are not only the product of its operation, but also maintain the necessary conditions for its further existence [11.11, 12].

The photographs reproduced in Fig.11.3 demonstrate once again that the determination of the cathode spot current density should be

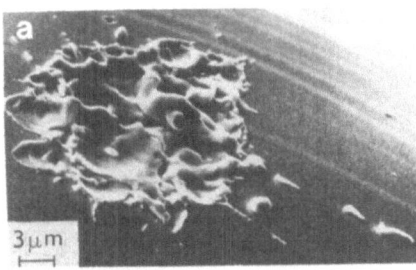

3 μm

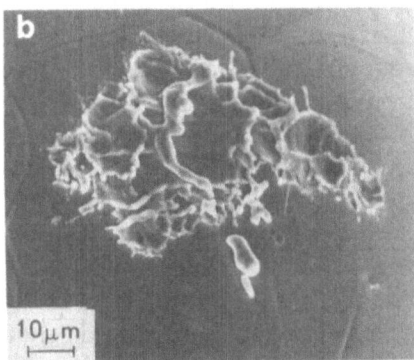

10 μm

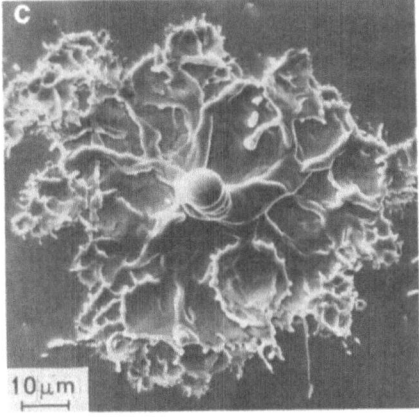

10 μm

Fig.11.3a–c. Cathode spot autographs on a tungsten (a) and molybdenum (b,c) [11.5] cathodes. t_p = 100 ns (a) and 500 ns (b,c); i = 30 A (a), 80 A (b), and 200 A (c)

approached carefully. If one divides the current by the area of the erosion trace, then the current density will be about $2 \cdot 10^7$ A/cm² at t_p = 100 ns and falls to about $2 \cdot 10^6$ A/cm² at t_p = 500 ns, although it is quite obvious that the true current density is significantly greater.

Summarizing the above discussion, it can be stated that at a certain stage of the arc operation the second-type spots are essentially nonstationary first-type spots grouped within a very small region and are incapable of scattering with a great velocity because of the absence of non-metallic films and inclusions.

11.1.2 The Influence of a Magnetic Field

A great number of investigations have been devoted to the effect of retrograde spot motion in an applied tangential magnetic field [11.2, 4, 13-18]. *Kesaev* [11.4] showed experimentally that the vacuum arc spot shifts to the region where the total magnetic field (the applied field plus the arc self-field) is maximum. In *Kesaev's* opinion, the plasma density maximum corresponds to the maximum of the spot's magnetic field as a consequence of the reduced level of diffusion of charged particles in the direction transverse to the magnetic field. Since new spots should appear where the plasma density is highest, this explanation ought to be considered justified. However, *Lyubimov* and *Rakhovsky* [11.2] believe that the theoretical proof of the "principle of a maximum magnetic field" given in [11.4] is unreasonable.

Some publications [11.16-18] suggested a mechanism of retrograde spot motion based on the Hall effect in a plasma, where the process of spot motion consisting of elemental acts of EC appearance and cessation is not involved. The following important experimental facts should be noted: (i) electrons and ions move from a vacuum arc cathode spot isotropically, as they would from an explosive electron emission center; the velocity of motion of the ions (plasma jets) is about 10^6 cm/s; (ii) the spot motion occurs due to the appearance of new ECs and the disappearance of old ones; (iii) new emission centers occur within a radius of not more than 10^{-2} cm from the spot center, i.e., where the plasma density is higher. The efficiency of their formation increases with the plasma potential.

If a magnetic field exists at a distance $r > 10^{-2}$ cm from the emission center (provided that the plasma density $n \propto r^{-2}$), Hall's parameter $\omega_e \tau_e$, where ω_e is the electron plasma frequency, τ_e is the time between the electron collisions, becomes larger than unity. The ion trajectories bend under the action of Hall's field and the plasma as a whole starts to expand in Ampere's direction. The existence of Hall's field results in that the cathode drop; the plasma density and consequently the density of ion current to the cathode turns out to be maximum in the "opposite" direction from the EC. That is, the conditions for more efficient formation of new ECs are created in the magnetic field at the spot "reverse" edge. Having occurred there, the new ECs are under more favorable energetic conditions than the old ones, which later disappear. In [11.16] it was shown in dynamic terms that after a rapid application of magnetic field the cathode spot fragments, initially spread randomly over the cathode surface, group along a magnetic line of force at the spot "reverse" edge. Judging by the light emission intensity, the maximum plasma density was right at the reverse side of the spot. The value of Hall's field was estimated by *Emtage* et al. [11.16] as 70 V/cm. *Lyubimov* and *Rakhovsky* [11.2] observed the appearance of cathode spot fragments in the Ampere direction with $di/dt \simeq 5 \cdot 10^7$ A/s but these fragments disappeared after a short time and the spot moved in the reverse direction.

It should be noted that the influence of Hall's effect on the retrograde motion is still unclear. Indeed, if one takes into account that a new EC forms at a distance $r \leq 10^{-2}$ cm, then even with a Hall's field of about 100 V/cm the difference in the cathode drop values at this distance is less than 1 V and therefore it is unreasonable to suppose that the additional field strongly affects the formation of new ECs. However, such a difference seems to be sufficient for the newly created ECs to be more stable. It is very likely that the enlarged plasma density produced by Hall's field at the cathode "reverse" edge (and, consequently, the higher density of current to the cathode) have a stronger influence on the efficiency of the formation of new ECs.

Another view has been expressed on the possible reason for the retrograde motion of cathode spots. *Litvinov* et al. [11.19] showed that at the EC periphery ($r < 10^{-3}$ cm) the contribution of fast plasma electrons moving toward the cathode exceeds those of the ions and the field-assisted-thermionic-emission electrons. This means that part of the total current flowing through the spot center is fed back to the cathode surface by the fast electrons of the plasma. It is obvious that the magnetic force for the "reverse" electrons, and consequently for the current-carrying plasma where the "reverse" electron current prevails, is in the same direction as the retrograde motion of the spot. Since the reason for the formation of new ECs might be related to the "reverse" electron current (Sect.8.1.2), the approach suggested in [11.19] opens new perspectives for understanding of the abnormal motion of a cathode spot in a magnetic field tangential to its surface [11.20].

We assumed in [11.21] that the trace left by a spot on a clean cathode placed into an ultrahigh vacuum under a sufficiently strong tangential magnetic field will be a continuous strip of micron-sized overlapping craters. This was confirmed by *Jüttner* [11.22] (Fig.11.4), who pointed out that only one EC operates at each given moment. The maximum velocity of the spot in an ideal one-directional motion can therefore be estimated by the known crater radius and the EC lifetime. Using the data of Fig.11.3a we found that for the current of an arc with a molybdenum cathode of several tens of Amperes the crater radius is of the order of 10^{-4} cm and the EC lifetime ranges from 5 to 10 ns. The maximum velocity of motion of the spot would then be of the order of 10^4 cm/s.

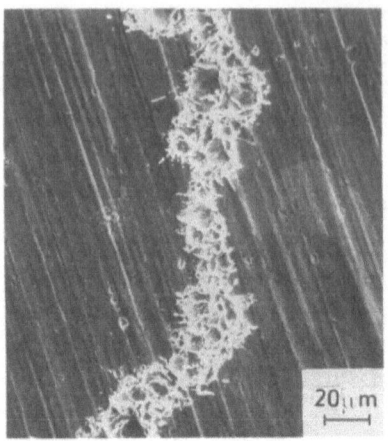

Fig.11.4. Fragment of a trace left by a cathode spot on a clean Mo cathode with arc current 20 A, and magnetic field 0.3 T

11.1.3 Spontaneous Formation of Cathode Spots in Pulsed Arc Discharges

In pulsed vacuum arc discharges, when the rate of current rise is 10^8 A/s and more, spontaneous formation of cathode spots can be observed [11.4, 23-25].

This process was studied carefully by *Cummings* [11.24] with the use of a high-voltage triggered spark-gap with a mercury cathode. It was shown that the radial propagation of the front of the region occupied by spots took place only in the stage of current rise, the maximum velocity of propagation being about $2 \cdot 10^6$ cm/s. The growth of current in the spark gap was accompanied by current bursts and formation of luminous strata in the electrode gap. *Cummings* suggested that oxide films on the mercury surface were responsible for the spontaneous appearance of cathode spots. However, because of the absence of sufficiently reliable data on the processes at the cathode and in the plasma of high-current discharges no convincing model of the phenomenon was suggested. We believe that the strata formation observed by *Cummings* might be identified with the motion from the ignition site to the cathode of the more dense plasma balls which form as a result of the irregular input of cathode material to the plasma. This leads to the formation of double electric layers with a high cathode potential drop. Comparing the data of [11.24, 25] with our results, it can be assumed that the spontaneous appearance of cathode spots in a spark discharge proceeds according to a mechanism associated with the breakdown of non-metallic inclusions and films. It is due to charging of the peripheral plasma layers up to a high potential. With this, the maximum velocity of propagation of the front of the region occupied by spots should be determined by the velocity of expansion of the cathode plasma, i.e., it must be not greater than $2 \cdot 10^6$ cm/s. As shown in Chap.8, application of a transverse magnetic field does not change the nature of the process of the spontaneous formation of cathode spots, but this process occurs only in the direction of the plasma drift in mutually perpendicular electric and magnetic fields.

11.2 Response of the Vacuum Arc to Current Transients

One of the methods for investigating nonstationary phenomena in an arc discharge is to detect transient processes during an abrupt change of the arc current. On studying the response of a vacuum arc to current transients ($di/dt \simeq 10^7$ A/s), *Kesaev* [11.4] arrived at the conclusion that the duration of the transient stage is entirely determined by processes in the cathode region, they necessitate the transition of the cathode spot to a new regime (change of position, spot division, etc.). With the help of bettter equipment the response of a vacuum arc to current transients was investigated in the range $10^6 \le di/dt \le 7 \cdot 10^8$ [A/s] [11.26]. For $di/dt \le 5 \cdot 10^6$ A/s no increase of the arc voltage above a usual level was observed. With $di/dt > 10^7$ A/s, a transient voltage burst occurred the amplitude of which increased with di/dt up to 500 to 1000 V, the duration of the transient stage ($0.4-0.8\,\mu s$) being dependent on the cathode material constants. *Paulus* et al. [11.26] suggested that the duration of the transient process is determined by the time it takes to heat up some cathode sections to boiling point. To show this they assessed the dependence of the transient stage duration on the cathode material constants.

However, the results of the studies of cathode processes at explosive electron emission question the validity of the conclusions in [11.4, 26]. As is shown above, emission centers can provide for high rates of current rise (10^9 to 10^{10} A/s) without any essential increase in the cathode drop (Chap.7). The time lag of the new EC formation does not exceed $\sim 10^9$ s (Chap.5). We assumed that the duration of the transient stage during a sudden current burst is determined by the time lag of the event occurring in the arc column. We shall now analyze the transient processes involved in an abrupt current rise with $di/dt = 2 \cdot 10^{10}$ A/s [11.27].

11.2.1 Experimental Equipment and Technique

The experiment was carried out in a discharge chamber with a low-inductance cathode (Fig.7.7). Copper electrodes were used. The electrode separation was varied in the range $0.5 \leq d \leq 8$ [cm]. In some tests the plane anode was replaced by a hollow one. The end surface of the latter was covered with a tungsten grid with a transparance of 95%. The cathode had a radial slit in which a needle-shaped trigger electrode had been mounted. A probe made of wire 0.2 mm in diameter and with an uninsulated section of 0.5 mm was also mounted in the slit. The triggering pulse of amplitude $V_2 = 17$ kV was applied to the cathode from a generator via a cable with $\rho_w = 750$ Ω. The current in the trigger circuit was limited to 20 A. After the gap had been filled up with plasma, the "pedestal" of a step-wise pulse ($V_1 = 1.5$kV) was applied to the anode via another cable, and an arc discharge was initiated in the main gap. In 1.3 μs the voltage pulse amplitude started to increase and after 10 ns it rose to $V_1+V_2 = 18.5$ kV (i.e., the current varied from a level $i_1 = 20$A to a level $i_2 = 220$A). Using a Langmuir probe operating in a floating regime the plasma potential was measured within an accuracy of $\simeq 10\%$ along both the cathode surface and the discharge axis, and the values obtained were compared with the voltage drop across the electrodes. Such measurements were performed only at distances $r \leq 4$ mm from the triggering site, since the probe could not satisfy the conditions for floating at $r \geq 4$ mm because of the low plasma density [11.28]. With $d \leq 1$ mm, the probe dimensions introduced a significant error into the measurement of the potential distribution between the electrodes. It should be noted that at distances $r \leq 0.5$ mm, when the absolute value of the plasma potential is small, it is necessary to take into account the difference between the floating potential and the true one. In our case, when the arc was operating in the copper electrode vapor, the plasma potential was about 7 kT_e in excess of the floating potential [11.29].

When the current changed from the level i_1 to the level i_2, a transient burst of voltage at the electrodes was observed. Typical current and voltatge waveforms are shown in Fig.11.5. At gap spacings $d = 0.5$ to 4 mm the arc current increased without any time lag for about 10 ns, the electrode voltage increasing up to 200 to 600 V (Fig.11.5a). With $d > 4$ mm, the current lag with respect to the leading edge of the applied voltage pulse increased as the gap was lengthened (Fig.11.5b). The behavior and the rise time to the level i_2 resembled the commutation characteristic of a vacuum spark. The transient voltage amplitude reached 2 to 10 kV. Given in Fig.11.6 are plots of the transient stage duration t_t with respect to the electrode separation d. It can be seen

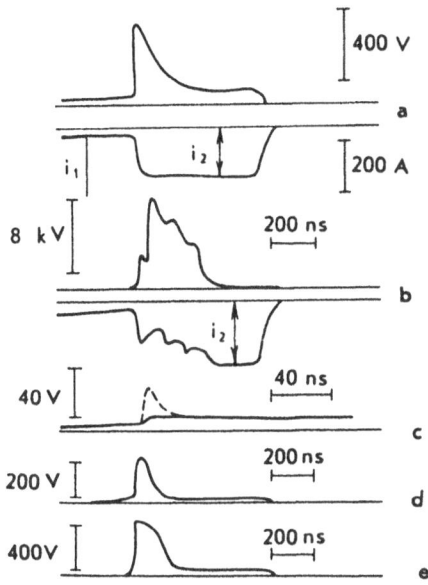

Fig.11.5a,b. Typical waveforms of the discharge current (lower traces) and the gap voltage (upper traces) characterizing the processes occurring at arc current transients, recorded with d = 4 mm (a) and 6 mm (b)

that when using a grid anode the plot of $t_t(d)$ is a straight line. Its slope gives the velocity of the gap closure $v_c = 2 \cdot 10^6$ cm/s. For a solid anode with d ≥ 4 mm the time of gap closure is essentially shorter.

The voltage drop across the electrodes at the current i_1 increases with gap length from ≃20 V at d = 0.5 mm to ≃80 V at d = 8 mm. The plasma "floating" potential within a radius $r \simeq 0.5$ mm from the site of triggering remains unchanged at 13 to 15 V. After the current transient the plasma potential within the radius $r \leq 0.5$ mm usually increased to 20 to 25 V which corresponds to a new value of the current. However, it sometimes reached 50 to 100 V and in 15 to 20 ns

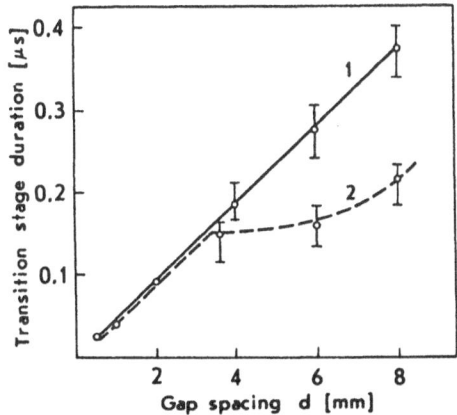

Fig.11.6. The transient process duration as a function of gap spacing

237

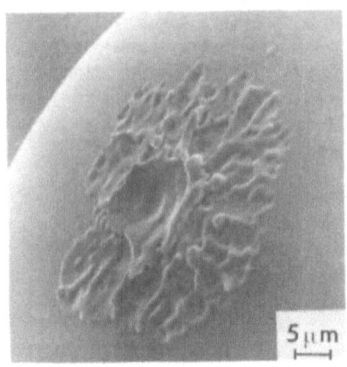

Fig.11.7. Erosion trace left on a tungsten sphere by a single arc discharge with a current transient $i_1 = 70$ A, $i_2 = 500$ A; $t_1 \simeq 30$ ns, $t_2 \simeq 60$ ns

fell to 20 to 25 V. As the probe was moved away from the triggering site both the amplitude and the duration of the plasma potential increased. At the time of current passage the voltage drop across the plasma was distributed approximately linearly. The duration of the probe pulse was practically equal to the transient stage duration t_t for the corresponding gap spacing ($r = d$). This correlation between the probe signal and the electrode voltage was observed up to $d = 4$ mm. For large gaps, when the probe was at a distance $r \geq 4$ mm from the site of triggering, probe measurements became incorrect because of the low initial plasma density and subsequent formation of a vacuum part of the gap in the probe region.

Figure 11.7 shows an erosion trace caused by a single arc discharge with a current jump. The total discharge duration was 100 ns, and after 30 ns the current rose from 70 A to 500 A in 10 ns. The appearance of the erosion trace supports the view that emission and erosion processes are of a nonstationary nature. Compared with the case illustrated in Fig.11.3a, the erosion trace diameter is considerably greater (40 μm) and the pattern is somewhat different. The average velocity of the radial expansion of the zone controlled by the cathode spot reaches $5 \cdot 10^4$ cm/s. Even under such extreme conditions ($di/dt \geq 5 \cdot 10^{10}$ A/s) the erosion trace is continuous, which is evidence supporting the field-assisted thermionic emission mechanism of the formation of new ECs on a clean cathode surface. The average current density measured throughout the erosion trace was about $4 \cdot 10^7$ A/cm.

11.2.2 Results

Since the transient process duration is dependent on the gap spacing (Fig.11.6), it can be assumed that the arc transition from one state to another is determined by the events occurring in the gap and is characterized by the velocity $v_t = 2 \cdot 10^6$ cm/s. This can be explained as follows: In a stationary regime the current in each cross-section of the plasma column is defined as $i_1 = j_1 S = e n_1 v_d S$, where v_d is the electron drift velocity. A new current level i_2 should be associated with a new value of plasma density n_2. The time necessary for attaining this value is defined as $t_t = d/v_t$.

However, as experiment shows, with $d \leq 4$ mm the new current value is reached in a time much shorter than t_t. This can be accounted for only by an increase in v_d, i.e., it is due to the growth of the plasma electric field. The dynamics of the transient process can be represented as follows: At the moment of the onset of current rise the cathode drop increases by a certain value, thus intensifying the emission processes in the existing centers and giving rise to the explosive appearance of new ECs. The value of the cathode drop remains lower than 100 V and the time of the new EC formation is not in excess of 10^{-9} s. The intensification of the emission processes is accompanied by generation at the cathode of a higher-density plasma which corresponds to the new current level. A double electrical layer forms between the plasma of the newly appeared centers and the initial plasma of a lower density. The voltage drop across this layer is of the order of kT_e. The layer moves with the velocity of cathode plasma motion $v_c = 2 \cdot 10^6$ cm/s. The interelectrode voltage decreases as the layer moves and the layer-anode gap is reduced. Probe measurements confirm the pattern of the layer motion and show that the main voltage drop across the plasma occurs between the layer and the anode.

With large gaps ($d \geq 4$mm) the initial plasma density is so small that at the moment of the abrupt current change a vacuum gap appears rapidly between the newly formed plasma and the anode. The current passage in the gap obeys the "3/2 power" law, and the commutation time is determined by the velocity of expansion of the new plasma. The vacuum gap formation seems to occur through the starvation of the initial plasma as a result of the withdrawal of ions from its boundary which leads to a rapid increase of the double layer thickness. The rate of increase of the layer thickness, according to [11.30], is expressed as follows:

$$\frac{dx}{dt} \simeq \frac{j_e}{n_e e} \sqrt{\frac{m_e}{m_i}}. \tag{11.2}$$

To obtain estimates, one can put $j_e \simeq 200$ A/cm, $n_e < 10^{12}$ cm^{-3} (for r > 3mm). Then we obtain $dx/dt > 10^7$ cm/s, which is much in excess of the velocity of cathode plasma expansion. As n_e decreases with distance, the rate dx/dt will increase nonlinearly with time. For instance, with $d = 6$ mm the time for the vacuum gap to form will amount to several tens of microseconds. On the current waveform shown in Fig.11.5 b this corresponds to the first current drop.

As a result of the formation of the vacuum gap between the anode and the new cathode plasma across which a high voltage is concentrated, the anode experiences an intense electron bombardment. At the anode plasma appears which diminishes the efficient gap spacing and hence also t_t as it moves toward the cathode. The use of a highly transparent grid as anode reduced the efficiency of the anode plasma formation and weakened its influence on the time t_t (Fig.11.6).

The results obtained explain the dependence of the transient stage duration t_t on the cathode material constants in [10.26]. For example, the time t_t was 0.4 μs and 0.7 μs for copper and lead cathodes, respectively, resulting in $v_t \simeq 1.2 \cdot 10^6$ and $7 \cdot 10^5$ cm/s, respectively. These velocities are smaller by a factor of 1.5 than the velocities of the cathode plasma motion for the corresponding materials observed at breakdowns of vacuum gaps. The lower value of v_t in [10.26] is possibly related to the fact that the newly formed plasma expands into the medium with a counterpressure (Sect.6.4). The lower limit of $(di/dt)_1$, at which the process of filling the gap with new plasma still has time to "follow" the rate of current rise and the transient voltage is not significantly different from the voltage across the arc gap, can be estimated form the expression:

$$\left(\frac{di}{dt}\right)_1 \simeq \frac{i_{th}}{t_t} \simeq i_{th} \frac{v_t}{d} , \qquad (11.3)$$

where i_{th} is the threshold arc current. In [11.26] ($d = 0.5$ mm, $v_t = 1.2 \cdot 10^6$ cm/s) $(di/dt)_1 = 6 \cdot 10^6$ A/s. *Kesaev* [11.4] believed that when the rates of current rise exceed the critical rate the cathode spots change qualitatively (spontaneous mulitplication of spots occurs, a spark spectrum appears). Therefore, he treated the discharge with $di/dt > (di/dt)_1$ as a spark. We think that at high values of di/dt the essence of the emission processes at the cathode is the same as at low di/dt values. The processes in the discharge column are essentially different due to the formation of the moving double layers.

The lower limit of $(di/dt)_2$, at which the time lag of the processes in a single cathode spot becomes noticeable, can be estimated as follows:

$$\left(\frac{di}{dt}\right)_2 \simeq j \frac{dS}{dt} \simeq 4\pi j \, v_c^2 \, t \, . \tag{11.4}$$

From (11.4) it follows that with an explosive emission current density of $5 \cdot 10^8$ A/cm^2 and a characteristic time scale of current change of $\sim 10^{-8}$ s, $(di/dt)_2 \simeq 10^{14}$ A/s. It may therefore be difficult to measure the time lag.

In conclusion, we wish to note the following: The vacuum arc discharge is characterized by a short, pulsed rise of the voltage across the electrodes. According to *Kesaev* [11.4], the voltage burst at the electrodes should prevent arc failure. The burst duration is determined by the cathode processes. Nevertheless, it can be concluded that the recovery of the initial state after the voltage burst is determined by the change of the plasma column conductivity rather than by the time for which the cathode spot returns to the previous regime. The correlation between the voltage bursts and the motion of plasma jets from the cathode in an arc discharge was established in [11.31].

11.3 Vacuum Arcs at Threshold Currents

11.3.1 The Threshold Current of a Vacuum Arc

Progress in studies of explosive emission phenomena have led to a need for refining some of the vacuum arc parameters known from literature. The spot lifetime and the cathode voltage drop characteristics at vacuum arc threshold currents are of a particular interest. As is well known [11.4], the arc threshold current is a minimum current at which the features of discharge are still noticeable. Since threshold currents have been determined experimentally with a time resolution of the order of 10^{-5} s [11.4] and the spot fragment liefetime is of the order of 10^{-6} s [11.2], the values of threshold currents given in [11.4] should be considered overestimated. *Bazhenov* and *Chesnokov* have shown that this is the case in measuring minimum EEE currents [11.32].

The experiment described in [11.32] was carried out in a vacuum diode with the cathode made of a thin copper wire and a plane anode ($d = 2$cm). A stepwise positive voltage pulse was applied to the anode ($V_1 = 35$kV, $t_1 = 10^{-8}$ to $3 \cdot 10^{-7}$s; $V_2 = 0$ to 4kV, $t_2 = 10^{-6}$s). The diode

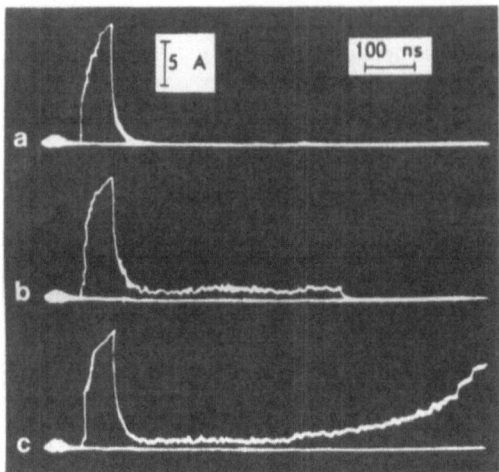

Fig.11.8. Minimum EEE current waveforms

current initially increased in accordance with (7.11), but after the voltage dropped to V_2 it became essentially dependent on the voltage. At low V_2 values the EEE current ceased simultaneously with the termination of the first pulse step. At high V_2 values the current increase at the second step recommenced. These two ranges of V_2 values are separated by a small range of intermediate values of V_2 where the outcome depends on the time-dependent behavior of V_2.

It is clear from Fig.11.8 that the current can either be absent or break off after a certain arc-operation time, or increase irreversibly up to a value defined as V_2/R at the same V_2 value. The time at which current breaks off is less than 10^{-8} s. Under controlled experimental conditions the values of the break-off current i_{min} and the time of arc operation vary from pulse to pulse, but the average time increases with i_{min} and t_1 (Fig.11.9). The minimum EEE currents and the discharge operation time are much lower than those determined for a copper cathode [11.4, 33].

An investigation of the parameters of short vacuum arcs has been carried out using a high temporal resolution set-up [11.34]. An arc was initiated under ultra-high vacuum conditions between closely placed wire electrodes using a short (10^{-8}s) trigger pulse. A pulse oscilloscope was connected to the vacuum gap to measure the voltage drop directly at the electrodes. Some typical arc voltage waveforms for successive discharge realizations are presented in Fig.11.10. By processing such oscilloscope records the dependence of the average time of arc operation on the arc current is obtained (Fig.11.9). Our data are in good agreement with the measurements of *Bazhenov* and *Chesnokov* [11.32] and show a roughly proportional increase of this time with arc current.

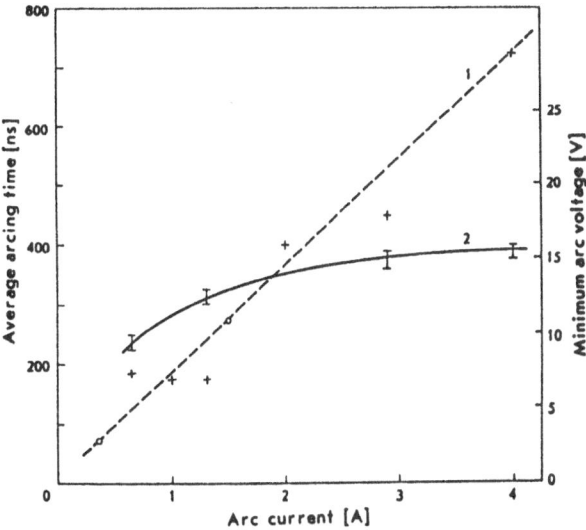

Fig.11.9. Average time of vacuum discharge (*1*) and the minimum arc voltage (*2*) as a function of current for copper electrodes (circles denote the data of [11.32])

The use of shorter time triggering led to a reduction of the arc shunting by the trigger discharge and resulted in a dependence of the time of arc operation on i_{min} different from that in [11.4]. This time became as low as a few hundreds of nanoseconds at currents comparable with i_{th}; the current i_{min} sank below i_{th}. A similar result was obtained for a tungsten cathode. For example, with i_{min} = 1 and 2 A the average arc operation time was 100 and 500 ns, respectively.

Attention should be paid to another important fact. Although the arc voltage is essentially variable and there is a pronounced minimum of this quantity [11.4], the variation time scale is two or three orders

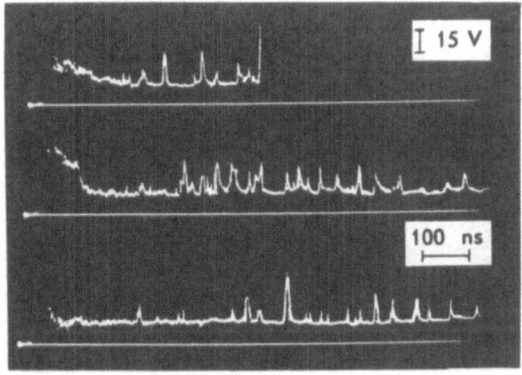

Fig.11.10. Typical waveforms of the voltage drop on short vacuum arcs. (Copper electrodes, i = 4A)

of magnitude less than that in [11.4]. The short duration of the bursts is explained by the use of an arc which was shorter than in [11.4] (d $\simeq$ 0.1 mm). The interval between the bursts indicates that the spot life-time is not more than a few tens of nanoseconds and it is much less than the average time of arc operation.

The minumum arc voltage as a function of arc current for copper electrodes obtained from voltage waveforms is given in Fig.11.9. At low arc currents it is less than that determined in [11.4] and only 2 V greater than the ionization potential for copper. The value of the mini-mum arc voltage for tungsten turned out to be close to that found in [11.3].

11.3.2 Cathode Spot Current Density

We carried out experiments [11.36, 37] which indicate that high curent densities at a low voltage drop across the arc gap occur. Experiments were modified in line with comments stated in [11.2, 35] about the val-idity of measuring the current density by erosion traces. Also taken into consideration was the observation [11.35] that explosive emission processes can only occur at a vacuum breakdown when a high voltage is applied between cathode and anode.

Classical field-electron emitters made of tungsten were used as cathodes. Employing such cathodes for measuring the current density has a number of advantages. These are the possibility of obtaining a clean surface, being able to localize a single cathode spot at the tip and, measuring the erosion rate at a sufficient accuracy, a relatively small plasma-cathode surface contact area, and the opportunity of using low supply voltages comparable in magnitude with the arc volt-age drop. A tungsten ball was obtained by pulsed heating of the wire tip to melting and was used as the anode. The cathode spot was initi-ated at the point tip by slowly closing the distance between the elec-trodes, whereas a dc voltage of 50 V was applied. The discharge cir-cuit ensured the duration of a short-circuit rectangular current pulse of 1 μs. The current was controlled in the range of 0.5 to 4.0 A by vary-ing the discharge circuit wave resistance.

Arc current and voltage waveforms are illustrated in Fig.11.11. At first there occurs a relatively stable stage of arc operation with a dura-tion of 50 to 250 ns. Then either the current breaks and the voltage recovers up to the level of the charging voltage (Fig.11.11a,b) or at the beginning of the current break a small increase of voltage results in several cycles of arc operation having different durations (Fig.11.11 c, d). During these cycles, a gradual decrease of current and increase of

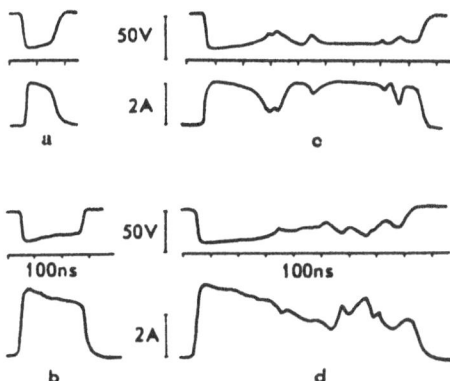

Fig.11.11. Combined waveforms of the arc current and operating voltage

voltage often take place, but for the last cycle the current break is pronounced. The arc operation voltage was 20 to 30 V.

Only those tests were taken into consideration in which a single cycle took place and the discharge time could be considered to be equal to the lifetime of the cathode spot on the point cathode t_0. Examination of the points in a Transmission Electron Microscope (TEM) after discharge indicated erosion and melting of the cathode tips. On examination under SEM no signs of cathode-spot plasma action on the side surface adjoining the molten tip, and no indications of new cathode spot formation were observed. Hence, one can believe that the discharge current passed mainly through the point tip which was destroyed during cathode-spot operation. This provided us with an opportunity to estimate a lower bound for the average current density in the cathode spot by the end of the its lifetime as $j_{min} = i/S$, where S is the cross-sectional area of the molten point tip. It turned out that j_{min} is in the range $(2 \text{ to } 10) \cdot 10^7$ A/cm^2 for different points. With this, the average velocity of propagation of the point destruction boundary is $(1 \text{ to } 5) \cdot 10^3$ cm/s and the average erosion rate is $(5 \text{ to } 20) \cdot 10^{-4}$ g/C. These data are in good agreement with the experimental results for increasing current pulses (Chap.5) typical for the high-voltage (spark) stage of a vacuum discharge. They do not contradict the cathode erosion model based on intense heat release in the cathode tip due to Joule's energy dissipation. It should be noted that to provide the above values of the propagation velocity of the evaporation boundary, the temperature of the evaporating surface according to [11.38], should be equal to $2 \cdot 10^4$ K. This agrees well with the results of a more rigorous numerical simulation of the explosive emission process [11.39].

It has therefore been shown experimentally that a vacuum arc cathode spot can produce a current density of the order of 10^8 A/cm^2 with a voltage drop across the arc gap of 20 to 30 V. A calculation of the voltage drop V_{np} across a section of nonideal plasma adjoining an exploding copper point was carried out by *Litvinov* et al. [11.40]. The following results were obtained. At an arc current i = 6 A and an emitting area radius r_{cr} = 10^{-4} cm ($j \simeq 2 \cdot 10^8$ A/cm^2), V_{np} amounts to 15 V. Reducing r_{cr} to $3 \cdot 10^{-5}$ cm ($j \simeq 2.5 \cdot 10^9$ A/cm^2) results in an increase in V_{np} up to 18 V. From the calculation it also follows that $V_{np} \propto i^{1/3}$. The voltage drop across the section of ideal plasma which can be approximated by V_{ip} = $i/(4\pi\sigma r_{ip})$ (r_{ip} is the radius originating from where the plasma can be considered ideal) is about 30 V at the same value of current and with $r_{ip} \simeq 3 \cdot 10^{-4}$ cm. About the same values of the voltage drop across both plasma sections are obtained from the expression V = ϵ_{pl}M/i discussed in Sect.6.4. Attention should also be drawn to [11.41,42] where it was shown that by including a diffusion term in the equation for the near-cathode plasma voltage drop it is possible to obtain lower voltage-drop values than those predicted by the model of purely resistive current passage.

Recently, *Litvinov* et al. carried out a numerical simulation of the expansion of the plasma jet ejected by a single EC of the vacuum-arc cathode spot [11.43]. It was confirmed that the cathode voltage drop observed in the experiment corresponds to a current density of $j > 10^8$ A/cm^2.

11.4 Numerical Simulation of Processes in an Explosive Emission Center

There is a great deal of experimental data on the properties of single explosive emission centers appearing on the cathode during vacuum discharges. Nevertheless, some questions concerning the mechanism of the operation of such centers under a plasma still remain unanswered. It is important to establish the emission current density, the mechanism of the electron emission from the cathode, the role played by ions as current carriers, the dominant factors in the lifetime of emission centers, etc. To do this it is necessary to understand the emissive, energetic and erosive processes characterizing the dynamics of emission-center operation. A numercial simulation of the processes occurring in a single emission center at vacuum discharge has been carried out [11.39,44-46].

As the system is spherically symmetrical, the EC can be assumed by a hemisphere on a plane in a simplified mathematical model. We suppose that a zone of evaporation exists on the cathode surface. It coincides with the emission zone, and a layer of molten metal is present behind it. The processes of heat release, heat conduction, evaporation and melting may be described by

$$\rho c \, \frac{T}{t} = \frac{\lambda}{r^2} \, \frac{\partial}{\partial r} \, r^2 \, \frac{\partial T}{\partial r} - \frac{ic_2}{2\pi er^2} \, \frac{\partial T}{\partial r} + \left(\frac{i}{2\pi r^2}\right)^2 \kappa_0 T \, , \tag{11.5}$$

$$T_{t=0} = T_0 \, , \, T_{r \to \infty} = T_0 \, , \, T(r_m, t) = T_m \, ,$$

$$\lambda \nabla T|_{\mathrm{s}} - \lambda \nabla T|_1 = \rho v_m \epsilon_m \, , \tag{11.6}$$

$$v_{ev} = v_{\mathrm{s}} \, \exp[-\epsilon_{\mathrm{s}} m_{\mathrm{a}}/kT(r_0, t)] \, , \tag{11.7}$$

$$-\lambda \nabla T_{r=r_0} = -\rho v_{ev} \epsilon_{\mathrm{s}} - (j/e) \, (2kT + \phi) + j_i i \epsilon_i/e \, . \tag{11.8}$$

Here, $c_e = \pi^2 k^2 T/(2\epsilon_F)$ is the electronic heat conduction; ϵ_F is the Fermi energy; ϵ_i is the energy delivered by an ion to the cathode surface; r_0, v_{ev}, ϵ_{s} are, respectively, the radius of the evaporation zone, the velocity of evaporation front, and the specific heat of evaporation; v_{s} is the transverse velocity of sound in the metal; the indexes $|_{\mathrm{s}}$ and $|_1$ indicate the boundary conditions at the solid-liquid phase boundary. The values of ρ, c, λ were considered to be independent of temperature and identical for liquid and solid phases. It was assumed that a fraction of evaporating atoms, f, returned back to the cathode as ions with energy ϵ_i, i.e.,

$$j_i = f\rho v_{ev} e/m_{\mathrm{a}} \, . \tag{11.9}$$

The values of λ, v_{s}, κ_0, f, ϵ_i were not known exactly and were therefore varied over a wide range. The EC current was calculated using formulae for field-assisted thermionic emission [11.47]. It was noted that the EC current did not exceed a certain peak value limited only by the external conditions. The electric field at the cathode was calculated using the well-known Langmuir-Mackeown formula. It was assumed that the EC had already been initiated (a micropoint may have exploded or a thin dielectric film broken down). The current flowing through the EC heats the cathode matter leading to its evaporation and melting. The phase-transition boundaries move deep into the cathode.

The main results of the calculation are as follows. The present model could answer the question about the mechanism of energy release in the EC. The cathode heating and erosion are affected solely by

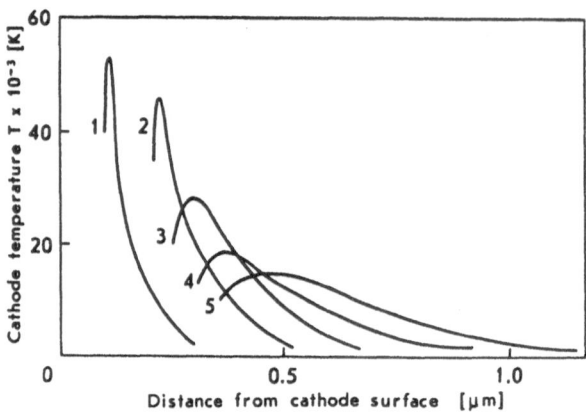

Fig.11.12. Calculated temperature distribution in a copper cathode at t=0.1 ns (*1*), 0.2 ns (*2*), 0.3 ns (*3*), 0.5 ns (*4*), and 1.0 ns (*5*), with EC current 20 A

the Joule heat release, and the contribution of the plasma ion bombardment of the cathode surface to the energy balance is not significant. Moreover, the energy flux removed by the emittted electrons from the surface is in excess of the maximum energy input from ion bombardment. Thus the surface cools and the temperature maximum is attained in the bulk of the cathode rather than at the surface (Fig. 11.12).

The EC current pulse predicted by the model has an almost rectangular form, i.e., the rise time and the fall time are noticeably shorter than the pulse duration. Rapid current rise is necessary for erosion and emission processes to develop, while rapid current fall is caused by a high rate of cooling of the cathode surface (Fig.11.13) which causes a decrease of its emissive power.

As erosion develops, the emission zone increases in size (Fig.11.14) and this leads to a decrease of the current density at a fixed current amplitude. This in turn yields to a less efficient Joule's heat release and a decrease of the cathode surface temperature; and, in the end, to the disappearance of the emissive power of the center. The EC lifetime is in the range 10^{-9} to 10^{-8} s, which is in good agreement with the measured times of crater formation. During the EC lifetime a melt zone of about the same size as the crater has time to form. Calculations have shown that the melt is formed not only due to the Joule heat released in the given cathode site, but also due to the heat which propagates deep into the cathode from the hottest surface layers. This noticeably speeds up the melting process (Fig.11.14). The formation of craters within nanoseconds remained unclear for a long time, since the Joule heat release model, which does not involve the effect of heat

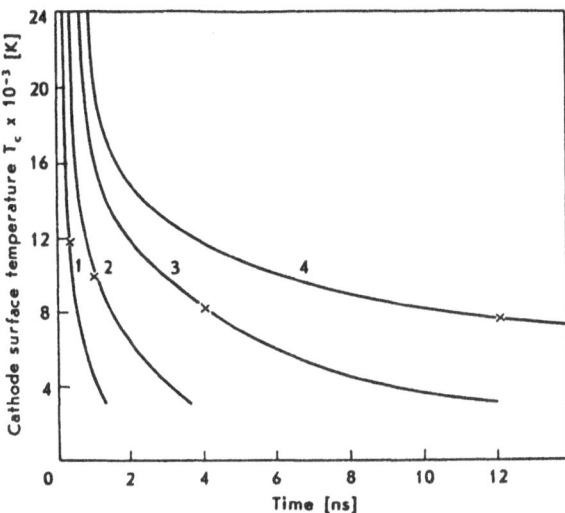

Fig.11.13. Surface temperature of a copper cathode as a function of time for EC currents of 10 A (*1*), 20 A (*2*), 50 A (*3*), and 100 A (*4*). The moment of the termination of emission is denoted by a cross

conduction [11.48] is not able to explain the observed crater depth. The same could be said about the other models which assume that the melting is due to surface heating of the cathode [11.22].

The calculation showed that a steady-state thermal regime within an EC is unattainable. The observation that the development of an EC inevitably results in its extinction, i.e., that the finite lifetime of an EC is its intrinsic feature, is the main result of [11.39, 44-46]. This conclu-

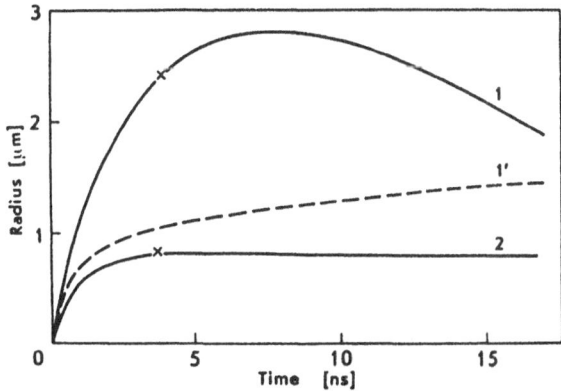

Fig.11.14. Time behavior of the melting zone radius with (*1*) and without (*1'*) taking into account heat conduction and the evaporation zone radius (*2*) in the EC on a copper cathode at a 50 A arc current. The moment of the termination of emission is denoted by a cross

Table 11.1. Charateristics of an emission center on a plane copper cathode at the moment of emission termination

i [A]	t [ns]	R_0 [μm]	T [K]	j [A/cm^2]	n [cm^2]	γ_c [g/C]
10	0.4	0.17	12000	$5.5 \cdot 10^9$	$4 \cdot 10^{19}$	$2.7 \cdot 10^{-5}$
20	1	0.33	10000	$3.0 \cdot 10^9$	$2 \cdot 10^{19}$	$3.4 \cdot 10^{-5}$
50	4	0.8	8400	$1.2 \cdot 10^9$	$1 \cdot 10^{19}$	$3.8 \cdot 10^{-5}$
100	12	1.5	7800	$0.7 \cdot 10^9$	$0.6 \cdot 10^{19}$	$4.5 \cdot 10^{-5}$

sion allows the observed cyclicity of the EC operation to be treated in a new way. The "jumping" of the emission zone now seems not to be an occasional but rather a necessary event. The cyclic nature of the EEE turns out to be caused by the finite lifetime of a single center.

Table 11.1 presents some results for a copper cathode [11.45, 46], namely, the EC lifetime t_0, the emission zone radius r_{cr}, the cathode surface temperature T, the e c c cmission current density, the "vapor" density by the end of EC operation, and the erosion rate γ_c. The current density at the moment of EC destruction is 10^8 to 10^9 A/cm^2, and is higher than in the beginning of its operation. The calculation predicts that the EC emission zone is several times smaller than the melt zone. Therefore, the current density obtained by the crater size is underestimated.

The metal "vapor" density at the surface was estimated from

$$n = \rho v_{ev} \, (T_v/m_a)^{-1/2} \, , \qquad (11.10)$$

where the "vapor" temperature T_v was assumed to be equal to the sublimation temperature. It follows from the calculation that in the beginning of the emission process the evaporation front velocity is high (of the order of magnitude of the velocity of sound in the metal) and the "vapor" density near the cathode surface is $\simeq 10^{22}$ cm^{-3}. Further on the evaporation front velocity is reduced; the "vapor" density at the time of emission termination is 10^{20} to 10^{21} cm^{-3}. *Litvinov* and co-workers [11.39, 44-46] suggested that in the initial period after the EC appearance there is no pronounced metal-plasma interface, but it becomes clearly visible by the end of EC operation.

This observation has lead to a new qualitative hypothesis about the mechanism of electron emission. *Litvinov* and co-workers [11.39, 44-46] proposed that there is a "smearing" of the potential barrier in the region of the metal-plasma transition, whereas its total height remains unchanged. Electrons, after transfering to the "vapor" region and undergoing heating, occupy increasingly higher levels, thus providing for field-assisted thermionic emission with a high current density. In the course of time, when discontinuities in the density of the material form, a barrier occurs at the metal-nonmetal interface. Its value may be estimated by

$$\Delta\phi = \phi\left[1 - \left(\frac{nm_a}{\rho}\right)^{1/3}\right]. \tag{11.11}$$

This resembles a semiconducting n-p junction. As the evaporation rate falls and the "vapor" density decreases to 10^{20} cm^{-3}, corresponding to the case of ideal plasma ($T_p \simeq 5eV$), the potential barrier for electrons at the metal boundary can be calculated using Shottky's correction.

Figure 11.15 is a plot of the calculated radii of the melting and evaporation zones at the end of EC operation versus the discharge current. This also illustrates the experimental relationships between current and crater radius on the cathode obtained by *Daalder* [11.49] and *Jüttner* [11.5]. It can be seen that the calculated melting zone radius agrees well with the crater radius. This indicates that the crater size is determined by the size of the zone of the molten metal pushed away by the vapor pressure. The metal splashing seems to occur either after the completion of EC operation or at the very end of its lifetime.

Analysis of the results shows that the emission mechanism has no principal importance in the model discussed. The EC emissive power was obtained with a large margin, so that *Litvinov* and co-workers were forced to assume the existence of a virtual cathode which returned the

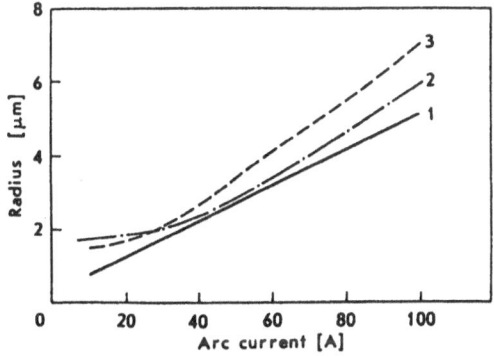

Fig.11.15. Calculated melting zone radius (*1*) and the experimentally obtained crater radius (*2*) [11.49], (*3*) [11.5] as a function of the discharge current for a copper cathode.

"extra" electrons. Emission formulae were used only to determine the moment when emission ceased. Since the cathode surface cools at a high rate, different assumptions concerning the emission mechanism did not introduce any essential difference into the final results. In other words, the energy concentration in the EC turned out to be so high that the EC current was determined by the external conditions (which were initiated by the prescribed current amplitude) rather than by the surface emissive power.

11.5 Explosive Electron Emission and the Vacuum Arc Cathode Spot

As stated above, the most interest up until now has been in the development of steady-state models of the cathode spot. However, none of the steady-state models was able to explain such experimental facts as the rapid motion of the spot, the arc voltage fluctuations, the discrete pattern of the erosion traces and the presence of multiply charged ions in the cathode plasma. The models did not take into account cathode surface microirregularities and the inhomogeneity of its chemical composition. Analysis of the steady-state energy balance in cathode spots of high-melting-point metals has shown [11.50, 51] that throughout the possible range of cathode surface temperatures the energy carried away by emitted electrons is much greater than the energy which can be brought to the cathode by ions. The conclusion was drawn [11.50, 51] that in a cathode spot a more powerful energy source exists, viz. a volume source. The volume heat source turns out to be more efficient when compared to the surface one at current densities $j \geq 10^8$ A/cm^2 [11.4, 52, 53]. At these current densities one can expect explosive non-stationary processes to occur [11.54-57] as was found in a study of the transition of field electron emission into a vacuum arc and the vacuum breakdown [11.58-60]. Moreover, numerous estimates of the electric field under the cathode spot are close to those values at which vacuum breakdown is initiated. Fast processes at vacuum breakdown and explosive emission therefore attracted attention of those investigators who are interested in the cathode spot phenomenon [11.2, 5, 11, 48, 61-64].

On the basis of the studies about vacuum breakdown and about EEE phenomena and by comparing them with the processes occurring in a cathode spot we suggested that the cathode flare represents the initial stage of the vacuum arc cathode spot [11.58-67]. The further studies described above made it possible to compare the EEE and the

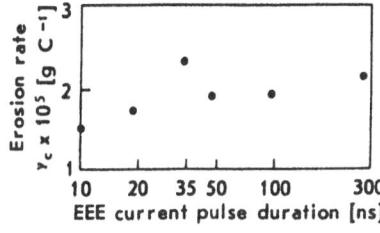

Fig.11.16. Erosion rate of copper wire cathodes as a function of the EEE current pulse duration

cathode spot properties in more detail. Below we shall give a general summary of this problem.

1) As is well known [11.4,68], an arc discharge can be initiated by a number of techniques, and it is easy to show that all of them are based on the creation of a high energy density in a cathode microvolume. This energy density can be reached in the initiation of EEE when field-assisted-thermionic current passes through a micropoint or when current flows at the moment of the break-off of a liquid drop from a micropoint.

2) A typical feature of both processes is that metal must be lost from the cathode in order to maintain a high density near the emitting surface. As an example, shown in Fig.11.16 is the average erosion rate as a function of the duration of the explosive emission current pulse for copper-wire cathodes. It is clear that γ_c is about $2 \cdot 10^{-5}$ g/C, i.e., it is close to the calculated values given in Table 11.1 and to the value of the "ion" erosion rate for a copper cathode at vacuum arcing ($\sim 4 \cdot 10^{-5}$ g/C) [11.69,70]. *Jüttner* also showed [11.5] that the erosion rate for wire cathodes made of Cu, Al, and W at $t_p = 10^{-8}$ to 10^{-6} s is comparable to the erosion rate for quasistationary vacuum arcs.

3) The parameters of the cathode flare plasma are comparable those of with the plasma of arc cathode spots. We may assume that near the emsision zone on the cathode the plasma density can reach a value of the order of 10^{20} cm^{-3} or more (measurements of the plasma density averaged over the spot give a value of the order of 10^{18} cm^{-3}). The electron temperature in the plasma, according to different estimates, ranges from 1 to 5 eV. The cathode plasma includes multiply charged ions of the cathode material.

4) At moderate values of current the cathode flare plasma and the cathode plasma of vacuum arcs expand isotropically from the emission zone. The velocity of plasma expansion is usually of the order of 10^6 cm/s for arc discharges and $(1$ to $3) \cdot 10^6$ cm/s for EEE. It is assumed that the energy necessary for plasma to expand with such a velocity is accumulated in a small region which covers the cathode emission zone and the adjoining plasma.

5) At explosive electron emission with a high rate of current rise the near-cathode voltage drop is not in excess of several tens of volts, i.e., it is comparable in magnitude with the cathode drop in vacuum arcs.

6) Typical for EEE processes are spontaneous stopage of emission as $di/dt \rightarrow 0$ and small currents commensurable with the arc threshold current.

7) Both the EEE and the vacuum arc cathode spot phenomena are characterized by a high current density in the region of the cathode connection. At EEE, the current is in excess of 10^8 A/cm^2. We have also shown that with an arc discharge duration of $t_p = 10^{-7}$ s, the current density on a tungsten cathode is of the same order of magnitude. The spot current density of the order of 10^8 A/cm^2 was also measured in a number of experiments with pulsed and quasistationary arcs.

8) The EEE erosion traces left on the cathode appear as craters of several micrometers in size having a specific substructure. Micropoints solidify on the crater rims. The craters left by a cathode spot are of similar appearance and size, which suggests that they have been formed in the same way.

9) During operation of a cathode spot and several explosive emission centers, liquid drops leave the cathode with a velocity of up to $5 \cdot 10^4$ cm/s. This indicates the existence of a high pressure on the cathode in the region of the emission zone.

10) The explosive emission and arcing plasma have a stimulating effect on the process of formation of new emission centers, which gives rise to a displacement of the cathode junction along the cathode surface.

11) Explosive emission is characterized by the cyclic nature of the associated processes. This is due to the finite lifetime of a single EC. The cycle duration is of the order of a nanosecond. The cyclic nature of the cathode-spot processes of approximately the same characteristic duration has been confirmed by recent research. Such a comparison of the characteristics of the two phenomena suggests that they are basically identical. We therefore suggested [11.65, 67] that the explosion of a micropoint accompanied by the generation of cathode plasma and by the current flowing from the cathode is an elementary phenomenon in the nonstationary cathode spot operation.

Further comprehensive studies of the individual emission centers and the peculiarities of the self-consistent cathode emission have provided a more complete description of the processes involved. It led to the conclusion that explosive electron emission is fundamental in cathode spot operation [11.10, 71].

12. Pulsed Electrical Discharge in Vacuum at Cryogenic Electrode Temperatures

The studies discussed in the preceding chapters have shown the fundamental role played by microexplosions on the cathode surface in the initiation and development of a vacuum discharge. Under conditions of high vacuum and a clean cathode surface, the main cause of these microexplosions is heating of cathode microprotrusions by field emission current.

Bearing in mind that Joule heating can be reduced by lowering the cathode temperature and even eliminated under superconducting conditions, one would expect the pulsed breakdown strength of vacuum gaps to increase when the electrodes are cooled. Moreover, the superconducting-to-normal transition requires a finite time [12.1], which could also promote vacuum gap strengthening under the conditions of pulsed breadkown. These considerations prompted us to investigate the pulsed electrical discharge at cryogenic temperatures [12.2].

12.1 Field Electron Emission at Low Cathode Temperatures

12.1.1 Effect of Superconductivity on FEE Current

We first consider field electron emission (FEE) at low cathode temperatures, particularly when the cathode is in the superconducting state.

In the late 1950s, *Bardeen*, *Cooper*, and *Schrieffer* developed a theory of superconductivity which promoted an understanding of the observations in their totality (the so-called BCS theory) [12.3]. First studies of FEE from superconducting metals sought further verification of the BCS theory, i.e., were aimed to demonstrate that when a field emitter goes from the normal to the superconducting state the emission current actually decreases by a certain value Δi due to the formation of a superconducting condensate [12.4].

Gomer and *Hulm* [12.5] investigated FEE from a tantalum point cathode at temperatures of 2.2 and 4.4 K (the transition temperature for tantalum is $T_s = 4.39 K$). For a point cathode with a tip radius 0.57

μm, a current density j = 80 A/cm² was obtained. Within the measurement error of 0.2%, *Gomer* and *Hulm* observed no difference in the emission current values measured for point temperatures below and above T_s. From these data they concluded that the normal-to-superconducting transition causes no change either in the potential barrier at the metal surface or in the chemical potential of conduction electrons.

Leger [12.4] attempted to measure Δi for a niobium point. The transition to the normal state was induced by heating the point by passing current through the niobium hoop which supported the point. The predicted increase in current is $3.8 \cdot 10^{-13}$ A according to BCS theory. In spite of the sufficiently high resolution of the apparatus used, *Leger* was unable to observe the expected effect. His conclusion was that either the emitter tip does not change to the superconducting state at all or the emitter surface changes to the normal state with the existence of a strong electric field and field emission. Application of magnetic field to induce the transition of a tantalum emitter to the normal state in attempts to measure Δi [12.6] gave no convincing results either.

The results of the studies [12.4-6] suggest that cooling of a superconducting emitter to temperatures below T_s does not lead to any noticeable change in the work function ϕ. This was confirmed by careful measurement of ϕ using the photoelectric effect [12.7]. The absence of any change in the field emission current, notwithstanding the predictions of BCS theory, may be associated with either the penetration of the electric field into the emitter surface layer [12.8] or the Nottingham effect. For an emitter made of a superconducting material the Nottingham effect is the sole heating factor at temperatures below T_s, as no Joule heating occurs.

12.1.2 The Nottingham Effect and Superconductivity

In Sects.2.4 and 5.1 we mentioned the contribution of the Nottingham effect to the energy balance in the heating of an emitter. The essence of this effect is as follows [12.9]. The electrons emitted due to the tunnel effect remove from the emitter a mean energy ϵ, which may either be equal to the Fermi energy ϵ_F or differ from it by a certain value $\Delta\epsilon$. At low temperatures $\epsilon = \epsilon_F - \Delta\epsilon$, i.e., the conduction electrons bring a mean energy approximately equal to the Fermi energy from the emitter bulk to the emission boundary. This means that the electrons that leave the cathode in the process of emission are replaced by "hotter" electrons coming from the external circuit. These electrons give up their energy to the lattice in the surface layer at a distance of the order of the electron mean free path from the surface. Thus, the thickness of the heated layer is usually only a few tens of Ångstroms. Hence, the Nottingham effect is purely a surface effect.

The more the lattice is heated, the larger the number of electrons leaving the cathode. This is because the number of electrons with energies above the Fermi energy increased, and so does the average energy of the emitted electrons. At a certain emitter temperature this energy becomes higher than the Fermi energy ($\epsilon \simeq \epsilon_F + \Delta\epsilon$). This results in heat absorption in the surface layer. The theory of the Nottingham effect based on Sommerfeld's model of a metal gives the following expression for the power deposited at the emitter surface with a field emission current i

$$ P = \frac{9.78 \cdot 10^{-9} \, Ei}{e\sqrt{\phi}} , \tag{12.1} $$

where E is the electric field at the emitter surface and ϕ is the work function of the emitter material. The temperature at which $P = 0$ is the inversion temperature T^* defined in Sect.2.5. If the emitter temperature is lower than T^*, the emitter is heated due to the Nottingham effect. If it is higher than T^*, the emitter is cooled. The inversion temperature can be calculated using (2.17).

The heating of a superconducting emitter by a Nottingham source at cryogenic temperatures was studied experimentally by *Bergeret* and *Septier* [12.10]. A niobium point cathode was supported by a niobium wire which could be cooled to liquid helium temperature. The electrodes were mounted in a UHV chamber immersed in a helium bath. The voltage applied to the diode was modulated at 0.25 Hz. Current was passed through the wire and, if the wire was in the superconducting state, the voltage across it was zero. The Nottingham effect caused both the wire and the emitter to go over to the normal state and the voltage across them was therefore non-zero.

Later *Bergeret* et al. [12.11] investigated the heating of a niobium emitter in a wider range of electric fields. They observed that as the electric field at the emitter tip was varied in the range (3 to 6.5)$\cdot 10^7$ V/cm, the value of the energy $\langle \epsilon \rangle$ left on the point surface by each electron encreased linearly from about 0.13 to 0.27 eV.

Alekseyevsky [12.12] investigated the effect of the emission current on the critical parameters of superconducting tantalum. Current was made to flow through a wire cathode made of superconducting tantalum and the voltage drop across the wire was measured. The voltage applied between the cathode and the anode induced emission current from the cathode. If the current was zero, the cathode was in the superconducting state. If the emission current reached a certain value, a voltage appeared across the wire, which indicated that the wire was in the normal state. Moreover, on increasing the voltage between the cathode and the anode the voltage across the wire cathode increased, tending to a value characteristic of the normal state of tantalum. These results were consistently explained in terms of the Nottingham effect by *Litvinov* et al. [12.13]. The heat released on the cathode surface due to this effect is removed through the leads, whose temperature is equal to that of the helium bath. As the emission current is increased, more heat is released and a part of the wire cathode goes over to the normal state. With increasing cathode-anode voltage and, correspondingly, emission current, a larger and larger part of the cathode wire will go over to the normal state and the voltage drop across it will increase.

Thus, the studies [12.10-13] have shown that the assumption that the Nottingham effect can cause a conductor to go from the superconducting to normal state is valid.

It is interesting to estimate the field emission current density at which the superconducting state of an emitter collapses. The current density j_n at which the transition to the normal state occurs due to the Nottingham effect can be found from [12.14]

$$j_n r_e \simeq \frac{e(T_c - T_0)\lambda}{2kT^*} , \tag{12.2}$$

where T_0 is the initial temperature of the emitter. For a niobium point with tip radius $r_e \simeq 10^{-6}$ cm, $T_c - T_0 = 5$ K, $\lambda \simeq 1$ W/cmK and $T^* \simeq 10^3$ K this current density will be of the order of 10^5 A/cm^2.

12.1.3 Other Emission Effects

There exist two further effects that promote the superconducting-to-normal transition. Firstly, there is the self magnetic field at the point tip. At a certain critical magnetic field a metal loses its superconducting properties. This occurs when

$$j_n r_e = 2H_{cr} , \tag{12.3}$$

where H_{cr} is the critical self magnetic field. For a niobium point with $r_e \simeq 10^{-6}$ cm and $H_{cr} \simeq 1300$ Oe the transition current density is $j_n \simeq 10^9$ A/cm^2.

Secondly, there exists a limitation on the current density associated with the onset of the decomposition of Cooper pairs. This occurs at a certain current density j_n. For niobium $j_n \simeq 10^7$ A/cm^2.

Thus, the main cause of the collapse of the superconducting state of a field emission point is the Nottingham effect. This has to be taken into accout in analyzing the initiation of pulsed vacuum discharge with superconducting electrodes. This effect appears to explain the absence of any peculiarities in the current-voltage characteristics and emission properties of superconducting point cathodes compared to those in the normal state [12.4-6, 10, 11].

12.2 Field Emission Current Preceding the Explosion of a Point

As shown in Chap.5, the initiation of a vacuum discharge occurs due to heating and explosion of micropoints present on the cathode under the action of field emission current. By cooling the cathode or making it change to the superconducting state one can hinder the heating of these micropoints and so increase the pulsed breakdown strength of the vacuum gap.

Aksyonov [12.15] investigated the critical field emission currents at which points start exploding for tungsten, niobium and tantalum point cathodes with temperatures of 300 and 4.2 K and applied voltage pulse durations of 5-100 ns. He observed that the critical current for a cathode cooled to 4.2 K is a factor of two or more higher than that for the same cathode at room temperature. *Aksyonov* believes that the emitting surface area remains unchanged and so the current density increases by the same factor. He explained this increase by the expansion of the region in the superconducting cathode where the Nottingham effect takes place.

Shkuratov [12.16] carried out experiments with niobium, tungsten, tantalum and lead point cathodes at four fixed temperatures, T_c = 4.2, 15, 80, and 300 K. He showed convincinly that the superconducting state itself does not affect the critical field emission current density. For example, increasing T_c from 4.2 to 15 K caused practically no change in the critical current of all the investigated metals, in spite of the fact that all the points were initially in the superconducting state and then, after heating, in the normal state. For explosion delay times of the point $t_d \leq 100$ ns the relationship $j^2 t_d$ = const (Chap. 5) was satisfied as a rule, although the right-hand side increased with a decrease in the initial temperature. It was concluded that the mechanism of explosion of cathode micropoints does not change with temperature. For niobium points with T_c = 4.2 K the critical current was a factor of about 1.5 higher than for the case T_c = 80 K for the same t_d. When T_c was increased to 300 K, the factor of critical current increase was more than two.

In order to establish which mechanism of heating is responsible for the explosion of micropoints at cryogenic temperatures it is necessary to evaluate the time taken for the point to be heated from helium to room temperature. As the total explosion delay time is $t_d = t_r + t_e$, where t_r and t_e are the times of heating the point from helium to room temperature and from room temperature to explosion, respectively, t_r can be easily estimated by measuring the total time in one test, then t_e in another test, and finally subtracting the latter from the former. According to [12.15,16], t_r for niobium points is estimated as (5 to 7)·10^8 /j^2, i.e., for $j \simeq 10^8$ A/cm^2 the time t_r amounts to a few tens of nanoseconds.

It is interesting to see how a variation in the temperature of a long, thin niobium conductor from helium to room temperature will change its explosion delay time. In fact, this means finding the temperature dependence of the "action integral" for niobium (Sect.5.1). Analysis of the heat balance assuming that the heat source is Joule heating gives

$$\int_0^t j^2 \, dt = Af(T) , \tag{12.4}$$

for the action integral, where A is the action integral at room temperature and f(T) is a temperature function which has the values 2.1, 1.6, 1.35, 1.15 and 1.0 for T = 5, 50, 100, 200, 300 K, respectively.

From estimates of the action integral for an exploding conductor it follows that increasing the conductor temperature from helium to room temperature results in an increase in the current density at which the conductor explodes by a factor of 1.5. However, a more accurate estimate of the increase of the critical current with a decrease in temperature requires consideration of the Nottingham effect, heat conduction, temperature dependences of thermophysical constants, etc.

Aksyonov [12.15] observed a spontaneous rise of the field emission current from niobium and tungsten points at the peak of the applied voltage pulse before the explosion of the point. He interpreted the observed effect as due to a change in the emitter geometry due to elastic deformation of the emitter tip under the action of the electric field ponderomotive forces. However, the results obtained by *Shkuratov* [12.16] disprove this because the waveforms showing a spontaneous rise of current were reproducible in a number of shots for the same cathode. Moreover, there was no hysteresis. Therefore, the idea of a change in point shape during the extraction of the field emission current is untenable.

The origin of the spontaneous current rise is analyzed in [12.17]. It is suggested that this current rise is related to an increase of the cathode emission are due to a high temperature wave. This wave propagates from the point tip (where the current density is maximum) to its periphery and gives rise to an increase in the field-assisted thermionic emission current.

12.3 Characteristics of the Vacuum Discharge at Cryogenic Temperatures

12.3.1 Experimental Conditions

Practically important characteristics of a pulsed vacuum discharge are the breakdown delay time t_d as a function of the electric field and the commutation time t_c as a function of the cathode temperature T_c . In an experimental study of these characteristics for a variety of metals [12.16] hemispherical cathodes of radius 3 mm and wire cathodes of thickness up to 200 μm were used. As the largest micropoints are typically not more than 1 μm in size, the cathode surface can be considered quasiplane. The cathode and the anode were made of the same material. Experiments were carried out in a vacuum of up to 10^{-9} Torr after a careful conditioning of the electrodes.

The metals used as electrode materials had different values of the ratio γ of the resitstivity at room temperature to that just before the transition to the superconducting state. This quantity is dependent on two factors, namely the degree of order of the metal structure and the percentage of impurities of other metals, i.e., the spectroscopic purity of the metal. A spectroscopically pure metal may have a highly imperfect structure and, hence, a low γ value. The electrodes used were made of spectroscopically pure niobium with γ = 500, commercially pure niobium with γ = 4, and spectroscopically and commercially pure tungsten with γ = 1000 and 10, respectively. Lead and tantalum electrodes were also used. Most of the results were obtained with breakdown delay times not longer than 100 ns.

12.3.2 Experimental Results

The first experiments examined how the conditions of electrode conditioning affect the breakdown delay time as a fuction of the applied voltage amplitude (breakdown voltage). Conditioning was accomplished with high currents (200 to 400 A) and low currents (2 to 4 A). The difference between the t_d (E) characteristics were measured for these two cases was unimportant. Conditioning of electrodes by breakdowns increased the breakdown voltage by a factor of 2-2.5. Conditioning erodes a cathode and removes the contaminated surface layer that is responsible for a decrease in the electric field initiating a vacuum breakdown . The breakdown voltage is affected by the γ and the purity of the cathode material. For example cooling a cathode made of a spectroscopically pure metal to 4.2 K resulted in a 20%-30% increase of the pulsed breakdown voltage. For commercially pure metals, cooling reduced the breakdown voltage insignificantly. Also, for a diode with the cathode made of a spectroscopically pure metal, while the anode was commercially pure, cooling did not cause any noticeable increase in the breakdown voltage after 10^3 discharges. This is related to an intense material transfer from the anode to the cathode.

An important result of the experiments [12.16] is a strong dependence of the breakdown voltage on the electrode temperature observed for all the materials studied. Plots of the function t_d (E) for electrodes of spectroscopically pure niobium are given in Fig. 12.1. Increasing the electrode temperature from 4.2 to 15 K did not noticeably change the breakdown voltage for any of the materials used. This is evidence that the superconducting state of a cathode in itself does not decisively affect the pulsed breakdown voltage. We must therefore conclude that the duration of the superconducting-to-normal state transition is not longer than 10^{-9} s. Such a fast transition is due, firstly, to the Notting-

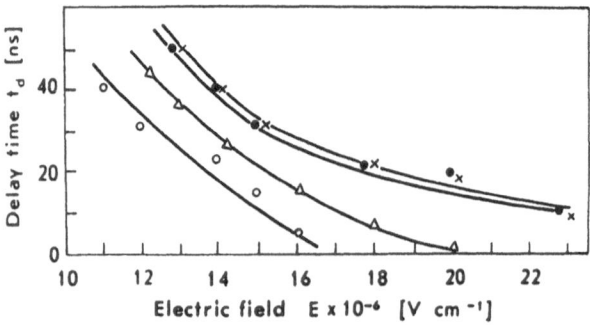

Fig.12.1. Breakdown delay time as a function of the electric field between Nb electrodes with γ = 300. $(\cdot)T_c$ = 4.2K; $(x)T_c$ = 15K; $(\Delta)T_c$ = 78K; $(o)T_c$ = 300K

ham heating of the emitter surface and, secondly, to the prebreakdown current density exceeding the critical current density of the superconductor [12.1].

Cooling a cathode does not affect the commutation time t_c or the velocity of expansion of the cathode plasma $v_c \simeq d/t_c$. For niobium and tungsten electrodes with $T_c = 4.2$ K this velocity is $3 \cdot 10^6$ and $2.6 \cdot 10^6$ cm/s, respectively. These values are in agreement with v_c measured at room temperature. The obtained data show that the explosive emission is independent of the cathode temperature T_c in the range 4.2 to 300 K.

12.4 Vacuum Discharge Between Electrodes Made of High-Temperature Superconductors

12.4.1 General Notions

Until recently, the transition temperatures of all known superconductors did not exceed 22 to 24 K. Therefore, the discovery of high-temperature superconductivity may reasonably be considered one of the outstanding scientific achievements of recent years. At the end of 1986 it was found [12.18] that the transition temperature of a mixture of copper oxides and rare-earth metals is 30 to 40 K.

At the beginning of 1987 a superconductor with a critical temperature of higher than 90 K was discovered [12.19]. This makes superconductivity at liquid nitrogen temperature possible. This superconductor is $YBa_2 Cu_3 O_y$ (y=7), an yttrium-barium ceramic with perovskite structure. Specimens of this ceramic are obtained by solid-phase synthesis from a mixture of superconducting oxides at temperatures of $900°$ -$1000°$ C in air or in an atmosphere with a controlled pressure of oxygen for tens of hours. A change in the sintering temperature, a deficit of oxygen, or for sintering too long, can cause specimens of this ceramic to change into semiconductors or dielectrics. Important features of the yttrium-barium ceramics are high resistivity, which in the normal state, is a few orders of magnitude higher than that of metals; low critical current densities, which are usually not in excess of 100 to 1000 A/cm^2; and low melting points ($\simeq 1600$ K). These properties make these ceramics promising as an eletrode material for vacuum discharges.

12.4.2 FEE from High-Temperature Superconducting Cathodes

The field electron emission from yttrium-barium high-temperature superconductors (HTSCs) was investigated on a set-up with a coaxial vacuum chamber made of stainless steel, which was impedance matched to the transmission line. The chamber design allowed the cathode temperature to be varied from 4.2 to 300 K. A pulsed generator produced voltage pulses with an amplitude of up to 50 kV and a rise time of ~10^{-9} s; the pulse duration was varied in the range of 10-1000 ns. Ceramic cathodes were manufactured mechanically; their tip radii were 5-50 μm. The transition temperature of the ceramic under investigation was 92 K with a transition width of 2-3 K. In addition, measurements for "classical" field emission points produced from the ceramic by electrochemical polishing were carried out. Part of these measurements were accomplished in a metal-glass electron-ion field emission microscope at a dc applied voltage. The energy distribution of field emission electrons was obtained using a dispersion energy analyzer with a resolution of 30 meV.

For fresh specimens, the field emission current from HTSC cathodes was unstable. After arc conditioning with currents not higher than 10^{-6} A for several hours, stable and reproducible current-voltage characteristics and stable field emission images of the emitting zone could be recorded. The current-voltage characteristics for T_c = 4.2, 80, 100, and 300 K were straight lines in Fowler-Nordheim coordinates. The number of emission centers was determined by the cathode size. On a cathode with a tip radius of about 10 μm a single emission center usually existed, while for a tip radius of about 50 μm two or three centers were observed. The energy distribution of field emission electrons was similar to that for tungsten cathodes. An abrupt fall in the high-energy region (0.09-0.11 eV), a width of 0.25-0.3 eV, a sharp increase in the width with cooling to 78 K and a Fermi level of 4.2 to 4.5 eV are characteristic of this distribution. All these data indicate that the HTSC ceramic has an electronic structure similar to that of metals.

However, there are some distinctions. For instance, the ceramic changes its emissive properties after being kept in vacuum for a prolonged time. The width of the energy distribution increases and no abrupt fall in the high-energy region is observed. This resembles the spectrum of electrons emitted by a semiconductor point. It can be stated that all the above changes are due to diffusion of oxygen from the ceramic's surface into vacuum, because a decrease of the content of lattice oxygen in the HTSC ceramic causes its electrophysical characteristics to become closer to those of semiconductors and dielectrics.

12.4.3 Vacuum Discharge

Let us now focus on the vacuum discharge between electrodes made of the HTSC ceramics. It is well-known that an increase of the electric field at cathode micropoints causes the field emission current to increase, and this leads to the explosion of the micropoints. The current density prior to the explosion as estimated from the prebreakdown current is an order of magnitude smaller than the corresponding value for niobium. From this observation it follows that a diode with HTSC ceramic electrodes should have low breakdown strength. The breakdown delay time t_d was measured as a function of the applied voltage, which was then converted into the macroscopic electric field E between the electrodes.

It should be noted that, firstly, discharge conditioning increased the pulsed breakdown strength of the HTSC ceramic cathodes by a factor of 3 to 4 and made it comparable with that for niobium and tungsten cathodes. Secondly, it was practically impossible to measure a dc current-voltage characteristic. With a very gradual increase of the voltage across the diode, current bursts on a level of 10^{-9} to 10^{-6} A were observed initially. Then the current sharply increased and a microdischarge occurred (the voltages were a factor of 8-10 smaller than the pulsed breakdown voltage). This may be due either to the mechanical properties of the cathode conditioned by dischargeds or to the influence of erosion products. The breakdown delay time was determined from the arrival of the pulse at the diode to the start of the current rise as a result of the formation and expansion of a CF plasma. Figure 12.2 shows the breakdown delay time as a function of the macroscopic electric field for different initial temperatures. On increasing E from $2 \cdot 10^6$ to 10^7 V/cm the time t_d decreased from tens to a few nanoseconds.

The velocity of expansion of the CF plasma into the interelectrode gap as determined by the commutation time from breakdown current oscilloscope traces, is $\sim 10^6$ cm/s, which is comparable with the

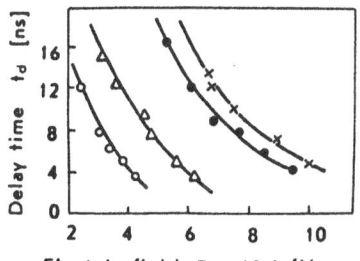

Fig.12.2. Breakdown delay time as a function of the electric field between ceramic electrodes. (x)T_c = 4.2K; (·)T_c = 78K; (Δ)T_c = 150K; (o)T_c = 300K

values for well-studied metals and is independent of the cathode temperature. The arc with HTSC ceramic electrodes is stable for hundreds of nonoseconds within current ranges of 300 to 400 A and 2 to 3 A. The mass loss from ceramic cathodes at pulsed arc discharges is much greater than that from a niobium cathode. This is due to a difference in the thermal and mechanical characteristics of these materials.

Ceramic high-temperature superconductors have a higher resistivity ($\kappa = 10^{-2} - 10^{-3}$ Ω/cm) and a lower thermal conductivity than niobium and tungsten. It is obvious that when the surface of a ceramic cathode remelts under the action of pulsed arc discharges in vacuum, a dielectric "crust" forms on the surface, which reduces the electric field at the protrusions underneath [12.20]. It is known that covering a cathode with a dielectric layer increases the breakdown strength of the diode. Cooling a cathode from 300 to 78 K results in an increase in the breakdown strength by a factor of 2 to 2.5 [12.14]. A more detailed investigation has shown that this increase is monotonic. This cannot be explained in terms of weaker Joule dissipation and stronger heat removal as in the case of a metal emitter, because the resistivity and thermal conductivity of the HTSC ceramic change insignificantly on cooling from 300 to 94 K. Such a noticeable effect of cooling on the pulsed breakdown strength seems to be associated with an increased critical current density of the dielectric crust. It is a factor of 10-20 smaller for semiconductors than for metals. This is in agreement with the following observation: with t_d = 300-500 ns the level of prebreakdown currents is 2-3 mA, which is a factor of 8-10 smaller than that for a niobium cathode.

The experiments have shown that the process of pulsed vacuum breakdown with HTSC ceramic electrodes has some differences from that with metal electrodes. The formation of a dielectric covering on the cathode surface makes it possible to use HTSC ceramic cathodes in high-voltage electrophysical devices.

As we mentioned above, the action integral plays an important role in the study of exploding conductors. This quantity is insensitive to a variation of the current density over a wide range. For most of the metals studied (copper, silver, iron, tungsten and others) it is of the order of 10^9 A$^2 \cdot$s$\cdot$cm^{-4}. Let us estimate the action integral for the yttrium-barium ceramic. If we restrict ourselves to small values of t_d ($\leq 10^{-8}$ s) and neglect heat conduction then the action integral for a thin, long wire can be calculated from (5.2). For the yttrium-barium ceramic the density ρ is 6.4 g/cm^3 and the melting point is 1600 K. Temperature dependences for heat capacity and resistivity can be found in [12.21, 22].

Our estimates show that the action integral for the HTSC ceramic under investigation is not more than 10^6 A$^2 \cdot$scm^{-4}, i.e., three orders of magnitude smaller than for many metals. This accounts for the low pre-explosion current densities and electric fields of a vacuum discharge with a HTSC ceramic cathode before it is conditioned with discharges.

References

Chapter 1

1.1 G.A. Mesyats, S.P. Bugaev, D.I. Proskurovsky: Pulse breakdown mechanism of short vacuum gaps in the nanosecond range. Proc. III Int'l Symp. on Discharges and Electrical Insulation in Vacuum (Paris, France 1968) pp.218-222

1.2 G.A. Mesyats, E.A. Litvinov, D.I. Proskurovsky: High-speed processes during pulse breakdown of vacuum gaps. Proc. IV Int'l Symp. on Discharges and Electrical Insulation in Vacuum (Waterloo, Ontario, Can. 1970) pp.82-95

1.3 G.A. Mesyats: The role of fast processes in vacuum breakdown. Proc. X Int'l Conf. on Phenomena in Ionized Gases (Oxford, UK 1971) pp.333-363

1.4 G.A. Mesyats, D.I. Proskurovsky: Pisma Zh. Eksp. Teor. Fiz. 13, 7-10 (1971) [Engl. transl.: JETP Lett. 13, 4 (1971)]

1.5 G.A. Mesyats: Electron explosive emission and electrical discharge in vacuum. Proc. VI Int'l Symp. on Discharges and Electrical Insulation in Vacuum (Swansea, UK 1974) Inv. Pap., pp. 21-47

1.6 S.P. Bugaev, E.A. Litvinov, G.A. Mesyats, D.I. Proskurovsky: Usp. Fiz. Nauk. 115, 101-121 (1975) [Engl. transl.: Sov. Phys.- Usp. 18, 51 (1975)]

1.7 G.A. Mesyats: Fundamental role of electron explosive emission during vacuum current commutation. Proc. XII Int'l Conf. on Phenomena in Ionized Gases (Eindhoven, Netherlands 1975)

1.8 E.A. Litvinov, G.A. Mesyats, D.I. Proskurovsky, E.B. Yankelevich: Cathode processes at electron explosive emission. Proc. VII Int'l Symp. on Discharges and Electrical Insulation in Vacuum (Novosibirsk, USSR 1976) pp. 55-69

1.9 G.A. Mesyats: Fast processes on the cathode in a vacuum discharge. Proc. X Int. Symp. on Discharges and Electrical Insulation in Vacuum (Columbia, SC 1982) pp.37-42

1.10 E.A. Litvinov, G.A. Mesyats, D.I. Proskurovsky: Usp. Fiz. Nauk 139, 265-302 (1983) [Engl. transl.: Sov. Phys.-Usp. 26, 138 (1983)]

1.11 E.A. Litvinov: IEEE Trans. EI-20, 683 (1985)

1.12 G.A. Mesyats: IEEE Trans. EI-20, 729 (1985)

Chapter 2

2.1 I.N. Slivkov, V.I. Mikhailov, N.I. Sidorov, A.I. Nastyukha: *Electrical Breakdown and Discharge in a Vacuum* (Atomizdat, Moscow 1966) [Engl. transl.: NTIS AD 745471]

2.2 R. Hawley, A. Maitland: *Vacuum as an Insulator* (Chapman & Hall, London 1967)

2.3 I.N. Slivkov: *Electrical Insulation and Discharges in Vacuum* (Atomizdat, Moscow 1972) (in Russian)

2.4 A.S. Denholm: A perspective of the first four symposia. Proc. V. Int'l Symp. on Discharge and Electrical Insulation in Vacuum (Poznan, Poland 1972) pp.21-33

2.5 P.A. Chatterton: Vacuum breakdown, in *Electrical Breakdown in Gases*, ed. by J.M. Meek, J.D. Craggs (Wiley, New York 1978) Chap.2, pp.129-208

2.6 G.A. Farrall: Electrical breakdown in vacuum, in *Vacuum Arcs*, ed. by J.M. Lafferty (Wiley, New York 1980) Chap.2, pp.20-80

2.7 R.V. Latham: *High Voltage Vacuum Insulation* (Academic, London 1981)

2.8 P.A. Jacquet: *Le Polissage Electrolitique et Chemique* (Editions Métaux, Saint-Germain-en-Laye, 1950)

2.9 H.W. Anderson: Rev. Sci. Instrum. 6, 309 (1935)

2.10 A.S. Pokrovskaya-Soboleva, V.V. Kraft, T.S. Borisova, L.I. Mazurova, V.M. Stuchenkov: Zh. Tekh. Fiz. 42, 1318 (1972) [Engl. transl.: Sov. Phys.-Tech. Phys. 17, 1047 (1972)

2.11 P.N. Chistyakov, A.L. Radionovsky, N.V. Tatarinova, N.E. Novikov, D.S. Treshchikova: Zh. Tekh. Fiz. 42, 821 (1972) [Engl. transl.: Sov. Phys.-Tech. Phys. 17, 646 (1972)

2.12 A.L. Radionovsky, D.S. Treshchikova: Complex conditioning method in the vacuum electrical breakdown studies. Proc. VII Int'l Symp. on Discharges and Electrical Insulation in Vacuum (Novosibirsk, USSR 1976) pp.446-449

2.13 G.E. Vibrans: J. Appl. Phys. 35, 2855 (1964)

2.14 P.A. Chatterton: Proc. Phys. Soc. (London) 88, 241 (1966)

2.15 F. Rohrbach: *Mechanisms Leading to the Formation of the Electric Spark at Very High Voltage and Ultra-High Vacuum* (Rep. CERN 71-28, Geneva 1971) [Engl. transl.: NTIS]

2.16 G.N. Fursey, P.N. Vorontsov-Vel'yaminov: Zh. Tekh. Fiz. 37, 1870 (1967) [Engl. transl.: Sov. Phys.-Tech. Phys. 12, 1370 (1967)

2.17 H.C. Miller: J.Appl. Phys. 55, 158 (1984)

2.18 R.H. Fowler, L. Nordheim: Proc. Roy. Soc (London) A119, 173 (1928)

2.19 D. Alpert, D.A. Lee, E.M. Lyman, H.E. Tomaschke: J. Vac. Sci. Technol. 1, 35 (1964)

2.20 H.E. Tomaschke, D. Alpert: J. Appl. Phys. 38, 881 (1967)

2.21 G.A. Farrall: J. Appl. Phys. 41, 563 (1970)

2.22 G.J. Bennete, R.W. Strayer, E.C. Cooper, L.W. Swanson: J. AIAA 3, 284 (1965)

2.23 G.P. Beukema: Physica 61, 259 (1972)

2.24 B.M. Cox, W.T. Williams: J. Phys. D 10, 15 (1977)

2.25 B.M. Cox: J. Phys. D 7, 143 (1974)

2.26 R.P. Little, W.T. Whitney: J. Appl. Phys. 34, 2430 (1963)

2.27 R.P. Little, S.T. Smith: IEEE Trans. ED-12, 77 (1965)

2.28 B.M. Cox: J. Phys. D 8, 2065 (1975)

2.29 G.A. Farrall, M. Owens, F.G. Hudda: J. Appl. Phys. 46, 610 (1975)

2.30 C.S. Athwal, K.H. Bayliss, R.Calder, R.V. Latham: IEEE Tans. PS-13, 226 (1985)

2.31 K.H. Bayliss, R.V. Latham: Proc. Roy. Soc. (London) A403, 285 (1986)

2.32 R.J. Noer, Ph. Niedermann, N.Sankarraman, O. Fischer: J. Appl. Phys. 59, 3851 (1986)

2.33 R.V. Latham: Vacuum 32, 137 (1982)

2.34 H.P.S. Powell, P.A. Chatterton: Vacuum 20, 419 (1970)

2.35 C.S. Walters, M.W. Fox, R.V. Latham: J. Phys. D 7, 911 (1974)

2.36 B. Jüttner, W. Rohrbeck, H. Wolf: Pressure dependence of prebreakdown current due to sorption processes. Proc. V Int'l Symp. on Discharges and Electrical Insulation in Vacuum (Poznan, Poland 1972) pp.65-69

2.37 G.N. Fursey, G.K. Kartsev: Zh. Tekh. Fiz. **40**, 310 (1970) [Engl. transl.: Sov. Phys.-Tech. Phys. **15**, 225 (1970)]

2.38 M. Goldman, A. Loubiere, M. Boutteau, P. Herion: Microdischarges. Proc. III Int'l Symp. on Discharges and Electrical Insulation in Vacuum (Paris, France 1968) pp.69-74

2.39 H.E. Tomaschke, D. Alpert: J. Vac. Sci. Tech. **4**, 192 (1967)

2.40 N.V. Tatarinova: Nature of microdischarges. Proc. XI Int'lSymp. on Discharges and Electrical Insulation in Vacuum (Berlin, GDR 1984) pp. 45-46

2.41 A. Maitland: J. Appl. Phys. **32**, 2399 (1961)

2.42 N.B. Rozanova, V.L. Granovsky: Zh. Tekh. Fiz. **26**, 489 (1956) [Engl. transl.: Sov. Phys.-Tech. Phys. **1**, 471 (1956)]

2.43 A.W. Hull, E.E. Burger: Phys. Rev. **31**, 1121 (1928)

2.44 L.B. Snoddy: Phys. Rev. **38**, 1678 (1931)

2.45 J.M. Meek, J.D. Craggs: *Electrical Breakdown of Gases* (Oxford Univ. Press, London 1953) pp. 152

2.46 D.H. Goodman, D.H. Sloan: Phys. Rev. **82**, 575 (1951)

2.47 I.N. Slivkov: Zh. Tekh. Fiz. **27**, 2081 (1957) [Engl. transl.: Sov. Phys.-Tech. Phys. **2**, 1928 (1957)]

2.48 W.S. Boyle, P. Kisliuk, L.H. Germer: J. Appl. Phys. **26**, 720 (1955)

2.49 W.J. Wijker: Appl. Sci. Res. B **9**, 1 (1961)

2.50 J.M. Somerville, C.T. Graiger: Br. J. Appl. Phys. **7**, 400 (1956)

2.51 W.P. Dyke, J.K. Trolan: Phys. Rev. **89**, 799 (1953)

2.52 D. Leader: Proc. IEE (London) **100**, 2A, 138 (1953)

2.53 L.V. Tarasova, V.G. Kalinin: Zh. Tekh. Fiz. **34**, 666 (1964) [Engl. transl.: Sov. Phys.-Tech. Phys. **9**, 514 (1964)]

2.54 A.S. Denholm: Can. J. Phys. **36**, 476 (1958)

2.55 G.A. Farrall, H.C. Miller: J. Appl. Phys. **36**, 2966 (1965)

2.56 H.C. Miller, G.A. Farrall: J. Appl. Phys. **36**, 1338 (1965)

2.57 I.I. Kalyatsky, G.M. Kassirov: Zh. Tekh. Fiz. **34**, 348 (1964) [Engl. transl.: Sov. Phys.-Tech. Phys. **9**, 274 (1964)]

2.58 I.I. Kalyatsky, G.M. Kassirov: Izv. Vyssh. Uchebn. Zaved. Fiz. **4**, 78 (1963)

2.59 N.F. Olendzskaya, M.A. Sal'man: Zh. Tekn. Fiz. **40**, 333 (1970) [Engl. transl.: Sov. Phys.-Tech. Phys. **15**,242 (1970)]

2.60 D. König, Hj. Schmidt: Impulse breakdown voltage and prebreakdown currents of prestressed circuit breaker vacuum tubes. Proc. XII Int'l Symp. on Discharges and Electrical Insulation in Vacuum (Shoresh, Israel 1986) pp. 59-64

2.61 I.D. Chalmers, B.D. Phukan: Vacuum **32**, 145 (1982)

2.62 J.A. Chiles: J. Appl. Phys. **8**, 622 (1937)

2.63 P.A. Chatterton, W.A. Smith, R. Cooper, C.T. Elliot, D.L. Pulfrey: An investigation into a Kerr cell system for photographing the prebreakdown phase of very short vacuum gaps. Proc. II Int'l Symp. on Insulation of High Voltage in Vacuum (MIT, Cambridge, Mass. 1966) pp.61-65

2.64 R.W. Wood: Phys. Rev. Ser. 1 **5**, 1 (1897)

2.65 G. Thomer: Pulsed X-ray engineering, in *High-Speed Physics I*, ed. by K. Vollrath, G. Thomer (Springer, Wien 1967) pp.336-381

2.66 V.S. Tsukerman, M.A. Manakova: Zh. Tekh. Fiz. **27**, 391 (1957) [Engl. transl.: Sov. Phys.-Tech. Phys. **2**, 353 (1957)]

2.67 M.F. Jamet: Compt. rend. **264**, 1186 (1967)

2.68 K.F. Zelensky, O.P. Pechersky, V.A. Tsukerman: Zh. Tekh. Fiz. **38**, 1581 (1958) [Engl. transl.: Sov. Phys.-Tech. Phys. **13**, 1284 (1969)]

2.69 W.P. Dyke, J.K. Trolan, E.E. Martin, J.P. Barbour: Phys. Rev. **91**, 1043 (1953)

2.70 W.W. Dolan, W.P. Dyke, J.K. Trolan: Phys. Rev. **91**, 1054 (1953)

271

2.71 V.A. Gor'kov, M.I. Elinson, G.D. Yakovleva: Radiotekh. Electron. 7, 1501 (1962) [Engl. transl.: Radio Eng. Electron. Phys. 7, 1404 (1962)]
2.72 D. Alpert, D.A. Lee, E.U. Lyman, H. Tomaschke: J. Vac. Sci. Tech. 1, 35 (1964)
2.73 I.L. Sokol'skaya, G.N. Fursey: Radiotekh. Electron. 7, 1474 (1962) [Engl. transl.: Radio Eng. Electron. Phys. 7, 1387 (1962)]
2.74 P. Kranjec, L. Ruby, J. Vac. Sci. Tech. 4, 94 (1967)
2.75 J. Brodie: J. Vac. Sci. Tech. 3, 222 (1966)
2.76 S.M. Bragin, A.V. Valter, N.N. Semyonov: Theory and Practice of the Breakdown of Dielectrics (Gosizdat, Moscow 1929) (in Russian)
2.77 M. Goldman, A. Goldman: Compt. rend. 253, 2654 (1961)
2.78 F.M. Charbonnier, C.J. Bennette, L.W. Swanson: J. Appl. Phys. 38, 627 (1967)
2.79 C.J. Bennette, L.W. Swanson, F.M. Charbonier: J. Appl. Phys. 38, 634 (1967)
2.80 T. Utsumi: J. Appl. Phys. 38, 2987 (1967)
2.81 D.K. Davies, M.A. Biondi: J. Appl. Phys. 39, 2979 (1968)
2.82 L. Cranberg: J. Appl. Phys. 23, 518 (1952)
2.83 N.B. Rozanova: Izv. Akad. Nauk. SSSR Ser. Fiz. 26, 1438 (1962) [Englisch transl.: Bull. Acad. Sci. USSR, Phys. Ser. Eng. Ed. 26, 1462 (1962)]
2.84 D.K. Davies, M.A. Biondi: J. Appl. Phys. 42, 3089 (1971)
2.85 D.K. Davies, M.A. Biondi: Dynamics of heating of anode particles in vacuum breakdown. Proc. VII Int'l. Symp. on Discharges and Electrical Insulation in Vacuum (Novosibirsk, USSR 1976) pp. 121-126
2.86 M. Steenbeck: Wiss. Veröff. Simens-Werken 17, 363 (1938)
2.87 K.H. Kingdon, H.E. Tanis: Phys. Rev. 53, 128 (1938)
2.88 E. Fünfer. Z. Angew. Phys. 5, 426 (1953)
2.89 P.T.G. Flynn: Proc. Phys. Soc. (London) 69B, 748 (1956)
2.90 G.M. Kassirov, G.A. Mesyats: Zh. Tekh. Fiz. 34, 1476 (1964) [Engl. transl.: Sov. Phys.-Tech. Phys. 9, 1141 (1964)]
2.91 I.G. Kesaev: Cathode Processes in Electric Arcs (Nauka, Moscow 1968) [Engl. transl.: SAND 78-6011, NTIS]
2.92 V.I. Rakhovsky: Physical Basis of the Commutation of Electric Current in a Vacuum (Nauka, Moscow 1970) [Engl. transl.: NTIS Rep. AD 773868 (1973)]
2.93 G.A. Farrall: Arc ignition processes, in Vacuum Arcs, ed. by J.M. Lafferty (Wiley-Interscience, New York 1980) pp. 81-119
2.94 G.A. Lyubimov, V.I. Rakhovsky: Usp. Fiz. Nauk 125, 665 (1978) [Engl. transl.: Sov. Phys.-Usp. 21, 693 (1978)]
2.95 L.P. Harris: Arc cathode phenomena, in Vacuum Arcs, ed. by J.M. Lafferty (Wiley-Interscience, New York 1980) pp.120-168
2.96 G. Ecker: Theoretical aspects of the vacuum arc, in Vacuum Arcs, ed. by J.M. Lafferty (Wiley-Interscience, New York 1980) pp.228-320
2.97 Yu.P. Rylov, Z.A. Pigulevskaya: Zh. Tekh. Fiz. 41, 2466 (1971) [Engl. transl.: Sov. Phys.- Tech. Phys. 16, 1957 (1971)]
2.98 C.C. Sanger, P.E. Secker: J. Phys. D. 4, 1939 (1971)
2.99 J.E. Daalder: J. Phys. D. 8, 1647 (1975)
2.100 A.A. Plyutto, V.N. Ryzhkov, A.T. Kapin: Zh. Eksp. Teor. Fiz. 47, 494 (1964) [Engl. transl.: Sov. Phys. - JETP 20, 328 (1965)]
2.101 J.D. Cobine, T.A. Vanderslice: AIEE Trans. Comm. Electr. 82, 240 (1963)
2.102 C.W. Kimblin: J. Appl. Phys. 44, 3074 (1973)
2.103 V.V. Kantsel, T.S. Kurakina, V.S. Potokin, V.I. Rakhovsky, L.G. Tkachev: Sov. Phys. - Tech. Phys. 13, 814 (1968)
2.104 B.N. Klyarfeld, N.A. Neretina, N.N. Druzhinina: Sov. Phys. - Tech. Phys. 14, 796 (1969)

2.105 I. Kutzner, Z. Zalucki: Electrode erosion in the vacuum arc. Proc. IV Int'l Conf. on Gas Discharges (London 1970) p.87

2.106 G.R. Mitchell: Proc. IEE 117, 2315 (1970)

2.107 M.P.. Reeece: Proc. IEE 110, 793 (1963)

2.108 W.G.J. Rondeel: J. Phys. D 6, 1705 (1973)

2.109 Z. Zalucki, J. Kutzner: Streams of neutral particles from the anode in vacuum arc. Proc. V Int'l Symp. on Discharges and Electrical Insulation in Vacuum (Poznan, Poland 1972) pp.275-281

2.110 W.D. Davis, H.C. Miller: J. Appl. Phys. 40, 2212 (1969)

2.111 D.T. Tuma, C.L. Chen, D.K. Davies: J. Appl. Phys. 49, 3821 (1978)

2.112 R. Tannberg: Phys. Rev. 35, 1080 (1930)

2.113 I.G. Nekrashevich, I.A. Bakuto: Inzh. Fiz. Zh. 11, 59 (1959)

2.114 J.A. Rich: J. Appl. Phys. 32, 1023 (1961)

2.115 V.E. Il'in, S.B. Lebedev: Zh. Tekh. Fiz. 32, 986 (1962) [Engl. transl.: Sov. Phys.-Tech. Phys. 7, 717 (1962)]

2.116 I.M. Tsinman: Radiotekh. Elektron. 8, 834 (1963)

2.117 J. Rothstein: The arc spot as a steady-state exploding wire phenomena, in *Exploding Wires*, Vol.3, ed. by W.G. Chace, H.K. Moore (Plenum, New York 1964) pp.115-124

2.118 G.W. McClure: J. Appl. Phys. 45, 2078 (1974)

2.119 T.H. Lee, A. Greenwood: J. Appl. Phys. 32, 916 (1961)

2.120 A.G. Goloveiko: Inzh. Fiz. Zh. 14, 478 (1968)

2.121 I.I. Beilis, G.A. Lyubimov, V.I. Rakhovsky: Dokl. Akad. Nauk SSSR 188, 202 (1969) [Engl. transl.: Sov. Phys.-Dokl. 14, 897 (1969)]

2.122 I.I. Beilis, G.A. Lyubimov, V.I. Rakhovsky: Dokl. Akad. Nauk SSSR 203, 71 (1972) [Engl. transl.: Sov. Phys.-Dokl. 17, 225 (1972)]

2.123 G. Ecker: Beitr. Plasma Phys. 11, 405 (1971)

2.124 G. Ecker: Z. Naturforsch. 26a, 935 (1971)

2.125 V.A. Nemchinsky: Zh. Tekh. Fiz. 49, 1373 (1979) [Engl. transl.: Sov. Phys.-Tech. Phys. 24, 764 (1979)]

2.126 V.A. Nemchinsky: Zh. Tekh. Fiz. 49, 1379 (1979) [Engl. transl.: Sov. Phys. - Tech. Phys. 24, 767 (1979)]

2.127 I.Kh. Bek-Bulatov, M.Yu. Borukhov, R.V. Nagaibekov: Zh. Tekh. Fiz. 43, 2211 (1973) [Engl. transl.: Sov. Phys.-Tech. Phys. 18, 1401 (1973)]

2.128 E. Hantzsche: Phys. Lett. 50A, 219 (1974)

2.129 G.N. Fursey, P.N. Vorontsov-Vel'yaminov: Zh. Tekh. Fiz. 37, 1880 (1967) [Engl. transl.: Sov. Phys.-Tech. Phys. 12, 1377 (1967)]

2.130 J. Mitterauer: Acta Phys. Austr. 37, 175 (1973)

2.131 G. Ecker: *Electrode Components of the Arc Discharge* (Springer, Berlin, Göttingen 1961) pp.1-104

2.132 G. Ecker: Z. Naturforsch. 28a, 417 (1973)

2.133 G. Ecker: IEEE Trans. PS-4, 218 (1976)

2.134 G. Ecker: Z. Naturforsch. 28a, 428 (1973)

Chapter 3

3.1 G.A. Vorob'yov, G.A. Mesyats: *High-Voltage, Nanosecond Pulse Forming Engineering* (Gosatomizdat, Moscow 1963) (in Russian)

3.2 G.A. Mesyats: *Generation of High-Power Nanosecond Pulses* (Sov. Radio, Moscow 1974) (in Russian)

3.3 V.L. Auslender, O.G. Il'in, A.M. Shenderovich: Prib. Tekh. Eksp. 2, 173-175 (1963)

3.4 D.I. Proskurovsky, E.B. Yankelevich: Prib. Tekh. Eksp. 5, 108-111 (1973)
3.5 R.H. Huddlestone, S.L. Leonard (eds.): *Plasma Diagnostic Techniques* (Academic, New York 1965)
3.6 M.M. Butslov, B.M. Stepanov, S.D. Fanchenko: *Electro-Optical Image Converters and Their Use in Scientific Research* (Nauka, Moscow 1978) (in Russian)
3.7 Z.A. Al'bikov, A.I. Veretennikov, O.V. Kozlov: *Pulsed-Ionizing-Radiation Detectors* (Atomizdat, Moscow 1978) (in Russian)
3.8 S.P. Bugaev, A.M. Iskol'dsky, G.A. Mesyats: Zh. Tekh. Fiz. 37, 1855-1860 (1967) [Engl. Transl.: Sov. Phys.-Tech. Phys. 12, 1358-1362 (1967)]
3.9 Ya.Ya. Yurike: Optical studies of vacuum breakdown development at steady voltage. Proc. V Int'l Symp. on Discharges and Electrical Insulation in Vacuum (Poznan, Poland 1972) pp.111-114
3.10 Ya.Ya. Yurike, V.F. Puchkarev, D.I. Proskurovsky: Izv. Vyssh. Uchebn. Zaved. Fiz. 3, 12-16 (1973) [Engl. transl.: Sov. Phys. J. 16, 293-297 (1973)]
3.11 H.R. Griem: *Plasma Spectroscopy* (McGraw Hill, New York 1964)
3.12 W. Lochte-Holtgreven (ed.): *Plasma Diagnostics* (North-Holland, Amsterdam 1968)
3.13 U. Ascoli-Bartoli, G. Benedetty-Michelangeli, L. Lovisetto: Appl. Phys. Lett. 8, 332 (1966)
3.14 D.I. Proskurovsky, Ya.Ya. Yurike: Izv. Vyssh. Uchebn. Zaved. Fiz. 9, 93-97 (1971) [Engl. transl.: Sov. Phys. J. 14, 1238-1242 (1971)]
3.15 W.P. Dyke, J.K. Trolan, W.W. Dolan, G. Barnes: J. Appl. Phys. 24, 570-576 (1953)
3.16 Yu.P. Danilov, M.P. Skvortsov: Prib. Tekh. Eksp. 3, 220-222 (1966)

Chapter 4

4.1 G.M. Kassirov, G.A. Mesyats: Zh. Tekh. Fiz. 34, 1476-1481 (1964) [Engl. transl.: Sov. Phys. Tech. Phys. 9, 1141-1145 (1964)]
4.2 G.M. Kassirov, B.M. Koval'chuk: Zh. Tech. Fiz. 34, 484-487 (1964) [Engl. transl.: Sov. Phys.-Tech. Phys. 9, 377-379 (1964)]
4.3 S. Dushman: *Scientific Foundations of Vacuum Technique*, 2nd. ed. (Wiley, New York 1962) Chap.6, pp.331-379
4.4 G.A. Mesyats. S.P. Bugaev, D.I. Proskurovsky, V.I. Eshkenazi, Ya.Ya. Yurike: Radiotekh. Elektron. 14, 2222-2230 (1969) [Engl. transl.: Radio Eng. Electron. Phys. 14, 1919-1925 (1969)]
4.5 G.A. Mesyats, S.P. Bugaev, D.I. Proskurovsky, V.I. Eshkenazi, Ya.Ya. Yurike: Study of initiation and development of short vacuum gaps pulse breakdown in nanosecond range. Proc. III Int'l Symp. on Discharges and Electrical Insulation in Vacuum (Paris, France 1968) pp.212-217
4.6 G.A. Mesyats, D.I. Proskurovsky: Izv. Vyssh. Uchebn. Zaved. Fiz. 1, 81-85 (1968) [Engl. transl.: Sov. Phys. J. 11, 49-51 (1968)]
4.7 S.P. Vavilov, G.A. Mesyats: Izv. Vyssh. Uchebn. Zaved. Fiz. 8, 90-94 (1970) [Engl. transl.: Sov. Phys. J. 13, 1058-1061 (1970)]
4.8 I.I. Kalyatsky, G.M. Kassirov, G.V. Smirnov, N.N. Frolov: Zh. Tekh. Fiz. 45, 1547-1550 (1975) [Engl. transl.: Sov. Phys.-Tech. Phys. 20, 988-989 (1975)]
4.9 H.C. Kärner, H.G. Bender: Breakdown of large vacuum gaps under lighting impulse stress. Proc. XII Int'l Symp. on Discharges and Electrical Insulation in Vacuum (Shoresh, Israel 1986) pp.54-58
4.10 S.P. Bugaev, A.M. Iskol'dsky, G.A. Mesyats, D.I. Proskurovsky: Zh. Tekh. Fiz. 37, 2206-2208 (1967) [Engl. transl.: Sov. Phys.-Tech. Phys. 12, 1625-1627 (1967)]

4.11   Ya.Ya. Yurike: Optical studies of vacuum breakdown development at steady voltage. Proc. V Int'l Symp. on Discharges and Electrical Insulation in Vacuum (Poznan, Poland 1972) pp.111-114

4.12   L.B. Snoddy: Phys. Rev. **38**, 1678 (1931)

4.13   J.A. Chiles: J. Appl. Phys. **8**, 622-626 (1937)

4.14   P.A. Chatterton, W.A. Smith, R. Cooper, C.T. Elliot, D.L. Pulfrey: An investigation into a Kerr cell system for photographing the prebreakdown phase of very short vacuum gaps. Proc. II Int'l Symp. on Insulation of High Voltage in Vacuum (MIT, Cambridge, Mass. 1966) pp.61-65

4.15   A. Maitland, R. Hawley: Vacuum **18**, 403-408 (1968)

4.16   R.K. Parker, R.E. Anderson, C.V. Duncan: J. Appl. Phys. **45**, 2463-2478 (1974)

4.17   I.D. Chalmers, B.D. Phukan: J. Phys. D **12**, 1285-1292 (1979)

4.18   J.D. Cross, B. Mazurek : Fast cathode processes in vacuum discharge development. Proc. XII Int'l Symp. on Discharges and Electrical Insulation in Vacuum (Shoresh, Israel 1986) pp.47-48

4.19   R.B. Baksht, G.M. Kassirov, G.V. Smirnov, F.G. Sekisov: Izv. Vyssh. Uchebn. Zaved. Fiz. 7, 130-132 (1975) [Engl. transl.: Sov. Phys. J. **18**, 1021-1023 (1975)]

4.20   G.M. Kassirov, F.G. Sekisov: Zh. Tekh. Fiz. **53**, 1279-1283 (1983)

4.21   I.I. Kalyatsky, G.M. Kassirov, F.G. Sekisov: IEEE Trans. EI-20, 701-703 (1985)

4.22   G.A. Mesyats, V.I. Eshkenazi: Izv. Vyssh. Uchebn. Zaved. Fiz. 2, 123-125 (1968) [Engl. transl.: Sov. Phys. J. **11**, 78-82 (1968)]

4.23   A. Maitland: J. Appl. Phys. **32**, 2399-2407 (1961)

4.24   L.V. Tarasova, A.A. Razin: Zh. Tekh. Fiz. **29**, 967-972 (1959) [Engl. Transl.: Sov. Phys.-Tech. Phys. **4**, 879-885 (1959)]

4.25   G.P. Bazhenov, G.A. Mesyats. D.I. Proskurovsky: Izv. Vyssh. Uchebn. Zaved. Fiz. **8**, 87-90 (1970) [Engl. transl.: Sov. Phys. J. **13**, 1054-1057 (1970)]

4.26   R. Hawley: Vacuum **11**, 32-35 (1961)

4.27   R.B. Baksht, G.A. Mesyats: Izv. Vyssh. Uchebn. Zaved. Fiz. 7, 144-146 (1970) [Engl. transl.: Sov. Phys. J. **13**, 969-971 (1970)]

4.28   D.I. Proskurovsky, Ya.Ya. Yurike: Izv. Vyssh. Uchebn. Zaved. Fiz. 9, 93-97 (1971) [Engl. transl.: Sov. Phys. J. **14**, 1238-1242 (1970)]

4.29   R.B. Baksht, S.P. Vavilov, M.N. Urbazaev: Izv. Vyssh. Uchebn. Zaved. Fiz. 2, 140-141 (1973) [Engl. transl.: Sov. Phys. J. **16**, 266-267 (1973)]

4.30   B. Bernstein, I. Smith: IEEE Trans. NS-20, 294-300 (1973)

4.31   G.A. Mesyats, S.P. Bugaev, D.I. Proskurovsky: Pulse breakdown mechanism of short vacuum gaps in nanosecond range. Proc. III Int'l Symp. on Discharges and Electrical Insulation in Vacuum (Paris, France 1968) pp.218-222

4.32   G.A. Mesyats, E.A. Litvinov, D.I. Proskurovsky: High-speed processes during pulsed breakdown of vacuum gaps. Proc. IV Int'l Symp. on Discharges and Electrical Insulation in Vacuum (Waterloo, Canada 1970) pp.82-91

4.33   G.A. Mesyats: The role of fast processes in vacuum breakdown. Proc. X Int'l Conf. on Phenomena in Ionized Gases (Oxford, UK 1971) pp.333-363

4.34   G.A. Mesyats, D.I. Proskurovsky: Pisma Zh. Eksp. Teor. Fiz. 13, 7-10 (1971) [Engl. transl.: JETP Lett. **13**, 4-6 (1971)]

4.35   G.N. Fursey, P.N. Vorontsov-Vel'yaminov: Zh. Tekh. Fiz. **37**, 1880-1888 (1967) [Engl. transl.: Sov. Phys.-Tech. Phys. **12**, 1377-1382 (1967)]

4.36   K.F. Zelensky, O.P. Pechersky, V.A. Tsukerman: Zh. Tekh. Fiz. **38**, 1581-1587 (1968) [Engl. transl.: Sov. Phys.-Tech. Phys. **13**, 1284-1289 (1968)]

4.37   M.F. Jamet: C. R. **264**, 1186-1188 (1967)

4.38   I.D. Chalmers, B.D. Phukan: Vacuum **32**, 145 (1982)

4.39 I.N. Slivkov: *Electrical Insulation and Discharge in Vacuum* (Atomizdat, Moscow 1972) pp.126-131 (in Russian)
4.40 N.B. Rozanova, V.L. Granovsky: Zh. Tekh. Fiz. **26**, 489-496 (1956) [Engl. transl.: Sov. Phys.-Tech. Phys. **1**, 471-478 (1956)]
4.41 J.M. Somerville, C.T. Grainger: Br. J. Appl. Phys. **7**, 400-405 (1956)

Chapter 5

5.1   G.K. Kartsev, G.A. Mesyats, D.I. Proskurovsky, V.P. Rotshtein, G.N. Fursey: Dokl. Akad. Nauk SSSR **192**, 309-312 (1970) [Engl. transl.: Sov. Phys.-Dokl. **15**, 475-477 (1970)]
5.2   G.A. Mesyats, V.P. Rotshtein, G.N. Fursey, G.N. Kartsev: Zh. Tekh. Fiz. **40**, 1551-1553 (1970) [Engl. transl.: Sov. Phys.-Tech. Phys. **15**, 1202-1204 (1970)]
5.3   B.A. Koval, D.I. Proskurovsky, V.P. Rotshtein: Prib. Tekh. Eksp. **4**, 243-247 (1979)
5.4   W.P. Dyke, J.K. Trolan: Phys. Rev. **89**, 799-808 (1953)
5.5   M. Drechler, E. Henkel: Z. Angew. Phys. **6**, 341-346 (1954)
5.6   I.L. Sokol'skaya, G.N. Fursey: Radiotekh. Elektron. **7**, 1474-1484 (1962) [Engl. transl.: Radio Eng. Electron. Phys. **7**, 1387-1394 (1962)]
5.7   V.A. Gor'kov, M.I. Elinson, G.D. Yakovleva: Radiotekh. Elektron. **7**, 1501-1510 (1962) [Engl. transl.: Radio Eng. Electron. Phys. **7**, 1404-1408 (1962)]
5.8   E.A. Litvinov, G.A. Mesyats, A.F. Shubin: Izv. Vyssh. Uchebn. Zaved. Fiz. **4**, 147-151 (1970) [Engl. transl.: Sov. Phys. J. **13**, 537-540 (1970)]
5.9   E.A. Litvinov, A.F. Shubin: Izv. Vyssh. Uchebn. Zaved. Fiz. **1**, 152-154 (1974) [Engl. transl.: Sov. Phys. J. **17**, 1549-1552 (1974)]
5.10 E.A. Litvinov: IEEE Trans. IE-20, 683 (1985)
5.11 G.N. Fursey: IEEE Trans. IE-20, 659 (1985)
5.12 G. Anderson, F. Neilson: Application of the "integral of action" notion to exploding wires studies, in *Exploding Wires*, Vol.1, ed. by W.G. Chace, H.K. Moore (Plenum, New York; Chapman & Hall, London 1959) pp.88-92
5.13 L.V. Tarasova, L.N. Khudyakova: Surface Phenomena and Electric Discharges in Vacuum and Gases at Pulse Voltages, Proc. III. Int'l Symp. on Discharges and Electrical Insulation in Vacuum (Paris, France 1968) pp.62-68
5.14 L.P. Babich, M.D. Tarasov: Izv. Vyssh. Uchebn. Zaved. Radiofiz. **23**, 1365-1372 (1980) [Engl. transl.: Radiophys. Quant. Electron. **23**, 914-920 (1980)]
5.15 S.I. Lobov, N.G. Pavlovskaya: Izv. Vyssh, Uchebn. Zaved. Radiofiz. **23**, 1373-1377 (1980) [Engl. transl.: Radiophys. Quant. Electron. **23**, 921-924 (1980)]
5.16 E.A. Litvinov, G.A. Mesyats, D.I. Proskurovsky: Usp. Fiz. Nauk **139**, 265-302 (1983) [Engl. transl.: Sov. Phys.-Usp. **26**, 138-159 (1983)]
5.17 G.A. Mesyats, G.P. Bazhenov, S.P. Bugaev, D.I. Proskurovsky, V.P. Rotshtein, Ya.Ya. Yurike: Izv. Vyssh. Uchebn. Zaved. Fiz. **5**, 153-154 (1969) [Engl. transl.: Sov. Phys. J. **12**, 688-689 (1969)]
5.18 W.P. Dyke, J.K. Trolan, E.E. Martin, J.P. Barbour: Phys. Rev. **91**, 1043-1053 (1953)
5.19 W.W. Dolan, W.P. Dyke, J.K. Trolan: Phys. Rev. **91**, 1054-1057 (1953)
5.20 E.A. Litvinov, G.A. Mesyats, D.I. Proskurovsky, E.B. Yankelevich: Cathode processes at electron explosive emission. Proc. VII Int'l Symp. on Discharges and Electrical Insulation in Vacuum (Novosibirsk, USSR 1976) pp.55-69

5.21 G.P. Bazhenov, E.A. Litvinov, G.A. Mesyats, D.I. Proskurovsky, A.F. Shubin, E.B. Yankelevich: Zh. Tekh. Fiz. **43**, 1255-1261 (1973) [Engl. transl.: Sov. Phys. - Tech. Phys. **18**, 795-798 (1973)]

5.22 V.M. Zukov, G.N. Fursey: Zh. Tekh. Fiz. **46**, 310-318; 319-327 (1976) [Engl. transl.: Sov. Phys.-Techn. Phys. **21**, 176-181; 182-187 (1976)]

5.23 G.P. Bazhenov, E.A. Litvinov, G.A. Mesyats, D.I. Proskurovsky, A.F. Shubin, E.B. Yankelevich: Zh. Tekh. Fiz. **43**, 1262-1268 (1973) [Engl. transl.: Sov. Phys.-Techn. Phys. **18**, 799-802 (1973)

5.24 E.A. Litvinov, G.A. Mesyats, D.I. Proskurovsky, E.B. Yankelevich: Zh. Tekh. Fiz. **48**, 541-545 (1978) [Engl. transl.: Sov. Phys.-Tech. Phys. **23**, 319-321 (1978)]

5.25 G.A. Mesyats, D.I. Proskurovsky, E.B. Yankelevich: Cathode surface microrelief formation at the explosive electron emission. Proc. VII Int'l Symp. on Discharges and Electrical Insulation in Vacuum (Novosibirsk, USSR 1976) pp.230-233

5.26 G.A. Mesyats, V.I. Eshkenazi: Izv. Vyssh. Uchebn. Zaved. Fiz. 2, 123-125 (1968) [Engl. transl.: Sov. Phys. J. 11, 79-82 (1968)

5.27 A.P. Komar, N.N. Syutkin: Dokl. Akad. Nauk SSSR **158**, 821-823 (1964) [Engl. transl.: Sov. Phys.-Dokl. **9**, 872-874 (1964)]

5.28 A.A. Plyutto, V.N. Ryzhkov, A.T. Kapin: Zh. Eksp. Teor. Fiz. **47**, 494-507 (1964) [Engl. transl.: Sov. Phys.-JETP **20**, 328-337 (1965)]

5.29 J.E. Daalder: J. Phys. D **9**, 2379-2395 (1976)

5.30 D.I. Proskurovsky, E.B. Yankelevich, G.A. Koval: Radiotekh. Elektron. **21**, 342-349 (1976) [Engl. transl.: Radio Eng. Electron. Phys. **21**, 100-107 (1976)]

5.31 B. Jüttner, V.F. Puchkarev, W. Rohrbeck: Nanosecond field emission. production and distruction of field emitting micro-tips by cathode flares. Preprint ZIE 75-3 (Akad. Wiss., Berlin, GDR 1975)

5.32 E. Hantzsche, B. Jüttner, V.F. Puchkarev, W. Rohrbeck, H. Wolff: J. Phys. D **9**, 1771-1781 (1976)

5.33 B. Jüttner: Beitr. Plasma Phys. **19**, 259-265 (1979)

5.34 B. Jüttner, W. Rohrbeck, H. Wolff: Nanosecond breakdown in ultra-high vacuum. Proc. III Int'l Symp. on Discharges and Electrical Insulation in Vacuum (Paris, France 1968) pp.209-211

5.35 D.I. Proskurovsky, E.B. Yankelevich: Radiotekh. Electron. 1, 132-137 (1979) [Engl. transl.: Radio Eng. Electr. Phys. 24, 99-103 (1979)]
B.A. Koval, D.I. Proskurovsky, E.B. Yankelevich: On the drop fraction of the cathode erosion at the explosive electron emission. Proc. VIII Int'l Symp. on Discharges and Electrical Insulation in Vacuum (Albuquerque, NM 1978) paper B5

5.36 T. Usumi, J.H. English: J. Appl. Phys. **46**, 126-131 (1975)

5.37 B.A. Koval, D.I. Proskurovsky, V.F Tregubov, E.B. Yankelevich: Pisma Zh. Tekh. Fiz. **5**, 603-606 (1979) [Engl. transl.: Sov. Tech. Phys. Lett. **5**, 246-247 (1979)

5.38 V.I. Rakhovsky, A.M. Yagudaev: Zh. Tekh. Fiz. **39**, 363-368 (1969) [Engl. transl.: Sov. Phys-Tech. Phys. **14**, 227-230 (1969)

5.39 N.A. Cokeen, G.T. Chang, T.M. Posten, D.J. Spencer: High Temp. Sci. **8**, 81-97 (1976)

5.40 G.A. Lyubimov, V.I. Rakhovsky: Usp. Fiz. Nauk **125**, 665-706 (1978) [Engl. transl.: Sov. Phys.-Ups. **21**, 693-718 (1978)]

5.41 E. Hantzsche: Beitr. Plasma Phys. **17**, 65-74 (1977)

5.42 L.D. Landau, E.M. Lifshits: *Mechanics of Continuous Media* (Fizmatgiz, Moscow 1954) [Engl. transl.: Pergamon, Oxford 1959] Chap.10, pp.35-47

5.43 E. Hantzsche: Droplet emission from vacuum arc spots. Proc. VII Int'l Symp. on Discharges and Electrical Insulation in Vacuum (Novosibirsk, USSR 1976) pp.324-327

5.44 E. Hantzsche: Cathode processes in vacuum arcs. Proc. XIII Int'l Conf. on Phenomena in Ionized Gases (Berlin, GDR 1977) pp.121-138

5.45 G.A. Mesyats: Nanosecond switching in vacuum and a cyclical explosive model of the cathode spot. Proc. XI Int'l Symp. on Discharges and Electrical Insulation in Vacuum (Berlin, GDR 1984) pp.93-100

5.46 G.A. Mesyats: Pisma Zh. Tekh. Fiz. 10, 593-596 (1984)

5.47 G. Ecker: Theoretical aspects of the vacuum arc, in *Vacuum Arcs*, ed. by J.M. Lafferty (Wiley, New York 1980) Chap.7, pp.228-320

5.48 E.A. Litvinov: A theory of explosive electron emission. Proc. XI Int'l Symp. on Discharges and Electrical Insulation in Vacuum (Berlin, GDR 1984) pp.9-18

Chapter 6

6.1 S.P. Bugaev, A.M. Iskol'dsky, G.A. Mesyats, D.I. Proskurovsky: Zh. Tekh. Fiz. 37, 2206-2208 (1967) [Engl.transl.: Sov. Phys.-Tech. Phys. 12, 1625-1627 (1967)]

6.2 G.A. Mesyats, V.P. Rotshtein, G.N. Fursey, G.K. Kartsev: Zh. Tekh. Fiz. 40, 1551-1553 (1970) [Engl. transl.: Sov. Phys.-Tech. Phys. 15, 1202-1204 (1970)]

6.3 Ya.Ya. Yurike, V.F. Puchkarev, D.I. Proskurovsky: Izv. Vyssh. Uchebn. Zaved. Fiz. 3, 12-16 (1973) [Engl. transl.: Sov. Phys. J. 16, 293-297 (1973)]

6.4 R.B. Baksht, S.P. Vavilov, A.P. Kudinov: Izv. Vyssh. Uchebn. Zaved. Fiz. 5, 145-146 (1974)

6.5 R.B. Baksht, G.A. Mesyats: Izv. Vyssh. Uchebn. Zaved. Fiz. 7, 144-146 (1970) [Engl. transl.: Sov. Phys. J. 13, 969-971 (1970)]

6.6 S.P. Bugaev, A.A. Kim, V.I. Koshelev, P.A. Khryapov: IEEE Trans. EI-18, 234-237 (1983)

6.7 G.P. Bazhenov, G.A. Mesyats, D.I. Proskurovsky, V.P. Rotshtein, S.P. Vavilov: Anode processes of vacuum pulse breakdown spark stage. Proc. IV Int'l Symp. on Discharges and Electrical Insulation in Vacuum (Waterloo, Canada 1970) pp.121-124

6.8 G.A. Mesyats, G.P. Bazhenov, S.P. Bugaev, D.I. Proskurovsky, V.P. Rotshtein, Ya.Ya. Yurike: Izv. Vyssh. Uchebn. Zaved. Fiz. 5, 153-154 (1969) [Engl. transl.: Sov. Phys. J. 12, 688-689 (1969)]

6.9 G.P. Bazhenov, R.B. Baksht, G.A. Mesyats, D.I. Proskurovsky, V.P. Rotshtein, S.P. Vavilov: Cathode flares during pulse breakdown in vacuum. Proc. IV Int'l Symp. on Discharges and Electrical Insulation in Vacuum (Waterloo, Canada 1970) pp.116-120

6.10 G.A. Mesyats, D.I. Proskurovsky: Pisma Zh. Tekh. Fiz. 13, 7-10 (1971) [Engl. transl.: JETP Lett. 13, 4-6 (1971)]

6.11 G.A. Mesyats: The role of fast processes in vacuum breakdown. Proc. X Int'l Conf. on Phenomena in Ionized Gases (Oxford, U.K. 1971) pp.333-363

6.12 G.A. Mesyats, E.A. Litvinov, D.I. Proskurovsky: High-speed processes during pulses breakdown of vacuum gaps. Proc. IV Int'l Symp. on Discharges and Electrical Insulation in Vacuum (Waterloo, Canada 1970) pp.82-91

6.13 R.B. Baksht, V.I. Manylov: Izv. Vyssh. Uchebn. Zaved. Fiz. 9, 148-150 (1971) [Engl. transl.: Sov. Phys. J. 14, 1297-1298 (1971)]

6.14 R.B. Baksht, A.P. Kudinov, E.A. Litvinov: Zh. Tekh. Fiz. 43, 146-151 (1973) [Engl. transl.: Sov. Phys.-Tech. Phys. 18, 94-97 (1973)]

6.15 S.P. Bugaev, R.B. Baksht, E.A. Litvinov, V.P. Stas'ev: Teplofiz. Vys. Temp. **14**, 1145-1150 (1976) [Engl. transl.: High Temp. **14**, 1024-1032 (1976)]

6.16 G. Yonas, J. Poukey, K. Prestwich, J. Freeman, A. Toepfer, M. Clauser: Nucl. Fusion **14**, 731-740 (1974)

6.17 R. Stinnett, M. Palmer, R. Spielman, R. Bengtson: Small gap experiments in magnetically insulated transmission lines. Proc. X Int'l Symp. on Discharges and Electrical Insulation in Vacuum (Columbia, SC, USA 1982) pp.281-285

6.18 R.B. Baksht, B.A. Kablambaev, G.T. Razdobarin, N.A. Ratakhin: Zh. Tekh. Fiz. **49**, 1245-1247 (1979)

6.19 B. Jüttner: Characterization of the cathode spot. Proc. XII Int'l Symp. on Discharges and Electrical Insulation in Vacuum (Shoresh, Israel 1986) pp.90-98

6.20 V.A. Ivanov, B. Jüttner, H. Pursch: IEEE Trans. PS-13, 334-336 (1985)

6.21 A.N. Pustovit, V.I. Zhila, G.G. Sikhorulidze: Zh. Tekh. Fiz. **56**, 813-815 (1986)

6.22 D.D. Hinshelwood: Explosive emission cathode plasmas in intense relativistic electron beam diodes. PhD Thesis, Cambridge University (1984)

6.23 R.B. Baksht, N.A. Ratakhin, B.A. Kablambaev: Zh. Tekh. Fiz. **50**, 487-491 (1980) [Engl. transl.: Sov. Phys.-Tech. Phys. **25**, 294-297 (1980)]

6.24 N.A. Ratakhin, B.A. Kablambaev: In *High-Current Emission Electronics*, ed. by G.A. Mesyats (Nauka, Novosibirsk 1984) pp.62-69 (in Russian)

6.25 R.W.P. McWhirter: In *Plasma Diagnostic Techniques*, ed. by R.H. Huddlestone, S.L. Leonard (Academic, New York 1965) Chap.5

6.26 A. Maitland, R. Hawley: Vacuum **18**, 403-408 (1968)

6.27 R.K. Parker, R.E. Anderson, C.V. Dunkan: J. Appl. Phys. **45**, 2463-2478 (1974)

6.28 Ya.B. Zel'dovich, Yu.P. Raizer: *Physics of Shock Waves and High-Temperature Hydrodynamic Phenomena* (Nauka, Moscow 1966) p.180 [Engl. transl.: Academic, New York 1966]

6.29 V.V. Loskutov, A.V. Luchinsky, G.A. Mesyats: Dokl. Akad. Nauk SSSR **271**, 1120-1122 (1983)

6.30 Yu.D. Bakulin, V.F. Kuropatenko, A.V. Luchinsky: Zh. Tekh. Fiz. **46**, 1963-1968 (1976)

6.31 S.I. Braginsky: In *The Plasma: Theoretical Aspects*, Vol.1, ed. by M.A. Leontovich (Atomizdat, Moscow 1963) pp.191-273 (in Russian)

6.32 L. Spitzer, Jr.: *Physics of Fully Ionized Gases*, 2nd. ed. (Wiley, New York 1962) Sect.5.4

6.33 H.W. Drawin: Z. Phys. **164**, 513 (1961)

6.34 A.A. Plyutto: Zh. Eksp. Teor. Fiz. **39**, 1589-1592 (1960) [Engl. transl.: Sov. Phys.-JETP **12**, 1106-1108 (1961)]

6.35 A.A. Plyutto, V.N. Ryzhkov, A.T. Kapin: Zh. Eksp. Teor. Fiz. **47**, 494-507 (1964) [Engl. transl.: Sov. Phys.-JETP **20**, 328-337 (1965)]

6.36 B.Ya. Moizhes, V.A. Nemchinsky: Zh. Tekh. Fiz. **50**, 78-86 (1980) [Engl. transl.: Sov. Phys.-Tech. Phys. **25**, 43-48 (1980)]

6.37 G.A. Lyubimov: Zh. Tekh. Fiz. **47**, 297-303 (1977) [Engl. transl.: Sov. Phys.-Tech. Phys. **22**, 173-177 (1977)]

6.38 R.B. Baksht, S.P. Vavilov, A.P. Kudinov, E.A. Litvinov, V.I. Manylov, M.N. Urbazaev: *Investigation of Cathode Flare Plasma Caused by Vacuum Breakdown*, Proc. V. Int'l Symp. on Discharges and Electrical Insulation in Vacuum (Poznan, Poland 1972) pp.139-144

6.39 S.P. Bugaev, E.A. Litvinov, G.A. Mesyats, D.I. Proskurovsky: Usp. Fiz. Nauk **115**, 101-120 (1975) [Engl. transl.: Sov. Phys.-Usp. **18**, 51-61 (1975)]

6.40 E.A. Litvinov: IEEE Trans. EI-20, 683 (1985)

6.41 E. Waisman, M. Chapman: J. Appl. Phys. **53**, 724-730 (1982)

7.1 A.B. Boim, E.M. Reikhrudel: Zh. Tekh. Fiz. **31**, 1127-1133 (1961) [Engl. transl.: Sov. Phys.-Tech. Phys. **6**, 821-826 (1961)]

7.2 P.T.G. Flynn: Proc. Phys. Soc. (London) **69** B, 748-762 (1956)

7.3 E.D. Korop, A.A. Plyutto: Zh. Tekh. Fiz. **40**, 2534-2537 (1970) [Engl. transl.: Sov. Phys.-Tech. Phys. **15**, 1298-1302 (1970)]

7.4 E.D. Korop, A.A. Plyutto: Zh. Tekh. Fiz. **41**, 1055-1057 (1971) [Engl. transl.: Sov. Phys.-Tech. Phys. **16**, 830-831 (1971)]

7.5 G.A. Mesyats: The role of fast processes in vacuum breakdown. Proc. X Int'l Conf. on Phenomena in Ionized Gases (Oxford, UK 1971) pp.333-363

7.6 D.I. Proskurovsky, V.P. Rotshtein: Izv. Vyssh. Uchebn. Zaved. Fiz. **11**, 142-144 (1973)

7.7 E.A. Litvinov, G.A. Mesyats, D.I. Proskurovsky: Usp. Fiz. Nauk **139**, 265-302 (1983) [Engl. transl.: Sov. Phys.-Usp. **26**, 138-159 (1983)]

7.8 L. Spitzer, Jr.: *Physics of Fully Ionized Gases*, 2nd rev. ed. (Wiley, New York 1962) p.45

7.9 M. Kaminsky: *Atomic and Ionic Impact Phenomena on Metal Surfaces* (Academic, New York 1964)

7.10 G.P. Bazhenov, O.B. Ladyzhensky, E.A. Litvinov, S.M. Chesnokov: Zh. Tekh. Fiz. **47**, 2086-2091 (1977) [Engl. transl.: Sov. Phys.-Tech. Phys. **22**, 1212-1215 (1977)]

7.11 D.I. Proskurovsky, V.P. Rotshtein, A.F. Shubin, E.B. Yankelevich: Zh. Tekh. Fiz. **45**, 2135-2143 (1975) [Engl. transl.: Sov. Phys.-Tech. Phys. **20**, 1342-1346 (1975)]

7.12 A.A. Plyutto, K.V. Suladze, E.D. Korop, V.N. Ryzhkov: On vacuum breakdown mechanism at the stage of cathode-side plasma formation. Proc. V Int'l Symp. on Discharges and Electrical Insulation in Vacuum (Poznan, Poland 1972) pp.145-149

7.13 I.I. Levintov: Dokl. Akad. Nauk SSSR **85**, 1247-1250 (1952)

7.14 G.P. Bazhenov, R.B. Baksht, G.A. Mesyats, D.I. Proskurovsky, V.P. Rotshtein, S.P. Vavilov: Cathode flares during pulse breakdown in vacuum. Proc. IV Int'l Symp. on Discharges and Electrical Insulation in Vacuum (Waterloo, Canada 1970) pp.116-120

7.15 G.A. Mesyats, D.I. Proskurovsky: Pisma Zh. Eksp. Teor. Fiz. **13**, 7-10 (1971) [Engl. transl.: Sov. Phys.-JETP Lett. **13**, 4-6 (1971)]

7.16 G.A. Mesyats, E.A. Litvinov: Izv. Vyssh. Uchebn. Zaved. Fiz. **8**, 158-160 (1972)

7.17 W.W. Dolan, W.P. Dyke, J.K. Trolan: Phys. Rev. **91**, 1054-1057 (1953)

7.18 A.F. Shubin, Ya.Ya. Yurike: Izv. Vyssh. Uchebn. Zaved. Fiz. **6**, 134-136 (1975) [Engl. transl.: Sov. Phys. J. **18**, 870-872 (1975)]

7.19 G.P. Bazhenov, S.P. Bugaev, G.A. Mesyats, S.M. Chesnokov: Pisma Zh. Eksp. Teor. Fiz. **2**, 462-465 (1976) [Engl. transl.: Sov. Phys. - JETP Lett. **2**, 180-181 (1976)]

7.20 G.A. Mesyats: Electron explosive emission and electrical discharge in vacuum. Proc. VI Int'l Symp. on Discharges and Electrical Insulation in Vacuum (Swansea, UK 1974) pp.21-47

7.21 G.P. Bazhenov, R.B. Baksht, G.A. Mesyats, D.I. Proskurovsky: Teplofiz. Vys. Temp. **13**, 184-186 (1975)

7.22 D.I. Proskurovsky, V.F. Puchkarev: Zh. Tekh. Fiz. **49**, 2611-2618 (1979) [Engl. transl.: Sov. Phys.-Tech. Phys. **24**, 1474-1478 (1979)]

7.23 E.D. Korop, A.A. Plyutto: Izv. Vyssh. Uchebn. Zaved. Fiz. **4**, 131-132 (1973) [Engl. transl.: Sov. Phys. J. **16**, 560-561 (1973)]

7.24 D.I. Proskurovsky, V.F. Puchkarev: Izv. Vyssh. Uchebn. Zaved. Fiz. **12**, 57-63 (1975)

7.25 G.P. Bazhenov, O.B. Ladyzhensky, S.M. Chesnokov, V.G. Shpak: Zh. Tekh. Fiz. **49**, 117-124 (1979) [Engl. transl.: Sov. Phys.-Tech. Phys. **24**, 67-71 (1979)

7.26 G.A. Mesyats, D.I. Proskurovsky: Izv. Vyssh. Uchebn. Zaved. Fiz. **1**, 81-85 (1968) [Engl. transl.: Sov. Phys. J. **11**, 49-51 (1968)

7.27 G.A. Mesyats, E.A. Litvinov, D.I. Proskurovsky: High-speed processes during pulse breakdown of vacuum gaps. Proc. IV Int'l Symp. on Discharges and Electrical Insulation in Vacuum (Waterloo, Canada 1970) pp.82-95

7.28 G.M. Kassirov, G.A. Mesyats: Zh. Tekh. Fiz. **34**, 1476-1481 (1964) [Engl. transl.: Sov. Phys.-Tech. Phys. **9**, 1141-1145 (1964)]

7.29 S.P. Bugaev, G.M. Kassirov, B.M. Koval'chuk, G.A. Mesyats: Pisma Zh. Eksp. Teor. Fiz. **18**, 82-85 (1973)

7.30 G.P. Bazhenov, S.P. Bugaev, G.A. Mesyats, S.M. Chesnokov: Pisma Zh. Tekh. Fiz. **2**, 462-465 (1976) [Engl. transl.: Sov. Tech. Phys. Lett. **2**, 180-181 (1976)]

7.31 S.P. Vavilov, G.A. Mesyats: Izv. Vyssh. Uchebn. Zaved. Fiz. **8**, 90-94 (1970) [Engl. transl.: Sov. Phys. J. **13**, 1058-1061 (1970)]

7.32 I.I. Kalyatsky, G.M. Kassirov, G.V. Smirnov, N.N. Frolov: Zh. Tekh. Fiz. **45**, 1547-1550 (1975) [Engl. transl.: Sov. Phys.-Tech. Phys. **20**, 988-989 (1975)]

7.33 R.K. Parker, R.E. Anderson, C.V. Duncan: J. Appl. Phys. **45**, 2463-2478 (1974)

7.34 K.F. Zelensky, O.P. Pechersky, V.A. Tsukerman: Zh. Tekh. Fiz. **38**, 1581-1587 (1968) [Engl. transl.: Sov. Phys.-Tech. Phys. **13**, 1284-1289 (1969)]

7.35 R.B. Baksht, S.P. Vavilov, M.N. Urbazaev: Izv. Vyssh. Uchebn. Zaved. Fiz. **2**, 140-141 (1973) [Engl. transl.: Sov. Phys. J. **16**, 266-267 (1973)]

Chapter 8

8.1 I.G. Kesaev: *Cathode Processes in Electric Arcs* (Nauka, Moscow 1968) [Engl. transl.: SAND 78-6011, NTIS]

8.2 G.A. Lyubimov, V.I. Rakhovsky: Usp. Fiz. Nauk **125**, 665-706 (1978) [Engl. transl.: Sov. Phys.-Usp. **21**, 693-718 (1978)]

8.3 L.P. Harrris: In *Vacuum Arcs*, Chap.4, ed. by J.M. Lafferty (Wiley, New York 1980) pp.120-168

8.4 G.N. Fursey, P.N. Vorontsov-Vel'yaminov: Zh. Tekh. Fiz. **37**, 1880-1888 (1967) [Engl. transl.: Sov. Phys.-Tech. Phys. **12**, 1377-1382 (1967)]

8.5 J. Mitterauer: Acta Phys. Austr. **37**, 175-192 (1973)

8.6 E. Hantzsche: Beitr. Plasma Phys. **17**, 65-74 (1977)

8.7 G. Ecker: In *Vacuum Arcs*, ed. by J.M. Lafferty (Wiley, New York 1980) Chap.7, pp.228-320
      G. Ecker: IEEE Trans. PS-4, 218-227 (1976)
      G. Ecker: Commun. Siberian Branch Academy of Sciences USSR **2**, 3 (1979)

8.8 M.A. Lutz: IEEE Trans. PS-2, 1-10 (1974)

8.9 E.A. Litvinov, G.A. Mesyats, D.I. Proskurovsky: Usp. Fiz. Nauk **139**, 265-302 (1983) [Engl. transl.: Sov. Phys.-Usp. **26**, 138-159 (1983)]

8.10 G.A. Mesyats: IEEE Trans. EI-20, 729-734 (1985)

8.11 E.A. Litvinov, G.A. Mesyats, A.G. Parfyonov, A.N. Fedosov: Zh. Tekh. Fiz. **55**, 2270-2273 (1985)

8.12 E. Hantzsche: Cathode processes in vacuum arcs. Proc. XIII Int'l Conf. on Phenomena in Ionized Gases (Berlin, DDR 1977) pp.121-138

8.13 I.M. Tsinman: Radiotekh. Elektron. **8**, 834-844 (1963)

8.14 G.A. Mesyats: Fundamental role of electron explosive emission during vacuum current commutation. Proc. XII Int'l Conf. on Phenomena in Ionized Gases (Eindhoven, Netherlands 1975) p.245

8.15 D.I. Proskurovsky, V.F. Puchkarev: Zh. Tekh. Fiz. **49**, 2611-2618 (1979) [Engl. transl.: Sov. Phys.-Tech. Phys. **24**, 1474-1478 (1979)]
8.16 G.A. Mesyats: IEEE Trans. EI-**18**, 218-225 (1983)
8.17 E.A. Litvinov: IEEE Trans. EI-**20**, 683-689 (1985)
8.18 B. Jüttner, V.F. Puchkarev, W. Rohrbeck: Nanosecond field emission. Production and destruction of field emitting micro-tips by cathode flares. Preprint ZIE 75-3 (Akad. Wiss. DDR, 1975)
8.19 G.A. Vorob'yov, V.A. Mukhachev: *Breakdown of Thin Dielectric Films* (Sov. Radio, Moscow 1977) (in Russian)
8.20 G.A. Mesyats: Electron explosive emission and electrical discharge in vacuum. Proc. VI Int'l Symp. on Discharges and Electrical Insulation in Vacuum (Swansea, UK 1974) pp.21-47
8.21 D.I. Proskurovsky, V.F. Puchkarev: Izv. Vyssh. Uchebn. Zaved. Fiz. **12**, 57-63 (1975)
8.22 G.P. Bazhenov, R.B. Baksht, G.A. Mesyats, D.I. Proskurovsky: Teplofiz. Vys. Temp. **13**, 184-186 (1975)
8.23 D.I. Proskurovsky, V.F. Puchkarev: Zh. Tekh. Fiz. **50**, 2120-2126 (1980) [Engl. transl.: Sov. Phys.-Tech. Phys. **25**, 1235-1240 (1980)]
8.24 D.I. Proskurovsky, V.F. Puchkarev: J. Phys. **40**, 411-412 (1979)
8.25 L. Reimer: *Scanning Electron Microscopy*, Springer Ser. Opt. Sci., Vol.45 (Springer, Berlin, Heidelberg 1985)
8.26 M.Yu. Borukhov, I.Kh. Bek-Bulatov, L.L. Lukashevich, R.B. Nagaibekov, N. Umurzakov: Zh. Tekh. Fiz. **42**, 1504-1507 (1972) [Engl. transl.: Sov. Phys.-Tech. Phys. **17**, 1199-1202 (1972)]
8.27 J.H. Holliday, G.G. Isaaks: Br. J. Appl. Phys. **17**, 1575-1585 (1966)
8.28 J.T. Maskrey: Br. J. Appl. Phys. **16**, 1583-1584 (1965)
8.29 G.A. Farrall, M. Owens, F.G. Hudda: J. Appl. Phys. **46**, 610-617 (1975)
8.30 B.M. Cox: J. Phys. D. **8**, 2065-2073 (1975)
8.31 B.M. Cox, W.T. Williams: J. Phys. D. **10**, 15-19 (1977)
8.32 R.E. Hurley, P.J. Dooley: J. Phys. D. **10**, 195-201 (1977)
8.33 P.C.L. Pfeil, L.B. Griffiths: Nature (London) **183**, 1481 (1959)
8.34 B. Jüttner: Beitr. Plasma Phys. **18**, 265-269 (1978)
8.35 S.Ya. Belomyttsev, S.D. Korovin, G.A. Mesyats: Pisma Zh. Tekh. Fiz. **6**, 1089-1092 (1980) [Engl. transl.: Sov. Tech. Phys. Lett. **6**, 466-467 (1980)]
8.36 E.N. Abdullin, V.G. Azarov, S.P. Bugaev: Zh. Tekh. Fiz. **46**, 2459-2461 (1976) [Engl. transl.: Sov. Phys.-Tech. Phys. **21**, 1455-1466 (1976)]
8.37 B.A. Koval, G.A. Mesyats, G.E. Ozur, D.I. Proskurovsky, E.B. Yankelevich: Pisma Zh. Tekh. Fiz. **7**, 1227-1230 (1981)
8.38 G.P. Bazhenov, G.A. Mesyats, D.I. Proskurovsky: Izv. Vyssh. Uchebn. Zaved. Fiz. **8**, 87-90 (1970) [Engl. transl.: Sov. Phys. J. **13**, 1054-1057 (1970)]
8.39 A.J. Topfer, L.P. Bradly: J. Appl. Phys. **43**, 3033-3036 (1972)
8.40 N.G. Pavlovskaya, T.V. Kudryavtseva, N.A. Dron, G.N. Sloeva, L.V. Tsvetkov: Prib. Tekh. Eksp. **1**, 22-24 (1973)
8.41 S.P. Bugaev, G.M. Kassirov, G.M. Koval'chuk, G.A. Mesyats: Pisma Zh. Eksp. Teor. Fiz. **18**, 82-85 (1973)
8.42 S. Singer, J. Ladish, M. Nutter: Cold cathode electron guns in the LASL high-power short-pulse CO_2 laser program. Proc. 2nd Int'l Top. Conf. on High Power Electron and Ion Beam Research and Technology (Cornell, New York 1977) pp.274-292
8.43 S.Ya. Belomyttsev, S.P. Bugaev, E.A. Litvinov, V.P. Il'in: To effect of "touches" at electron explosive emission. Proc. VI Int'l Symp. on Discharges and Electrical Insulation in Vacuum (Swansea, UK 1974) pp.83-85

Chapter 9

9.1 G.A. Mesyats: The role of fast processes in vacuum breakdown. Proc. X Int'l Conf. on Phenomena in Ionized Gases (Oxford, UK 1971) pp.333-363

9.2 D.I. Proskurovsky, V.P. Rotshtein: Izv. Vyssh. Uchebn. Zaved. Fiz. 11, 142-144 (1973)

9.3 S.I. Anisimov, Yu.A. Imas, G.S. Romanov, Yu.V. Khodyko: *Effects of High Power Radiation on Metals* (Nauka, Moscow 1970) (in Russian)

9.4 L.I. Mirkin: *Laser Treatment of Materials: The Physical Basis* (Izd. MGU, Moscow 1975) (in Russian)

9.5 J.F. Ready: *Effects of High Power Laser Radiation* (Academic, New York 1971)

9.6 M. von Almen: *Laser-Beam Interactions with Materials*, Springer Ser. Mat. Sci., Vol.2 (Springer, Berlin, Heidelberg 1987)

9.7 H.S. Carslow, J.C. Jaeger: *Conduction of Heat in Solids*, 2nd ed. (Clarendon, Oxford 1951)

9.8 Yu.D. Deniskin: Zh. Tekh. Fiz. 38, 1961-1965 (1968)

9.9 R.D. Birkhoff: The passage of fast electrons through matter. *Handbuch der Physik*, Vol.34 "Korpuskeln und Strahlung in Materie" (Springer, Berlin, Heidelberg 1958) pp.128-131

9.10 S.P. Bugaev, R.B. Gaksht, E.A. Litvinov, V.P. Stas'ev: Teplofiz. Vys. Temp. 14, 1145-1150 (1976) [Engl. transl.: High Temp. 14, 1027-1032 (1976)]

9.11 R.B. Baksht, N.A. Ratakhin, B.A. Kablambaev: Zh. Tekh. Fiz. 52, 1778-1782 (1982) [Engl. transl.: Sov. Phys.-Tech. Phys. 27, 1091-1093 (1982)]

9.12 G.A. Mesyats, S.P. Bugaev, D.I. Proskurovsky, V.I. Eshkenazi, Ya.Ya. Yurike: Radiotekh. Elektron. 14, 2222-2230 (1969) [Engl. transl.: Radio Eng. Electron. Phys. 14, 1919-1925 (1969)]

9.13 G.A. Mesyats, V.I. Eshkenaszi: Izv. Vyssh. Uchebn. Zaved. Fiz. 2, 123-125 (1968) [Engl. transl.: Sov. Phys. J. 11; 2, 79-82 (1968)]

9.14 A. Maitland: J. Appl. Phys. 32, 2399-2407 (1961)

9.15 J.M. Somerville, C.T. Graiger: Br. J. Appl. Phys. 7, 400-405 (1956)

9.16 R. Hawley: Vacuum 11, 32-35 (1961)

9.17 I.L. Gufel'd: Zh. Tekh. Fiz. 38, 304-305 (1968) [Engl. transl.: Sov. Phys.-Tech. Phys. 13, 221-224 (1968)]

9.18 G.P. Bazhenov, G.A. Mesyats, D.I. Proskurovsky: Izv. Vyssh. Uchebn. Zaved. Fiz. 8, 89-90 (1970) [Engl. transl.: Sov. Phys. J. 13, 1054-1057 (1970)]

9.19 G.P. Bazhenov, G.A. Mesyats, D.I. Proskurovsky, V.P. Rotshtein, S.P. Vavilov: Anode processes of vacuum pulse breakdown spark stage. Proc. IV Int'l Symp. on Discharges and Electrical Insulation in Vacuum (Waterloo, Canada 1970) pp.121-124

9.20 K. Vogel, P. Backlund: J. Appl. Phys. 36, 3697-3699 (1965)

9.21 L.I. Mirkin: Dokl. Akad. Nauk SSSR 189, 528-530 (1969)

9.22 D.I. Proskurovsky, V.P. Rotshtein, A.F. Shubin, E.B. Yankelevich: Zh. Tekh. Fiz. 45, 2135-2143 (1975) [Engl. transl.: Sov. Phys. - Tech. Phys. 20, 1342-1346 (1975)]

9.23 A.F. Shubin, V.P. Rotshtein, D.I. Proskurovsky: Izv. Vyssh. Uchebn. Zaved. Fiz. 7, 50-53 (1974)

9.24 V.P. Rotshtein, L.S. Bushnev, D.I. Proskurovsky: Izv. Vyssh. Uchebn. Zaved. Fiz. 3, 130-131 (1975)

9.25 A. Maitland: J. Appl. Phys. 33, 248-249 (1962)

9.26 S.P. Bugaev, V.I. Koshelev, M.N. Timofeev: Izv. Vyssh. Uchebn. Zaved. Fiz. 2, 57-61 (1974) [Engl. transl.: Sov. Phys. J. 17, 193-196 (1974)]

9.27 S.P. Bugaev, V.I. Koshelev, M.N. Timofeev: Izv. Vyssh. Uchebn. Zaved. Fiz. 2, 35-37 (1975) [Engl. transl.: Sov. Phys. J. 18, 173-175 (1975)]

9.28 R.B. Baksht, G.M. Kassirov, G.V. Smirnov, F.G. Sekisov: Izv. Vyssh. Uchebn. Zaved. Fiz. 7, 130-132 (1975) [Engl. transl.: Sov. Phys. J. 18, 1021-1023 (1975)]

9.29 E.D. Korop, A.A. Plyutto: Zh. Tekh. Fiz. 40, 2534-2537 (1970) [Engl. transl.: Sov. Phys.-Tech. Phys. 15, 1298-1302 (1970)]
E.D. Korop, A.A. Plyutto: Zh. Tekh. Fiz. 41, 1055-1057 (1971) [Engl. transl.: Sov. Phys.-Tech. Phys. 16, 830-831 (1971)]

9.30 R.K. Parker, R.E. Anderson, C.V. Duncan: J. Appl. Phys. 45, 2463-2478 (1974)

9.31 E.N. Abdullin, G.P. Bazhenov: Zh. Tekh. Fiz. 51, 1969-1971 (1981)

9.32 E.N. Abdullin, G.P. Bazhenov, S.P. Bugaev, G.P. Erokhin, S.M. Chesnokov: Generation of quasi-stationary electron beams on the basis of a vacuum discharge. Proc. XI Int'l Symp. on Discharges and Electrical Insulation in Vacuum (Berlin, GDR 1984) pp.393-397

9.33 R.B. Baksht, V.A. Kokshenev, V.I. Manylov: Zh. Tekh. Fiz. 45, 1678-1682 (1975)

9.34 S.P. Bugaev, A.M. Iskol'dsky, G.A. Mesyats, D.I. Proskurovsky: Zh. Tekh. Fiz. 37, 2206-2208 (1967) [Engl. transl.: Sov. Phys.-Tech. Phys. 12, 1625-1627 (1967)]

9.35 G.A. Mesyats, S.P. Bugaev, D.I. Proskurovsky: Dokl. Akad. Nauk SSSR 186, 1067-1069 (1969) [Engl. transl.: Sov. Phys.-Dokl. 14, 605-607 (1969)]

9.36 Ya.B. Zel'dovich, Yu.P. Raizer: *Physics of Shock Waves and High Temperature Hydrodynamic Phenomena* (Nauka, Moscow 1966) [Engl. transl.: (Academic, New York 1966)]

9.37 R.B. Baksht, N.A. Ratakhin, M.N. Timofeev: Pisma Zh. Tekh. Fiz. 1, 922-926 (1975) [Engl. transl.: Sov. Tech. Phys. Lett. 1, 399-402 (1975)

9.38 J.A. Chiles: J. Appl. Phys. 8, 622-626 (1937)

9.39 I.D. Chalmers, B.D. Phukan: J. Phys. D. 12, 1285-1292 (1979)

9.40 G. Tomer: In *High-Speed Physics*, Vol.1, ed. by K. Vollrath, G. Tomer (Springer, Wien 1967) pp.342-385

9.41 V.A. Tsukerman, L.V. Tarasova, S.I. Lobov: Usp. Fiz. Nauk 103, 319-337 (1971)

9.42 F. Jamet, G. Tomer: *Flash Radiography* (Elsevier, Amsterdam 1976)

9.43 G.A. Mesyats, D.I. Proskurovsky: Izv. Vyssh. Uchebn. Zaved. Fiz. 1, 81-85 (1968) [Engl. transl.: Sov. Phys. J. 11, 49-51 (1968)]

9.44 G.A. Mesyats: Zh. Tekh. Fiz. 44, 1521-1527 (1974) [Engl. transl.: Sov. Phys.-Tech. Phys. 19, 948-951 (1974)]

9.45 R.B. Baksht, S.P. Vavilov, M.N. Urbazaev: Izv. Vyshh. Uchebn. Zaved. Fiz. 2, 140-141 (1973) [Engl. transl.: Sov. Phys. J. 16, 266-267 (1973)]

9.46 J.P. Freitag: Extract Rev. Techn. Tomson CSF 1, 533-595 (1969)

Chapter 10

10.1 I.N. Slivkov: *Electrical Insulation and Discharge in Vacuum* (Atomizdat, Moscow 1972) (in Russian)

10.2 R.V. Latham: *High Voltage Vacuum Insulation* (Academic, London 1981)

10.3 G.A. Mesyats: The Role of fast processes in vacuum breakdown. Proc. X Int'l Conf. on Phenomena in Ionized Gases (Oxford, UK 1971) pp.333-357

10.4 D.I. Proskurovsky, Ya.Ya. Yurike: The spark current increase during steady voltage breakdown in vacuum. Proc. XI Int'l Symp. on Discharges and Electrical Insulation in Vacuum (Waterloo, Ontario, Canada 1970) pp.78-81

10.5 D.I. Proskurovsky, Ya.Ya. Yurike: Izv. Vyssh. Uchebn. Zaved. Fiz. **9**, 93-97 (1971) [Engl. transl.: Sov. Phys. J. **14**, 1238-1242 (1971)]

10.6. Ya.Ya. Yurike: Optical studies of vacuum breakdown development at steady voltage. Proc. V Int'l Symp. on Discharges and Electrical Insulation in Vacuum (Poznan, Poland 1972) pp.111-114

10.7 Ya.Ya. Yurike, V.F. Puchkarev, D.I. Proskurovsky: Izv. Vyssh. Uchebn. Zaved. Fiz. **3**, 12-16 (1973) [Engl. transl.: Sov. Phys. J. **16**, 293-297 (1973)]

10.8 Ya.Ya. Yurike: Izv. Vyssh. Uchebn. Zaved. Fiz. **11**, 140-141 (1974) [English transl.: Sov. Phys. J. **17**, 1604-1606 (1974)]

10.9 A.F. Shubin, Ya.Ya. Yurike: Izv. Vyssh. Uchebn. Zaved. Fiz. **6**, 134-136 (1975) [Engl. transl.: Sov. Phys. J. **18**, 870-872 (1975)]

10.10 D.K. Davies, M.A. Biondi: J. Appl. Phys. **41**, 88-93 (1970)

10.11 D.K. Davies, M.A. Biondi: Growth of radiation intensity emitted during the early phases of DC breakdown in vacuum. Proc. XI Int'l Symp. on Discharges and Electrical Insulation in Vacuum (Swansea, UK 1974) pp.45-50

10.12 D.K. Davies, M.A. Biondi: J. Appl. Phys. **48**, 4229-4233 (1977)

10.13 P. Kranjec, L. Ruby: J. Vac. Sci. Tech. **4**, 94-96 (1967)

10.14 D.K. Davies, M.A. Biondi: J. Appl. Phys. **39**, 2979-2990 (1968)

10.15 D.K. Davies, M.A. Biondi: J. Appl. Phys. **42**, 3089-3107 (1971)

10.16 D.K. Davies, M.A. Biondi: Dynamics and heating of anode particles in vacuum breakdown. Proc. XI Int'l Symp. on Discharges and Electrical Insulation in Vacuum (Novosibirsk, USSR 1976) pp.121-126

10.17 Y.I. Yen, D.T. Tuma, D.K. Davies: J. Appl. Phys. **55**, 3301-3307 (1984)

10.18 D.A. Eastham, P.A. Chatterton: Vacuum **32**, 151-155 (1982)

10.19 P.A. Chatterton: Vacuum breakdown, in *Electrical Breakdown of Gases*, ed. by J.M. Meek, J.D. Craggs (Wiley, New York 1978)

10.20 B. Jüttner, P. Siemroth: Beitr. Plasma Phys. **21**, 233-245 (1981)

10.21 A.A. Litvinov, G.O. Mesyats, A.G. Parfyonov, A.N Fedosov: Zh. Tekh. Fiz. **55**, 2270-2273 (1985)

10.22 W.W. Dolan, W.P. Dyke, J.K. Trolan: Phys. Rev. **91**, 1054-1057 (1953)

10.23 G.E. Vibrans: J. Appl. Phys. **35**, 2855-2857 (1964)

10.24 P.A. Chatterton: Proc. Phys. Soc. (London) **88**, 231-245 (1966)

10.25 F.M. Charbonnier, C.J. Bennette, L.W. Swanson: J. Appl. Phys. **38**, 627-633 (1967)

10.26 D.W. Williams, W.T. Williams: J. Phys. D **5**, 280-290 (1972)

10.27 G. Ecker: IEEE Trans. PS-4, 218-227 (1967)

10.28 E.A. Litvinov, A.F. Shubin: IZV. Vyssh. Uchebn. Zaved. Fiz. **11**, 90-93 (1974) [Engl. transl.: Sov. Phys. J. **17**, 1549-1552 (1974)]

10.29 D. Alpert, D.A. Lee, E.U. Lyman, H. Tomaschke: J. Vac. Sci. Tech. **1**, 35-50 (1964)

10.30 J. Brodie: J. Vac. Sci. Tech. **3**, 222-223 (1966)

10.31 A. van Oostrom: Influence of absorbates and electric fields on the nucleation and growth of microtips in a vacuum gap. Proc. IX Int'l Symp. on Discharges and Electrical Insulation in Vacuum (Swansea, UK 1974) pp.49-70

10.32 S.Y. Ettintger, E.M. Lyman: Effects of gas conditioning on cathode surfaces, field emission and electrical insulation. Proc. III Int'l Symp. on Discharges and Electrical Insulation in Vacuum (Paris, France 1968) pp.128-133

10.33 A. Maitland: J. Appl. Phys. **32**, 2399-2407 (1961)

10.34 G.N. Fursey, P.N. Vorontsov-Vel'yaminov: Zh. Tekh. Fiz. **37**, 1870-1879 (1967) [Engl. transl.: Sov. Phys.-Tech. Phys. **12**, 1377-1382 (1967)]

10.35 E.E. Martin, J.K. Trolan, W.P. Dyke: J. Appl. Phys. **31**, 50-57 (1960)

10.36 L.I. Pivovar, V.M. Tubaev, V.I. Gordienko: Zh. Tekh. Fiz. **27**, 997-1000 (1957) [Engl. transl.: Sov. Phys.-Tech. Phys. **2**, 909-912 (1957)]

10.37 G.N. Fursey, G.K. Kartsev: Zh. Tekh. Fiz. **40**, 310-319 (1970) [English transl.: Sov. Phys.-Tech. Phys. **15**, 225-232 (1970)]

10.38 B. Jüttner, W. Rohrbeck, H. Wolff: Pressure dependence of prebreakdown current due to sorption processes. Proc. V Int'l Symp. on Discharges and Electrical Insulation in Vacuum (Poznan, Poland 1972) pp.65-69

10.39 H. Wolff, B. Jüttner, W. Rohrbeck: Steady high current field emission from extended metal surfaces. Proc. V Int'l Symp. on Discharges and Electrical Insulation in Vacuum (Poznan, Poland 1972) pp.165-169

10.40 B. Jüttner, V.F. Puchkarev, W. Rohrbeck: Behaviour of micropoints during high voltage vacuum discharges. Proc. XII Int'l Symp. on Discharges and Electrical Insulation in Vacuum (Novosibirsk, USSR 1976) pp.189-192

10.41 W. Ermrich, A. van Oostrom: Solid Stat. Commun. **5**, 471-474 (1967)

10.42 F.G. Zheleznikov: Zh. Tekh. Tiz. **48**, 1224-1227 (1978) [English transl.: Sov. Phys.-Tech. Phys. **23**, 684-686 (1978)]

10.43 M. Kaminsky: *Atomic and Ionic Impact Phenomena on Metal Surfaces* (Academic, New York 1964)

10.44 N.B. Rozanova: Izv. Akad. Nauk SSSR Ser. Fiz. **26**, 1438-1440 (1962) [English transl.: Bull. Acad. Sci. USSR Phys. Ser. Eng. Ed. **26**, 1462-1464 (1962)]

10.45 R.P. Little, S.T. Smith: Investigations into the source of sharp protrusions which appear on flat cathode surfaces as a result of the application of high electric fields. Proc. II Int'l Symp. on the Insulation of High Voltages in Vacuum (MIT, Cambridge, Mass. 1966) pp.41-49

10.46 L.D. Landau, E.M. Lifshits: *Mechanics of Continuous Media* (Fizmatgiz, Moscow 1954) [English transl.: (Pergamon, Oxford 1959)]

10.47 P.A. Chatterton, M.M. Menon, K.D. Srivastava: J. Appl. Phys. **43**, 4536-4542 (1972)

10.48 A.K. Chakrabarti, P.A. Chatterton: J. Appl. Phys. **47**,. 5320-5328 (1976)

10.49 N.B. Rozanova, V.L. Granovsky: Zh. Tekh. Fiz. **26**, 489-496 (1956) [English transl.: Sov. Phys.-Tech. Phys. **1**, 471-478 (1956)]

10.50 P.V. Poshekhonov, M.M. Pogorel'sky: Zh. Tekh. Fiz. **39**, 1080-1085 (1969) [English transl.: Sov. Phys.-Tech. Phys. **14**, 1567-1572 (1969)]

10.51 V.A. Nevrovsky, V.I. Rakhovsky: J. Appl. Phys. **60**, 125-129 (1986)

Chapter 11

11.1 V.A. Nemchinsky: Zh. Tekh. Fiz. **49**, 1373-1378 (1979)

11.2 G.A. Lyubimov, V.I. Rakhovsky: Usp. Fiz. Nauk **125**, 665-706 (1978) [Engl. transl.: Sov. Phys.-Usp. **21**, 693-718 (1978)]

11.3 D.I. Proskurovsky, V.F. Puckarev: Zh. Tekh. Fiz. **49**, 2611-2618 (1979) [English transl.: Sov. Phys.-Tech. Phys. **24**, 1474-1478 (1979)]

11.4 I.G. Kesaev: *Cathode Processes in Electrical Arcs* (Nauka, Moscow 1968) [Engl. transl.: SAND 78-6011, NTIS]

11.5 B. Jüttner: Beitr. Plasma Phys. **19**, 25-48 (1979)

11.6 E. Schmidt: Ann. Phys. **4**, 246 (1949)

11.7 E. Hantzsche: Beitr. Plasma Phys. **17**, 65-74 (1977)

11.8 J. Achtert, B. Altrichter, B. Jüttner, P. Pech, H. Pursch, H.-D. Reiner, W. Rohrbeck, P. Siemroth, H. Wolff: Beitr. Plasma Phys. **17**, 419-431 (1977)

11.9 A.I. Bushik, B. Jüttner, H. Pursch: Beitr. Plasma Phys. **19**, 177-188 (1979)

11.10 E.A. Litvinov, G.A. Mesyats, D.I. Proskurovsky: Usp. Fiz. Nauk **139**, 265-302 (1983) [Engl. transl.: Sov. Phys.-Usp. **26**, 138-159 (1983)]

11.11 E. Hantzsche, B. Jüttner, V.F. Puchkaryov, W. Rohrbeck, H. Wolff: J. Phys. D **9**, 1771-1781 (1976)

11.12 G.A. Mesyats, D.I. Proskurovsky, E.B. Yankelevich: Cathode surface microrelief formation at the explosive electron emission. Proc. VII Int'l Symp. on Discharges and Electrical Insulation in Vacuum (Novosibirsk, USSR 1976) pp.230-233

11.13 A.E. Robson: J. Phys. D 11, 1917-1923 (1978)

11.14 M.G. Drouet: IEEE Trans. PS-13, 235-241 (1985)

11.15 B. Jüttner: Characterization of the cathode spot. Proc. XII Int'l Symp. Discharges and Electrical insulation in Vacuum (Tel-Aviv, Israel 1986) pp.90-98

11.16 P.R. Emtage, J.G. Gorman, J.V.R. Heberlein, F.A. Holmes, C.W. Kimblin, P.G. Slade, R.E. Voshall: The interaction of vacuum arcs with transverse magnetic fields. Proc. XIII Int'l Conf. on Phenomena in Ionized Gases (Berlin, GDR 1977) pp.673-674

11.17 V.I. Alfyorov, O.N. Vitkovskaya, G.I. Shcherbakov: Zh. Tekh. Fiz. 47, 102-111 (1977)

11.18 V.Ya. Moizhes, V.A. Nemchinsky: Pisma Zh. Tekh. Fiz. 5, 197-200 (1979) [English transl.: Sov. Tech. Phys. Lett 5, 78-80 (1979)]

11.19 E.A. Litvinov, G.A. Mesyats, A.G. Parfyonov, A.N. Fedosov: Zh. Tekh. Fiz. 55, 2270-2273 (1985)

11.20 H.O. Schrade: Arc cathode spots, their mechanism and motion. Proc. XIII Int'l Symp. on Discharges and Electrical Insulation in Vacuum (Paris, France 1988) pp.172-174

11.21 D.I. Proskurovsky, V.F. Puchkarev: Zh. Tekh. Fiz. 50, 2120-2126 (1980) [English transl.: Sov. Phys.-Tech. Fiz. 25, 1235-1239 (1980)]

11.22 B. Jüttner: J. Phys. D 14, 1265-1275 (1981)

11.23 K.D. Froome: Proc. Phys. Soc. (London) 60, 424-435 (1948)

11.24 D.B. Cummings: IEEE Trans. CE-68, 514 (1963)

11.25 A.M. Arsh, V.P. Andronov, Yu.D. Khromoy: Pisma Zh. Tekh. Fiz. 1, 86-88 (1975) [Engl. transl.: Sov. Tech. Phys. Lett 1, 38-40 (1975)]

11.26 I. Paulus, R. Holmes, H. Edels: J. Phys. D 5, 119-132 (1972)

11.27 D.I. Proskurovsky, V.F. Puchkarev: Zh. Tekh. Fiz. 51, 2277-2282 (1981) [English transl.: Sov. Phys.-Tech. Fiz. 26, 1342-1345 (1981)]

11.28 E.E. Yushmanov: Zh. Tekh. Fiz. 43, 1086-1089 (1973)

11.29 F.F. Chen: Electrical probes in Plasma Diagnostic Techniques, ed. by R.H. Huddlestone, S.L. Leonard (Academic, New York 1965) Chap.4

11.30 C.W. Mendel, S.A. Goldstein: J. Appl. Phys. 48, 1004-1006 (1977)

11.31 W.M. de Cock, J.E. Daalder: Ion velocities and the correlation between HF fluctuations of the arc voltage and the ion current in vacuum arcs. Proc. VII Int'l Symp. on Discharges and Electrical Insulation in Vacuum (Novosibirsk, USSR 1976) pp.288-292

11.32 G.P. Bazhenov, S.M. Chesnokov: Izv. Vyssh. Uchebn. Zaved. Fiz. 4, 133-135 (1976) [Engl. transl.: Sov. Phys. J. 19, 1500-1501 (1976)]

11.33 G.A. Farrall: Current zero phenomena, in Vacuum Arcs, ed. by J.M. Lafferty (Wiley-Interscience, New York 1980) Chap.6, pp.184-227

11.34 V.F. Puchkarev, D.I. Proskurovsky: Nonstationary processes in a vacuum arc at threshold current. Proc. XVI Int'l Conf. on Phenomena in Ionized Gases (Düsseldorf, FRG 1983) pp.268-269

11.35 M.P. Zektser, V.I. Rakhovsky: Dokl. Akad. Nauk SSSR; 78, 86- (1984)

11.36 G.A. Mesyats, A.M. Murzakaev, D.I. Proskurovsky, V.F. Puchkarev: Pisma Zh. Tekh. Fiz. (1985)

11.37 V.F. Puchkarev, D.I. Proskurovsky: IEEE Trans. PS-5, 257-260 (1985)

11.38 S.I. Anisimov, Ya.A. Imas, G.S. Romanov, Yu.V. Khodyko: Effects of High Power Radiation on Metals (Nauka, Moscow 1970) (in Russian)

11.39 E.A. Litvinov, G.A. Mesyats, A.G. Parfyonov: Dokl. Akad. Nauk SSSR, 269, 343-345 (1983)

11.40 E.A. Litvinov, G.A. Mesyats, A.Z. Nemirovsky, A.A. Starobinets: Digest of the V All-Union Symp. on Cold Cathodes (Tomsk 1985) pp.9-11 (in Russian)

11-41 B.Ya. Moizhes, V.A. Nemchinsky: Zh. Tekh. Riz. 50, 78-86 (1980) [English transl.: Sov. Phys.-Tech. Phys. 25, 43-48 (1980)]

11.42 B.E. Djakov: Ohm's law and ion acceleration in the cathode spot plasma of metal vapour arcs. Proc. XVI Int'l Conf. on Phenomena in Ionized Gases (Düsseldorf, FRG 1983) pp.270-271

11.43 E.A. Litvinov, G.A. Mesyats, A.G. Parfyonov, N.B. Volkov: An explosive emission model of the vacuum arc cathode spot. Proc. XIII Int'l Symp. on Discharges and Electrical Insulation in Vacuum (Paris, France 1988) pp.158-160

11.44 E.A. Litvinov, A.G. Parfyonov: Numerical simulation of cathode processes in a vacuum discharge. Proc. X Int'l Symp. on Discharges and Electrical Insulation in Vacuum (Columbia, USA 1982) pp.138-141

11.45 E.A. Litvinov, G.A. Mesyats, A.G. Parfyonov: Dokl. Akad. Nauk SSSR 279, 864-866 (1984)

11.46 E.A. Litvinov: IEEE Trans. EI-20, 683-689 (1985)

11.47 M. Kaminsky: *Atomic and Ionic Impact Phenomena on Metal Surfaces* (Academic, New York 1964)

11.48 J.E. Daalder: J. Phys. D 11, 1667-1682 (1978)

11.49 J.E. Daalder: IEEE Trans. PAS-93, 1747-1757 (1974)

11.50 I.Kh. Bek-Bulatov, M.Yu. Borukhov, R.V. Nagaibekov: Zh. Tekh. Riz. 43, 2211-2213 (1973) [Engl. transl.: Sov. Phys.-Tech. Phys. 18, 1401-1402 (1973)]

11.51 E. Hantzsche: Phys. Lett. A 50, 219-220 (1974)

11.52 J.A. Rich: J. Appl. Phys. 32, 1023-1031 (1961)

11.53 G.W. McClure: J. Appl. Phys. 45, 2078-2084 (1974)

11.54 I.G. Nekrashevich, I.A. Bakuto: Inzh. Fiz. Zh. 11, 59-65 (1959)

11.55 V.E. Il'in, S.B. Lebedev: Zh. Tekh. Fiz. 32, 986-992 (1962) [Engl. transl.: Sov. Phys.-Tech. Phys. 7, 717-721 (1962)]

11.56 I.M. Tsinman: Radiotekh. Electron. 8, 834-844 (1963)

11.57 J. Rotstein: The arc spot as a steady-state exploding wire phenomenon, in *Exploding Wires*, Vol.3, ed. by W.G. Chace, H.K. Moore (Plenum, New York 1964) pp.115-124

11.58 S.P. Bugaev, A.M. Iskol'dsky, G.A. Mesyats, D.I. Proskurovsky: Zh. Tekh. Fiz. 37, 2206-2208 (1967) [Engl. transl.: Sov. Phys. 12, 1625-1627 (1967)]

11.59 G.K. Kartsev, G.A. Mesyats, D.I. Proskurovsky, V.P. Rotstein, G.N. Fursey: Dokl. Akad. Nauk SSSR 192, 309-312 (1970) [Engl. transl.: Sov. Phys.-Dokl. 15, 475-477 (1970)]

11.60 G.A. Mesyats, D.I. Proskurovsky: Pisma Zh. Eksp. Teor. Fiz. 13, 7-10 (1971) [English transl.: Sov. Phys.-JETP Lett. 13, 4-6 (1971)]

11.61 G.N. Fursey, P.N. Vorontsov-Vel'yaminov: Zh. Tekh. Fiz. 37, 1870-1888 (1967) [Engl. transl.: Sov. Phys-Tech. Phys. 12, 1370-1382 (1967)]

10.62 J. Mitterauer: Acta Phys. Austr. 37, 175-192 (1973)

10.63 E. Hantzsche: Cathode processes in vacuum arcs. Proc. XIII Int'l Conf. on Phen9omena in Ionized Gases (Berlin, GDR 1977) pp.121-138

11.64 G. Ecker: Theoretical aspects of the vacuum arc, in *Vacuum Arcs*, ed. by J.M. Lafferty (Wiley-Interscience, New York 1980) Chap.7, pp.228-320

11.65 G.A. Mesyats: The role of fast processes in vacuum breakdown. Proc. X Int'l Conf. on Phenomena in Ionized Gases (Oxford, UK 1971) pp.333-363

11.66 G.A. Mesyats: Electron explosive emission and electrical discharge in vacuum. Proc. VI Int'l Symp. on Discharges and Electrical Insulation in Vacuum (Swansea, UK 1974) pp.21-47

11.67 S.P. Bugaev, E.A. Litvinov, G.A. Mesyats, D.I. Proskurovsky: Usp. Fiz. Nauk
115, 101-120 (1975) [English transl.: Sov. Phys.-Usp. 18, 51-61 (1975)]
11.68 G.A. Farrall: Arc ignition processes, in *Vacuum Arcs*, ed. by J.M. Lafferty
(Wiley-Interscience, New York 1980) Chap.3, pp.81-119
11.69 C.W. Kimblin: J. Appl. Phys. 44, 3074-3081 (1973)
11.70 J.E. Daalder: J. Phys. D 8, 1647-1659 (1975)
11.71 G.A. Mesyats: IEEE Trans. EI-20, 729-734 (1985)

Chapter 12

12.1 M. Tinkham: *Introduction to Superconductivity* (McGraw-Hill, New York 1975)
12.2 G.A. Mesyats, E.A. Litvinov, D.I. Proskurovsky, V.F. Puchkarev, S.I. Shkuratov: Dokl. Akad. Nauk SSSR 262, 598-600 (1982) [Engl. transl.: Sov. Phys. -
Dokl. 27, 73-75 (1982)]
12.3 W. Buckel: *Supraleitung. Grundlagen und Anwendungen* (Physik, Weinheim
1972)
12.4 A. Leger: J. de Phys. 29, 646-654 (1968)
12.5 R. Gomer, J. Hulm: J. Chem. Phys. 20, 1500 (1952)
12.6 R. Klein, L. Leder: Phys. Rev. 124, 1050-1052 (1961)
12.7 M.H. Cobourne, W.T. Williams: Physica 104C, 50-55 (1981)
12.8 H. Brechna: *Superconducting Magnet Systems* (Springer, Berlin, Heidelberg
1973)
12.9 W.B. Nottingham: Phys. Rev. 59, 907-908 (1941)
12.10 H. Bergeret, A. Septier: C.R. Acad. Sci. B 281, 405-408 (1975)
12.11 H. Bergeret, A. Septier, D. Drechsler: Phys. Rev. B 31, 149-153 (1985)
12.12 N.E. Alekseyevsky: Dokl. Akad. Nauk SSSR 242, 816-818 (1978) [in Russian]
12.13 E.A. Litvinov, G.A. Mesyats, A.A. Starobinets: Dokl. Akad. Nauk SSSR 249,
352-354 (1979) [in Russian]
12.14 G.A. Mesyats: Pulsed electrical discharge in vacuum at cryogenic electrode temperatures. Proc. XIII Int'l Symp. on Discharges and Electrical Insulation in
Vacuum (Paris, France 1988)
12.15 M.S. Aksyonov: Investigation of the critical current density of field electron
emission and explosive emission of metals cooled to ultra-low temperatures.
Cand. Sci. Thesis, Leningrad (1982) [in Russian]
12.16 S.I. Shkuratov: Investigation of pulsed vacuum breakdown at cryogenic electrode temperatures. Cand. Sci. Thesis, Tomsk (1986) [in Russian]
12.17 V.G. Mesyats, S.I. Shkuratov: Pisma Zh. Tekh. Fiz. 13, 756-758 (1987)
12.18 J.G. Bednorz, K.A. Müller: Z. Phys. B 64, 69 (1986)
12.19 M.K. Wu, J.R. Ashburn, C.J. Torng, P.H. Hor, R.L. Meng, L. Gao, Z.J. Huang,
Y.O. Wang, C.W. Chu: Phys. Rev. Lett. 58, 908-911 (1987)
12.20 J.N. Smith: J. Appl. Phys. 60, 1490-1492 (1986)
12.21 M. Gurvitch, A.T. Fiory: Phys. Rev. Lett. 59, 1337-1340 (1987)
12.22 A. Junod, A. Bezinge, T. Graf, J.L. Jorda, J. Muller, L. Antognazza, D. Cattani,
J. Cors, M. Decroux, F. Fischer, M. Ranovski, P. Genoud, L. Hoffmann, A.A.
Manuel, M. Peter, E. Walker, M. Francois, K. Yvon: Europhys. Lett. 4,
247-252 (1987)

Subject Index

Anode erosion
- craters 18,27,78,186,191
- material transport 13-15,26,70,71
- metallographic studies 187-190
- microprotrusions 32
- "touches" 71,72,178
- transmission electron microscopy (TEM) study 190,191
- vapor 18,20,27,77,158,196,213-219
Anode glow 18-20,65-69
Anode temperature 32,183-186,194,221

β-factor, see Electric field enhancement factor
Bardeen-Cooper-Schrieffer (BCS) theory 255
Boyle-Kisliuk-Germer (BKG) hypothesis 30
Breakdown
- anode initiated 19,26-31
- arc stage 35-42,78
- cathode initiated 19,24-31,75,76,88
- characteristic times, see Characteristic times of breakdown
- criteria 225,226
- direct current, see dc breakdown
- electric field 14,25,28,29,76,88,165
- electric field strength 60,266
- microparticle initiated 30-32, see also Microparticles
- spark stage 33-35,78,136-144, 154-156,181-202
- voltage 14-17,25,31,262

Cathode craters, see Microcraters
Cathode erosion, see also Microcraters, Micropoints, Micropoints
- erosion rate 36,37,253
- mass loss 89-96
- pressure in the emission zone 113,114

- scanning electron microscopy (SEM) study 104-107
- surface microstructure 6,7,114-117
- vapor 32,37,38,213-219
Cathode flare (CF), see also Cathode glow, Cathode spot
- brightness 62,77
- distribution density 177
- emission instability 150-154
- expansion velocity 64-67,118-121, 128-135,146,147,157,211,253,263,265
- multiplication 72
- plasma composition 37-39,124-128
- plasma density 121-124,137,138, 151,152,253
- plasma expansion models 128-135
- potential 147-151
- temperature 124-126,253
Cathode glow 18-20,62-69,77,126-128, 170,171, see also Cathode flare
Cathode plasma, see Cathode flare
Cathode plasma jets 38,39,241
Cathode spot (CS)
- autographs, see CS erosion traces
- current density 37,41,244-246
- division 36
- electron temperature 10,11,39,41
- erosion traces 230,231,234,238
- formation 160,234,235
- models 40-42
- motion 36,227-235
- plasma density 39
- retrograde motion 232,233
Cathode temperature 41,248-250, 256-264
Cathode voltage drop 41,227,245,246
Characteristic times of breakdown
- commutation time 15,61,62,74,75, 155,156,263
- delay time 15,58-62,75,87-89, 174,262,265

291

- gap voltage decay time 16,58,59
Child-Langmuir law ("3/2 power" law)
    87,138,144,154,155
Clusters 13
Conditioning
- "breakdown" 60,205
- "glow discharge" 6
- high-temperature heating 6
Current-voltage characteristic (CVC)
    25,80,81,144-146, *see also* Fowler-
    Nordheim plots

Dark current, *see* Prebreakdown current
Dashman formula 40, *see also* Richard-
    son-Dashman relationship
dc breakdown
- commutation time 207-209
- current rise time 207-210
- electric field 206,220
- electrical studies 204-210
- electrode erosion 210
- explosive emission initiation 219-226
- optical studies 210-219
- voltage 16,17,205-207
Debye length 137
Discharge circuit resistance 17,63,156
Drechsler-Henkel formula 80

Electric field enhancement
    factor 7-11,25,28-30,88,107-110,206
Electrical study of vacuum discharge
- using voltage dividers and Rogowski
    coils 46,47
- using voltage pulse generator
    (VPG) 43-46
Electrochemical pickling, *see* Electrodes,
    surface preparation
Electrochemical polishing, *see* Elec-
    trodes, surface preparation
Electrodes
- conditioning, *see* Conditioning
- critical temperature 30,32
- erosion studies 70-72, *see also* Anode
    erosion, Cathode erosion
- material 14,61,88,263-267
- surface examination 55-57
- surface microrelief 6,24, *see also*
    Microcraters, Microprotrusions
- surface preparation 5-7,55-57
- work function 7-9,137
Electron beam structure in a
    diode 178-180

Electronic flux perveance 144-146,157
Electro-optical image converter (EOIC),
    see Light emission
Emission centers (ECs)
- distribution density 104
- effect of transverse magnetic
    field 168-175
- lifetime 233,248-252
- mechanisms of formation 106,107,
    159-164,166,165,172-175
- multiplication 92,99-102,170-172
- numerical simulation studies 246-252
- "reverse motion" 175,232,233, *see also*
    Cathode spot, motion
- velocity 229
Emission sites, *see* Emission centers
Emitting area, *see* Microprotrusions
"Erosion method" for measuring current
    density, *see* Explosive electron
    emission
Explosive electron emission (EEE), *see
    also* dc breakdown, Point emitter
- centers, *see* Emission centers
- current behavior 82-84,139-144
- current density 96-112,140-144
- cyclic nature 117,254
- initiation 75,79-89
- pressure in the emission zone 113,114

Field-assisted thermionic emission
    22,86,163,164
Field electron emission (FEE)
- current 21-23,31,82-84
- current density 21-27,76,79-81,86,87,
    255,256
- from HTSC cathodes 264
- from monocrystal point emitters 21,80
- from non-metallic inclusions, *see* Non-
    metallic inclusions and films
- transition to breakdown 21-23,79
Field-electron emitter 56,76,83,244, *see
    also* Point emitter
Field-emission cathode 26,87, *see also*
    Field-electron emitter, Point emitter
Field enhancement factor, *see* Electric
    field enhancement factor
Fowler-Nordheim equation 7-10
Fowler-Nordheim plots 8-10,80,81

Gas desorption 14,24,88,174,192-195
Grounded grid and collector method, *see*
    Cathode flare, plasma expansion models

292

Hall effect, *see* Cathode spot retrograde motion
High-temperature superconducting (HTSC) cathodes 263-267
Hot electron current 10

Integral of action 86,92,260
Inversion temperature 27,28,257
Ion bombardment 6,21,22,31,174

Joule heating 22,27-29,86,93-95

Light emission, *see also* Anode glow, Cathode glow
- "cut-off discharge" method of investigation 69
- electro-optical recording 48-50,62-69
- photoelectrical recording 50-52
- spectral methods 52,53

Mechanical treatment, *see* Electrode surface preparation
Microcraters 6,37,104-107,172, *see also* Cathode erosion
Microdischarges 12-14
Microdrops 7,32,110-112,116,117, 163,164, *see also* Cathode erosion, Microparticles
Microparticles 7,13,24,31,32,105, 223-225, *see also* Cathode erosion
Micropoints 6-9,105,161,162,209,220, *see also* Cathode erosion, Microprotrusions
Microprotrusions 6,13

Non-metallic inclusions and films 9-12, 24,88,165,172-174,223
Nottingham effect 22,27-29,86,256-259

Particles, *see* Microparticles
Photoelectric method, *see* Cathode flares plasma expansion models
Photomultiplier tube (PMT), *see* Light emission
Point emitter, *see also* Cathode erosion
- avalanche-like destruction 22,26
- calculation of heating 85,86,219,220
- erosion, *see* Cathode erosion
- erosion rate 92
- explosion delay time 84,85
- mass loss 91-99
Polarity effect 224

"Polishing effect" 110
Ponderomotive forces 24,224
Power density at the anode 27,30, 77,181-183
Prebreakdown current 12,14,205-207, *see also* Breakdown
Prebreakdown current density 21,93, *see also* Breakdown
Prebreakdown FEE current, *see* Prebreakdown current
"Principle of maximum magnetic field" 36
Probe measurements, *see* Cathode flare, potential
Protrusions, *see* Microprotrusions
Pulse factor 16

Richardson-Dashman relationship 10,40
Richardson-Schottky plot 12

Scanning electron microscopy (SEM) 10,90,104-113,172
"Screening effect" 159,175-180
Secondary emission 24
Space charge effects 22,33,40,174
Spark stage of discharge, *see* Breakdown, spark stage
"Strata", *see* Electron beam structure
Superconductivity 255,256

Thin wires 86
Time-voltage characteristic 16
"Total voltage" effect 14,25,31,221-225
Tracer method, *see* Electrodes, erosion studies
Transverse magnetic field method, *see* Cathode flare, plasma expansion methods
Tunnelling electrons 10

Vacuum arc
- cathode spot, *see* Cathode spot
- threshold current 241-244
- transient processes 235-241
Vacuum equipment 54,55

X-radiation 19,53,54,74,75,198-202,210
X-ray tubes 19,20,33,158,198,200-202

Widdington formula 29,195
Work function, *see* Electrodes, work function